THE WETLANDS OF

Greater Manchester

D Hall
C E Wells
E Huckerby

With contributions by

A Mayer
C Cox

GIS mapping by

J T Dodds

LANCASTER IMPRINTS

1995

Published by
Lancaster University Archaeological Unit
Storey Institute
Meeting House Lane
Lancaster
LA1 1TH
(Phone: 01524 848666; *Fax:* 01524 848606*)*
(E-Mail: C.Wells@lancs.ac.uk; J.Dodds@lancs.ac.uk*)*
(Internet —World Wide Web URL: http://www.lancs.ac.uk/*)*
Further details of the North West Wetlands Survey can be obtained from the LUAU Home Page at the above address

Distributed by
Oxbow Books
Park End Place
Oxford
OX1 1HN
(Phone: 01865 241249; *Fax:* 01865 794449)

Printed by
Kent Valley Colour Printers, Kendal, Cumbria

ISBN 0 901800 80 5
ISSN 1343-5205

NWWS Series editor
Rachel Newman
Copy editor
Martin Lister
Indexer
Peter B Gunn
Design
R Middleton
Layout and production
R Middleton and R A Parkin

Front Cover: Carrington Moss from the south showing the River Mersey, Irlam, and Chat Moss *(top left)* (Photo: Sky Comm 1/5/1994 (neg no CCN 000 700))

Lancaster Imprints is the publication series of Lancaster University Archaeological Unit. The series covers work on major excavations and surveys of all periods undertaken by the Unit and associated organisations.

Contents

List of illustrations

Figures

A key for Figures 6, 29, 34, 45, 46, 55—60, and 62—66 is located on a flap on the inside of the back cover

Plates

Tables

Abbreviations

EH	English Heritage
ESA	Environmentally Sensitive Area
GIS	Geographic Information System
GMAU	Greater Manchester Archaeological Unit
LRO	Lancashire Record Office
LUAU	Lancaster University Archaeological Unit
NERC	National Environmental Research Council
NWWS	North West Wetlands Survey
OS	Ordnance Survey
OD	Ordnance Datum
PRO	Public Record Office
SMR	Sites and Monuments Record
SSSI	Site of Special Scientific Interest

Contributors

Chris Cox
Air Photo Services, 7 Edward Street, Cambridge CB1 2LS

John Dodds
Archaeological Unit, Lancaster University, Storey Institute, Lancaster LA1 1TH

David Hall
Fenland Project, Department of Archaeology, University of Cambridge, Downing Street, Cambridge CB2 3DZ

Elizabeth Huckerby
NWWS, Lancaster University Archaeological Unit, IEBS, Lancaster University, Bailrigg, Lancaster, LA1 4YW

Adele Mayer
Greater Manchester Archaeological Unit, The University of Manchester, Oxford Road, Manchester M13 9PL

Colin Wells
NWWS, Lancaster University Archaeological Unit, IEBS, Lancaster University, Bailrigg, Lancaster, LA1 4YW

Acknowledgements

The authors would like to thank the following for help, criticism and advice during the preparation of this volume:

Dr Charles French, University of Cambridge, and Dr Peter Moore, University of London. Bryony Coles, Chair NWWS, and Gerry Friell and Tony Wilmott of English Heritage. The staff of the Greater Manchester Archaeological Unit including its former Director, Phil Mayes, and Robina McNeil, Norman Redhead, and Dr Michael Nevell. Barri Jones, Department of Archaeology, University of Manchester, and Chris Perkins of Manchester University Geography Library. Greater Manchester Geological Survey.

Dr GT Cook and staff at the Scottish Universities' Research and Reactor Centre, Dr Alex Bayliss (Ancient Monuments Laboratory, English Heritage) for advice on Nook Farm radiocarbon dates, Mike Peace (stratigraphic survey), Ruth Parkin, and Bob Middleton (illustrations). The farmers to whom we are obliged for their friendly co-operation are too numerous to mention individually, but in particular we thank those at Carrington Moss, Chat Moss, Hollins Moss (Chew Moor), Red Moss and Siddall Moor. Thanks are due to the landholders of these areas, who include Peel Holdings, Land Improvement Group, Lancashire Wildlife Trust and Fisons plc. For Hopwood, to members of the Healey Dell environment team who escorted us across Hopwood Moss. Shell UK, Property Division, for access to land at Carrington, and John Headley and the staff of Local Peat Products Ltd., for co-operation and interest in our activities at the Nook Farm site, Chat Moss. Michelle Young and Selina Hill of English Nature, North-West Region, gave permission to survey at Red Moss, and Bolton Metropolitan Borough Council gave access to Red Moss.

Rog Palmer (Air Photo Services), staff of the Record Offices and Local History Study Units at Manchester Central Library, Chetham's Library, and Salford Art Gallery and Museum, Andrew Cross (Irlam Archive Office), Nicholas Webb (Wigan Archives Office), Lancashire Record Office, Cheshire Record Office, Trafford Local History Library, Hannah Haynes (Middleton and Heywood Local Studies Library), Sandy Roydes (Hopwood College), Alan Davies (Salford Mining Museum), Dr John Prag (Manchester Museum), Angela Thomas (Bolton Museum), Alan Leigh (Warrington Museum), and Judith Sandling (Salford Art Gallery).

Foreword

For many people, wetland archaeology has strong associations with lake villages, trackways, and bog bodies, but readers of the present volume will find few references to these traditional categories of evidence. Instead, they will be introduced to an industrialised wetland landscape and perhaps to the unexpected survival of deep peats and of palaeoenvironmental sequences alongside the fast disappearing physical remains of peatland industrial archaeology. Fieldwork in these conditions demands ingenuity, experience, and determination, and only those readers who have worked in similar conditions will perhaps fully appreciate the efforts put in by the two principal authors, David Hall and Colin Wells, to assemble the evidence presented here.

The wetness of the Greater Manchester region from the later prehistoric period onwards, and its subsequent key role in the early modern industrialisation of Britain, have conspired to make traditional archaeological survey difficult because of the low level of human activity from the Mesolithic to the medieval times has been masked rather than revealed by the recent and current land-use. The challenge of working in these conditions has stimulated the NWWS team and their helpers to explore every possible means of recovering evidence for past conditions and human activity, and we can be confident that the picture given here is representative.

One intriguing question raised by the results of the survey is that of the inter-relationship between the Manchester wetlands and the inception of industrialisation. Did the mires make a positive contribution, in the form of resources required by the developing industries? Or did they simply, as large marginal areas not claimed by agriculture, provide a setting for development once the technological difficulties of transport and drainage has been mastered? These questions are not the normal concern of wetland archaeologists, but they are proper to the Manchester context, and they underline the individuality of the human response to wetlands in this region. Neither in prehistory nor in historic times have the Greater Manchester wetlands been perceived and exploited quite according to our general models which are based largely on evidence from the Fens and the Somerset Levels.

In the long run, one of the most important results of the current survey will be the conclusive demonstration of the diversity of human responses to wetland environments in the past, responses influenced by the prevailing environmental conditions and compounded by the economic, technological, and social demands and capabilities of the local people.

Bryony Coles
Chair, North West Wetlands Survey

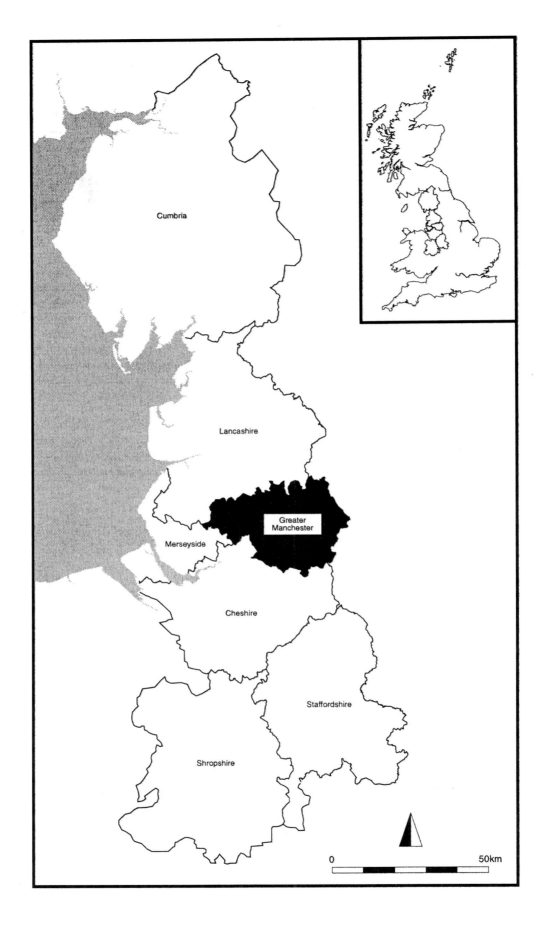

Fig 1 The North West Wetlands Survey region

1

THE NORTH WEST WETLANDS SURVEY

The most extensive wetlands of lowland England lie near major rivers and estuaries, where rising sea-levels during the last ten millennia have caused major interference with drainage. The largest of these wetlands in England are found in the Solway Firth, South Cumbria, the Mersey Basin, West Lancashire, the Somerset Levels, the Wash Fenlands, and Humberside.

Wetlands are characterised by drowned or paludified land surfaces, submerged by rising water-tables. The resultant marine clays, freshwater deposits and peats have buried and preserved archaeological remains of many periods. In recent centuries, peat digging and drainage of wetlands have much altered their character, exposing artefacts and sites. These finds gradually came to the attention of antiquarians; Dugdale and Stukeley wrote of Roman remains discovered in the Wash Fens during the seventeenth and eighteenth centuries, although most finds probably came from dried-out marine deposits rather than peat. The best known of the submerged wetland sites is the Glastonbury Iron Age lake village excavated by Bulleid in the Somerset Levels from 1892 to 1907. Archaeological interest in wetlands has been continuous since the nineteenth century, with studies large and small published from time to time; some of them are summarised in Coles and Coles (1989).

One of the most important properties of wetlands is that organic material is often well preserved by anaerobic conditions. Objects of wood, bone, and leather, which would not survive on dry-land sites, can be preserved intact on wetland sites. Equally important are biological remains of animals and plants, which can reveal the nature of past environments in the wetland and its surrounding landscape. Peats have the advantage of being datable by radiocarbon techniques, and dendrochronology can be used to determine the age of large samples of some species of wood.

Most of Britain's former wetlands are now no longer wet in an ecological sense, and the continued effects of drainage and industrialisation present major threats to fragile archaeological remains. In response to these threats, English Heritage and its predecessors have supported several work programmes to investigate and record remains in four of the major wetland regions: the Somerset Levels (1973—89), the Wash Fenlands (survey stage 1981—9), the North West Wetlands (1987—) and the Humber Wetlands (1993—).

In 1987 English Heritage began a three-stage programme to assess surviving wetland archaeology in the lowland regions of North West England. The study area comprises, in topographical order from north to south, the counties of Cumbria, Lancashire, Merseyside, Greater Manchester, Cheshire, Shropshire, and Staffordshire (Fig 1). The peatlands of the North West have a considerable range of types, with large raised mires, valley and floodplain mires, and basin peats. Pennine blanket peats are not studied in the present programme. In all, there are about 36,700 hectares of lowland peat according to the recent survey made by Burton and Hodgson (1987, appendix). The distribution of mire types varies across the region; large areas of raised mire peat lie in West Lancashire, the Solway lowlands, and the Mersey Basin, whilst valley mire and basin peats predominate on the Cheshire Plain and in Shropshire.

The first stage of work was to make an assessment of existing archaeological sites and finds in and around the wetlands (frequently called 'mosses'), and to this end a Gazetteer summarising previous palaeoecological studies was prepared with information of current land-use and present and future threats. The resulting study, *Peat and the past* (Howard-Davis *et al* 1988), confirmed the need for more detailed work. It was recommended that fieldwork was required to identify sites and define areas of archaeological potential. Palaeoecological fieldwork was also considered essential, making borehole sections of mosses to ascertain their stratigraphy, state of preservation, ontogeny, and to obtain samples for detailed analysis.

The second phase began in 1990 with an eight-year work programme — Lancashire and Merseyside being studied first, followed by Manchester (1992), and

Cheshire, Shropshire, and Staffordshire (1993—5). Work in Cumbria continued in parallel. Interim results have been published in *NWWS Annual Reports* (Middleton (ed) 1990, 1991, 1992, 1993; Middleton and Newman (eds) 1994). The third phase of the programme, which may overlap with later stages of field survey, will be an evaluation leading to the conservation of selected wetlands and their associated archaeological remains.

Previous wetland studies have been made of several types of terrain such as lowland peat, upland peat, coastal marsh, and alluviated river valleys. In the current project, the meaning of 'wetland' was limited to peats occurring in raised mires, floodplain mires, valley mires and basin mires, along with their catchment areas.

The scope of the work was determined by the landscape zones under examination. Each basin, valley, floodplain, and raised mire was surveyed in turn, as well as the surrounding organic soils derived from them and a sample of the hinterland. Lowland peats were selected for study because many are under imminent threat, and because of their importance in preserving a record of past human settlement and activity. The rate of destruction of mosses has increased during the last decade and it is urgently necessary to record surviving archaeological remains and peat stratigraphies before they are lost.

Palynology and mire stratigraphy can provide evidence of the development and exploitation of the landscape from prehistory to historically recent times. Since lowland mires occupied a considerable proportion of the land surface in some parts of the region (the Lancashire lowlands, the Mersey Basin, parts of the Solway Plain, and North Shropshire), they hold important evidence of the environmental and human history of the North West. The archaeology of the region is not well known, particularly prehistory, and information is needed to discover the extent and nature of lowland occupation. Even though a relatively large amount of palaeoecological research has been undertaken in the region, there are many gaps in the pollen record, especially in relation to archaeological sites.

Peat formation is controlled by hydrological, topographical, and climatic factors. Commonly, peat first formed in basins or valleys that had become permanently waterlogged. Such conditions led to the exclusion of oxygen which slows down decay, resulting in the accumulation of peat. Depending upon local conditions, a variety of different mire types evolved. In some cases peat surfaces rose above the water-table, leading to the development of raised mires which then became dependent on atmospheric precipitation as their major source of nutrients. Such mires are known as ombrotrophic systems and are characterised by deep peats and a flora associated with acid soils. So long as waterlogged conditions are maintained, organic materials, such as plant macrofossils, pollen, and the remains of insects and other fauna, are likely to be preserved.

Wooden trackways are among the best known man-made organic finds occurring in wetlands. Settlements and ceremonial sites on land surfaces near to mires may have been inundated by peat growth, preventing further disturbance. Inorganic materials such as flint implements and pottery are well-preserved, but, more significantly, the preservation of organic remains allows the recovery of artefacts not found on dry-land sites, as well as providing direct association between archaeological and contemporary environmental remains.

Threats to the archaeological resource

The work programme was necessitated by threats to fragile lowland peatlands and consequently to their unknown archaeological potential. Most threats involve some form of drainage which leads to oxidation of peat and irreversible drying. Wastage occurs as water is lost and peat dries and shrinks. The nature of the threat varies considerably; agricultural drainage may take centuries to desiccate a peat deposit completely, whereas exploitation for refuse disposal will poison and bury a mire within a decade. Peat extraction causes rapid destruction, although it does expose buried land surfaces systematically. A dramatic discovery during peat extraction of a lowland mire was made in 1984, with the exposure of bog bodies in Lindow Moss, Cheshire (Stead *et al* 1986). The major threats posed to the wetlands of the North West are listed below.

Agriculture

During the late eighteenth and nineteenth centuries, many mires in Lancashire, Cheshire, Merseyside, and Greater Manchester were drained and improved for agriculture. This process continues and it has been estimated that one-fifth of the peat in Lancashire was lost between 1964 and 1985, based on the current wastage rate of *c* 10 mm yearly (Burton and Hodgson 1987, 50—1). The same or higher levels of damage occur elsewhere in the region, and agricultural improvements form the greatest threat to lowland peats.

Refuse disposal

Many mosses have deep peats unsuitable for building purposes and have remained as open spaces within large conurbations. The wetlands are therefore convenient as landfill sites for refuse disposal, and such mires are amongst those most seriously threatened in the region. Their archaeological and palaeoecological potential urgently needs assessment and recording.

Peat extraction

Most of the mosses in the North West have been subject to peat extraction, usually for domestic fuel, although some have been cut for animal and poultry bedding and, more recently, for horticultural use. Domestic cutting is now very limited. Total peat extraction rates vary; estimates for three sites in Cumbria indicate that 250,000 cubic metres are removed annually (Burton and Hodgson 1987).

Construction and development

Urban expansion and road construction threaten several areas of wetland. This threat is considerable in Greater Manchester where mosses form the only space through which new roads and motorways can pass. Peat is also destroyed when underlying minerals are extracted.

Aims and objectives of the survey

A major objective of the survey is to map the extent and nature of settlement in and around the lowland peat of the North West from the prehistoric period to the Middle Ages. This is achieved by ground survey, recording evidence of all periods from the mires and their catchment areas. In many areas the present fieldwork is the first systematic survey to take place and the results help to interpret the nature of past settlement and economy in the region as a whole.

By studying lowland peat as a single landscape type, regional comparisons and contrasts will show if there is variability between contemporary settlement systems relating to different mire types. It will also be possible to compare settlement in different periods, subject to sufficient evidence being available. Settlement and exploitation of the many varieties of wetlands characteristic of the North West are likely to be different from those in other parts of the country, and regional results can be compared.

Previous studies have indicated the sensitivity of mires to climatic and other environmental changes (Barber 1981), and the sensitivity of human settlement to changes in wetland environments (Pryor 1984, 197—255). This is also true in several coastal areas of the North West, where mossland development has been greatly affected by relatively small sea-level fluctuations, with commensurate effects on both the immediate and wider environments.

For each individual mire system the aim of the palaeoecological studies has been to record peat depth, its degree of preservation, and the palaeoecological research potential, along with a record of mire ontogeny and, where appropriate, detailed palynological study to aid interpretation of outstanding archaeological discoveries. The results complement the archaeological survey, and provide the only means of determining early human impact on mosslands and their surroundings. The opening up of the countryside from woodland to pasture or agriculture can be identified by changes in pollen spectra, while hydrological changes, possibly caused by human activity, may be reflected in peat stratigraphy. Palaeoecological studies detail the condition of surviving peat deposits, and reveal the extent of disturbance. Intact profiles are the most useful for palaeoecological analysis, and survey enables location of such deposits for future work.

Geographic Information System

A number of new techniques have been developed and implemented which now play a key role in the achievement of the project objectives. Principal amongst these developments is the implementation of a Geographic Information System (GIS) which permits the graphic display of project data taken from a central database. Experience has shown that the GIS has three major uses within the project. Firstly, it enables large amounts of information of many different types, both archaeological and palaeoecological, to be stored and manipulated in a much more efficient manner than is otherwise possible. It also allows data manipulation in ways not possible manually. In addition, and perhaps most importantly, it permits the construction of models predicting the fate of the peatlands, based on all the data recovered during the project. The reconstruction of past landscapes plays a key role here, particularly where the impact of modern threats and developments needs to be assessed.

The GIS also allows the production of publication-quality maps directly from the same database at a variety of scales and formats, which saves time in the production of reports and archive material. The detailed maps in this volume were produced on the system.

Lastly, the computer-based system is also the main vehicle for the project archive; it holds, both textually and graphically, the major elements of the archaeological and palaeoecological datasets as well as final output for publication. The use of this system as the main archiving medium permits flexible output to the county SMRs and National Monuments Record in a variety of formats to suit local circumstances. The use of a single, centrally-based system also ensures the use of standard data structures throughout the work of the project. Storage of project data in this form will permit the archive to be continuously updated and maintained so that it provides an up-to-date database upon which long-term management strategies can be based.

The GIS now in place is based on a Sun SPARCstation platform with a large accelerated graphics monitor and almost 3 Gb of hard disk storage which is used to run the ARC/INFO GIS together with the ORACLE RDBMS (Relational Database Management System) (Dodds and Middleton forthcoming). This system provides the power to amass, store, and effectively analyse the vast body of data required for the project and gives the capacity to handle very detailed small-scale mapping of a large geographic area. It also has compatibility, in terms of input and output, with as wide a variety of data formats and platforms as possible, to allow data transfers to and from other bodies so that multi-variate analyses can be performed with data incorporated into the project database.

Future management of the wetlands

The archaeological field survey will map all individual sites and landscapes, and the palaeoecological work will record the condition and depth of remaining peat deposits. This information is required to identify areas of archaeological importance that need management, and to complement existing data for peat distribution (the reports published by the Soil Survey, the British Geological Survey, and place-name evidence) which currently provide an incomplete picture of wetland distribution.

The field-survey phase of the project will be followed by action to encourage management of the most significant and best-preserved wetland areas, aiming to arrest the processes of decay. In many cases the most appropriate action will be liaison with other bodies involved in the conservation of wetlands, especially English Nature. Some sites may achieve protection through local planning control, and others may have to be preserved by record, that is, excavated before destruction.

The project management of the North West Wetland Survey is based at the Lancaster University Archaeological Unit. Archaeological survey of Cumbria and Lancashire is undertaken from Lancaster (R Middleton), Merseyside work was based at Liverpool Museum (R Cowell), the Manchester survey was undertaken from Greater Manchester Archaeological Unit at Manchester University (D N Hall and Adele Mayer) and the southern counties from Cheshire County Council offices in Chester and English Heritage premises at Wroxeter (M Leah). Palaeoecological work for the whole region is undertaken by staff of Lancaster University Archaeological Unit (C Wells and E Huckerby) based at Lancaster University.

The results of the survey work in the lowland wetlands of North West England will be published in a series of monographs, one or more for each of the seven counties. This second volume describes the county of Greater Manchester.

2

THE WETLANDS OF GREATER MANCHESTER

The area studied in this volume, Greater Manchester County (Figs 2, 3), incorporates parts of the historic counties of Cheshire and Lancashire, and small parts of Derbyshire and the West Riding of Yorkshire. Its boundary straddles the upland Pennines and the Mersey Basin. The boundary between Lancashire and Cheshire until the late nineteenth century lay along the River Mersey as far as Stockport, and then followed the River Tame up the west flank of the Pennine ridge. The Pennine Hills are a boundary between Cheshire and Lancashire on the west and Yorkshire and Derbyshire on the east.

The Municipal Reform Act of 1835 formalised the developing urban status of Manchester (Carter 1962, 136). During the early years of the twentieth century the City of Manchester expanded in population, and housing estates had developed beyond the municipal boundary in Worsley, Wythenshawe, and elsewhere. By 1951 the area of Greater Manchester had been politically recognised as the South East Lancashire Conurbation, although it included parts of North East Cheshire (Kidd 1993, 198; Harris 1979, 2, 80, map).

In 1972 the Local Government Act (taking effect in 1974) created the County of Greater Manchester, the third largest metropolitan county after Greater London and the West Midlands (Kidd 1993, 198). The act also designated new metropolitan authorities based on the boroughs of Wigan, Bolton, Oldham, Bury, Rochdale, and Stockport, together with the city of Salford. Two districts were created based on artificial boundaries: Trafford, stretching out from the industrial centre of Old Trafford and taking in Altrincham and Hale; and Tameside incorporating Ashton-under-Lyne, reaching eastwards to Mottram in Longdendale (Figs 2, 3).

The topography of Greater Manchester is dominated by the sweeping curve of the Rossendale Hills along the north side, and the Pennines to the east. Three major rivers, the Mersey, Irwell, and Douglas, pass through the county, rising from the hills, and fed by a complex network of tributary rivers and streams. The Douglas, at the extreme west, begins in the Rossendale Hills, travels through the Lancashire low-lands, generally lying below 135 m OD, and flows north-east towards the Ribble estuary. The Irwell flows from the north, passes between the cities of Manchester and Salford, and although now redirected through the Manchester Ship Canal, opened in 1894 (Kidd 1993, 103), it was once a fluviant to the River Mersey, joining just south of Irlam. The River Mersey is formed by a confluence of the Rivers Tame and Goyt, on the north side of Stockport town centre.

Organisation of the Manchester data

This chapter outlines the geology of the region, refers to previous studies, and describes the techniques and methodology used in the current programme.

Chapters 3, 4, and 5 describe details of archaeological survey and palaeoecological work undertaken on all the mosses that still have surviving peat. Only five mosses are now extensive and only seven of the smaller mosses have any organic soils. The remaining 34 wetlands have been destroyed by industry or urbanisation and were not available for study; details of their present condition are to be found in the Gazetteer (*Appendix 1*). These three chapters contain the whole of the detailed palaeoecological results.

The final chapter (6) consists of a summary of the archaeology of the region, with particular reference to evidence connected with wetlands. The summary uses evidence from the present NWWS archaeological and palaeoecological fieldwork study, supplemented with information from published works and historical sources where relevant, and is integrated with data held in the Greater Manchester Archaeological Unit (GMAU) Sites and Monuments Record (SMR), fully sourced and referenced in the Gazetteer (*Appendix 1*).

Appendix 1 consists of a gazetteer of all the Manchester mosses in alphabetical order, listing the mosses arranged in project number order and by the modern districts. Each moss has a note of its location and

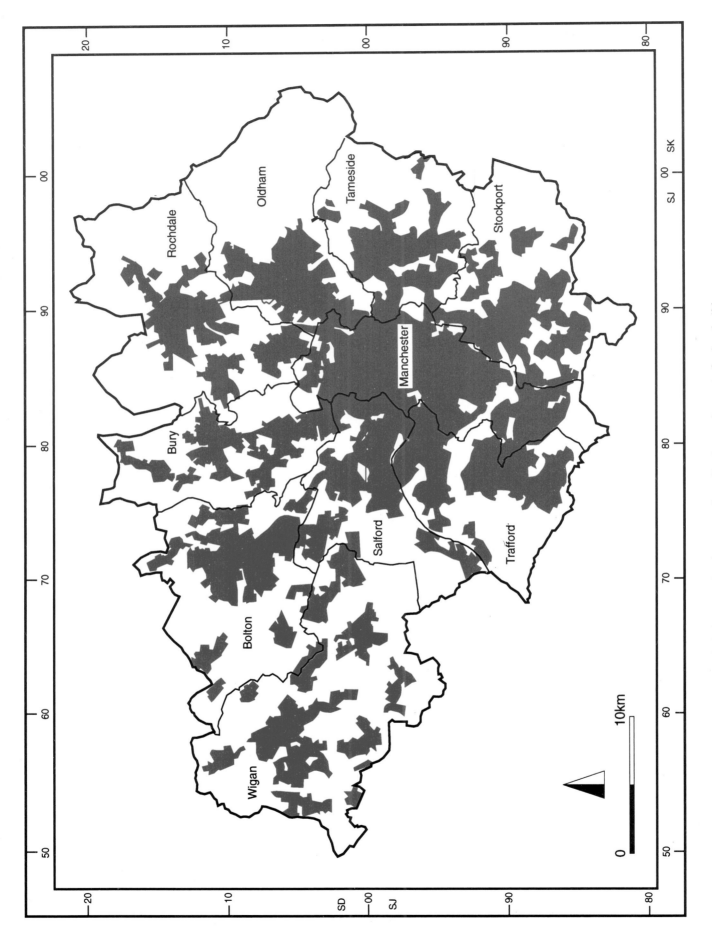

Fig 2 Greater Manchester: modern districts and extent of urban build-up

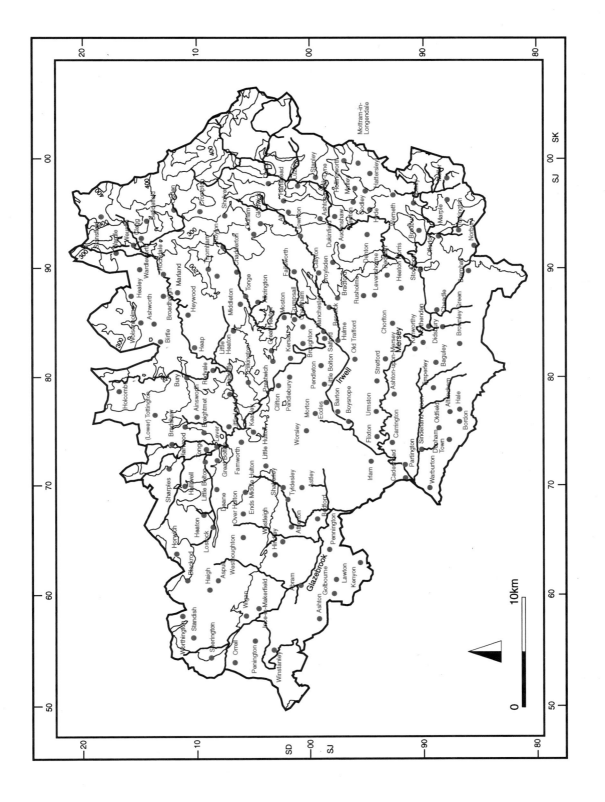

Fig 3 Greater Manchester: historic settlements and topography

7

geology, followed by the archaeological and documentary history, and comments and the results of mapping by fieldwork in 1992—3. The Gazetteer, with data from Chapters 3—5, represents the full record of the archaeological survey, on which other conclusions are based. Appendix 3 (by Chris Cox) gives details of aerial photographic interpretation. Some of the sites were subsequently visited and a note (by D Hall) of the ground evidence is appended. Appendix 4 lists the wetland sites, and Appendix 5 is a list of exotic artefacts from the region.

Geology of the region

The geology of a region determines, in part, the development of its wetland landscapes. The Manchester region of the North West is dominated by hills of Carboniferous rocks, consisting of gritstones, shales, mudstones, and Coal Measures, which have been much exploited by mining and quarrying. The lowland to highland change is marked by the west Pennine tectonic line, forming faulted scarps in the foothills. The east—west ridge at the bottom of the Rossendale Hills, visible at Horwich, was formed by folding and erosion of the Carboniferous Coal Measures and sandstones at a height of 100—122 m OD, later thinly covered with till (Johnson 1985, 239). In the southern Pennines, elevated areas are made of Carboniferous rocks: coarse Namurian sandstone/gritstone and mudstone rock, over which high moorland has developed, itself eroded and unstable, exposing the underlying gritstone. Limestone is exposed farther east, emerging from beneath the Kinder Scout gritstone in the Peak District (Johnson 1985, 4).

Carboniferous rocks in the region are composed of fine-grained sedimentary material: shales, mudstones and also sandstones, grits, coal seams, seatearths, and limestones (Carter 1962, 66), commonly known as the Millstone Grit Series. These bands of rock are overlain by Coal Measures, differentiated from the Grit by the greater development of the coal seams, and the occurrence of sandstones in place of gritstones. It is probable that coal seams developed from peat beds, rooted into a layer of seatearth (soil-bed), and grading down into mudstone or sandstones. Subsequent ground movement formed folds and basins within these deposits. Some seatearths, particularly the ganister, have been used as fireclay. Coal beds have been worked extensively in Lancashire (Carter 1962, 67—8).

Finally, the Permo-Triassic series of rocks is found, consisting mainly of New Red Sandstone. The lowest beds are Collyhurst Sandstone, a stone utilised in some of the buildings in central Manchester, which are overlain by Manchester Marls, including thin bands of mudstones, siltstone and limestones. Above this, the younger Triassic rocks consist of Bunter sandstone with Keuper sandstone above. Triassic beds extend beyond the boundary of the Permian deposits, and may be found overlying pre-Permian rocks (Carter 1962, 69). Keuper sandstone seems to have formed spasmodically; it is generally harder than the layers above and below, and is characterised by mineral seams of copper and lead which have been mined at Alderley Edge (north-east Cheshire, Carter 1962, 70). Overlying the sandstone are beds of sand and mud-stones called Waterstones, as well as Keuper Marls, and mottled mudstones which contain beds of rock salt reaching a maximum depth in mid-Cheshire at Middlewich and Northwich.

Folding and faulting occurred in the Carboniferous rocks; one of the best-known faults in Manchester is the Irwell Valley Fault, running north-west to south-east (visible in the coal mines at Wet Earth Colliery, Clifton). In the western part of the Pennines, close to Manchester, the upthrust of the rock has formed folded structures on a north—south axis cut by the drainage system. Subsequent action by Devensian ice sheets re-formed the landscape. Some valleys were filled with glacial sediment, which can be found in the Irwell valley north of Bury and in the Tame valley at Mossley, others filled with stagnating ice which prevented infill, but in both cases drainage may have been altered and recreated (Johnson 1985, 10).

The most extensive drift deposits, overlying the solid geology (the Northern Drift), are Pleistocene boulder clays of glacial origin, and sands, gravels, and clays of fluviatile/lacustrine origin, which cover most of the North West region. In the lowlands these deposits conceal older rocks. The exact sequence of ice flows, and thus drift origin, has not yet been determined. Ice-floes left a series of morainic deposits, found along the south Cheshire border, particularly at the division between the two water catchment areas of the Dee and Mersey and the Severn and Trent. Another area where morainic features have moulded the landscape is along the foothills of the Pennines, for example in the arc between Middleton, Oldham, and Rochdale. Above the Pennine hillslopes, rock has been eroded into a network of gullies and meltstreams which drain into this 'hummocky drift' (Johnson 1985, 18).

The major drainage alignments had developed in the Tertiary period, and were reformed in the Pleistocene, cutting into the glacial and peri-glacial deposits of the Mersey and Cheshire Basin. The River Douglas was created when glacial meltwater channels linked and formed a major drainage route; the Mersey, too, was etched into the glacial deposits of the plains. Around the Manchester conurbation the Mersey drainage system has been much altered in some areas, where upland streams have been dammed and redirected for industrial use and for reservoirs (Johnson 1985, 123).

Soil formation recommenced at the end of the last glaciation, and by the end of Flandrian I, forest covered the landscape up to a height of 800 m. A change may have occurred when clearance activities of Mesolithic people in the upland areas upset a fragile pedological balance leading to degradation of slowly permeable soils, and encouraged the growth of acidic blanket peats (Johnson 1985, 80). Soils are derived from the glacial till; on the upland Pennine drift, lying at c 60—90 m OD, they are mainly grits, sandstone, and shales. Glacial drift and soils of the lowlands consist mainly of red-brown, slightly stony, and calcareous material.

Ground drainage in the area of the Northern Drift is severely hampered by the sub-surface impermeability of clays, resulting in a perched water-table that has encouraged development of peat. Initially, fen-carr peats predominated, and later climatic and hydroseral changes influenced the subsequent dominance of raised mosses (Johnson 1985, 316).

The early post-glacial topography probably consisted of an ill-drained landscape and collection of pools and lakes, particularly at valley heads and in the Pennine foothills. Some of the low areas formed postglacial lakes, which developed a wetland vegetation. The remainder of the landscape was probably a mixture of forest, bog, marsh, and scrub.

Previous research and preparation for the present study

Previous palaeoecological studies have been made at Chat and Red Mosses (Birks 1964—5; Hibbert *et al* 1971); for more detail see the description of those mosses below. Peats of the region were studied, as part of a national survey, by Burton and Hodgson (1987). Initial work by the North West Wetlands Survey included preparation of a gazetteer summarising previous palaeoecological studies (Howard-Davies *et al* 1988). No systematic archaeological work in the region has been undertaken except of Ashton Moss (Nevell 1991b).

As part of the preparation for the fieldwork stage of the Manchester wetland survey, it was necessary to identify locations of all the former wetlands. Although some large mosses, such as Chat

and Red Moss, are well known, being major topographical features and having received detailed palaeoecological study, there were many smaller mosses existing before industrialisation, recorded only on estate and other maps. All these were potentially of interest to the North West Wetlands Survey and needed proper identification. Archaeological data relating to finds in and around the mosses were collected to provide as much background information as possible ready for fieldwork.

The main body of archaeological data derives from the Sites and Monuments Record (SMR) for Greater Manchester County. This contains the normal site and artefact information, some of it from secondary sources. Exotic finds of axes etc relevant to the mosses were relocated; of the 29 archaeological artefacts recorded, many pieces have no currently known location and two were rediscovered. The remainder are conserved in museums at Bolton, Manchester, Stockport, Warrington, and Cambridge (*Appendix 5*).

The SMR contained little historical or landscape data relevant to the mosses, and it was necessary to search antiquarian records and historical sources for information. Comments made by Charles Roeder in c 1888 are of interest, in view of the apparent paucity of known finds. When searching through the excavation records of the Manchester Ship Canal, he noted 'I have looked so far in vain, from Old Trafford onwards for any stone hammers, flint tools, whorls, or skulls and bones...' (Roeder 1889—90, 205). Other reports describe the stratigraphy revealed by the canal cutting (Thomson 1889, 217—19), who refers to a bed of green leaves 6.3 m down and horns of red deer.

Most of the information used to locate the many former mosses in the Greater Manchester region was obtained from maps of various dates. The sources are briefly discussed below with their relevance to Manchester. Specific details of the history of map-making in the region are available (Harley 1968; Harley and Laxton 1974).

The earliest maps covering an appreciable area are mid-sixteenth-century county maps of Cheshire and Lancashire, by Saxton and others. These are very schematic in their indication of open wasteland or forest, only Chat Moss is so dominant as to recur as a feature on any of them. More useful are maps of the late eighteenth century; Cheshire was published by Peter Burdett in 1777 at a scale of 1" inch to a mile (1:63,360), and Lancashire by William Yates in 1786 at the same scale. Both these show detail of several wetlands.

Private estate maps drawn at a large scale often mark topographical features. For the Manchester mosslands the earliest map is of White Moss, the 'theyle moor platt', a sketch dated 1556, which gives information on boundary points, although the moss itself is marked only schematically. Some late eighteenth century estate maps provide detail as, for example, those of Warburton (Arley Estate), Kearsley, and Worsley (Bridgwater Estate). The Warburton plan of *c* 1757 shows the extent of the moss and intakes at that date.

The other chief sources of large-scale maps in the pre-Ordnance Survey period are Parliamentary enclosure plans, of which there are 13 relating to mosslands in Manchester, Cheshire, and Lancashire, ranging in date from 1763 (Astley and Lowton, Wigan Archives; Astley map 1764 (UD/TY), and Lowton map 1763 (UD/GO)) to 1902 when Aspull Moor and Pennington Green were enclosed (Tate 1978, 148—52). Tithe maps made from 1836 onwards supplement the enclosure maps for areas of old enclosure (pre-1730). Apart from mapping the then surviving extent of moss, tithe map schedules give field-names, many of which indicate the extent of peat before reclamation ('moss', 'moor' etc). In areas such as Chew Moor, Bolton, the place name 'Moss Hall' approximately located a former moss extent and showed that the existing bog area lying away from these sites was only a remnant moss. At Red Moss, a farm site called Barkers lying to the west of the moss, and documented as a moss farm in the sixteenth century, probably marks the early medieval moss extent. Plate 1 shows Bryn Moss, south of Wigan, on a tithe map of 1843. The first Ordnance Survey 6" (1:10,560) scale maps were made for Lancashire between 1841 and 1854, and for Cheshire between 1870 and 1875. Subsequent editions show the fate of the mosses to the present day. The mosses identified from all these sources are shown on Figure 4. They are considerably more numerous and extensive than previously recognised in any recent publication.

All mosses, as far as they could be identified, were mapped on copies of the 1:50,000 and 1:10,000 scale maps. The smaller scale was used for general location and the large-scale plans showed details of the boundaries of each particular moss at different dates, if known. The 1:10,000 scale maps were marked with the location of all known sites and monuments, and with cropmark information. The preparative work identified 84 areas of former wetland from historical sources. Some mosses could no longer be identified in the present-day landscape, and others, like Chat Moss and the Kearsley complex, were formerly divided or classified into many parts. Counting the eight parts of Chat Moss, for example, as a single moss, and leaving out wetlands no longer identifiable, the total of free-standing mosses available for study becomes 46. For each moss a file was prepared containing printouts of SMR data, copies of relevant maps, and published literature.

Archaeological field survey

The survey area consisted of all the mosses of Greater Manchester, whether likely to survive or not, in order to record their present condition, extent, and potential. They are shown on Figure 4.

Techniques

The chief objectives of the field survey were to map moss boundaries and to identify sites of archaeological interest with wetland potential, particularly those buried by peat, or lying at a moss periphery. Sites in such locations are likely to leave evidence in the pollen record. The exact distance implied by 'periphery' has varied; all mosses that still have peat in open land have been surveyed up to about 0.5 km away from organic soil, and often farther if conditions and potential warranted it. Urban mosses have only been investigated as far as the boundaries suggested by their historical records.

First, wetlands had to be located by searching the areas indicated from historical sources, to record their type, survival, extent, and land-use. Three wetland types occur: basin mires, exhibiting small surface areas but containing relatively deep peat deposits, that often began developing in late-glacial or immediately post-glacial times; raised mires, which grew over time to dominate large shallow basins in glacial boulder clays, sands, and gravels; and blanket mires, which are topographically unconstrained peatlands occupying the Pennine uplands. As explained in the original NWWS research design, the survey does not consider the latter systems in the present study, in order to allow research to be concentrated on the large areas of more directly threatened peatlands still occupying the lowland zone.

Sites were identified by archaeological fieldwork prior to ecological confirmation. Where any significant mossland was found, plans and information were passed to the palaeoecological team for further study.

Site visits were made to each moss, recording data on a copy of the 1:10,000 scale map sheets, marked with SMR data and the approximate extent of all known wetlands, as described. Different techniques were required for the locally variable arable, pasture, and urban areas.

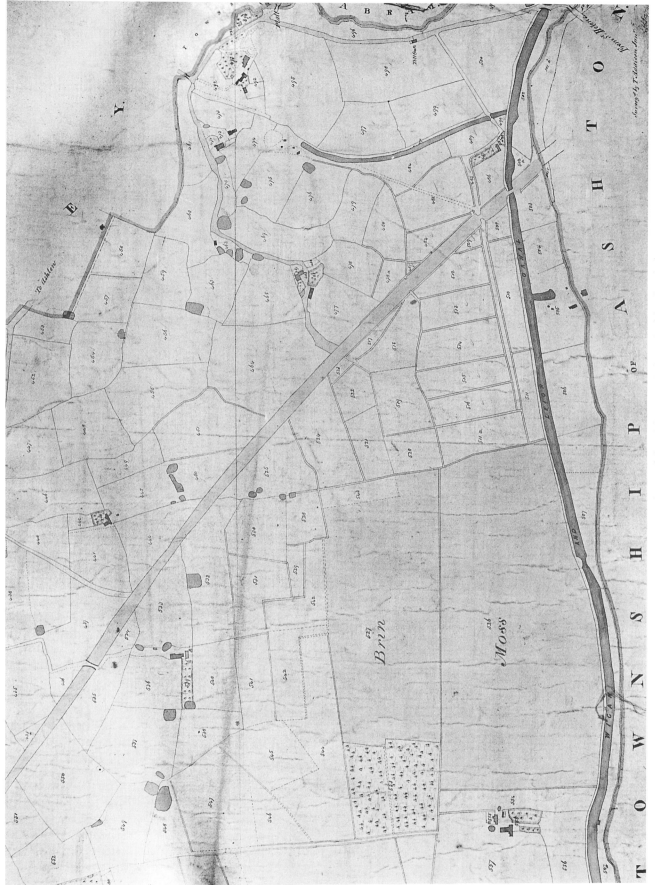

Plate 1 Bryn Moss in 1843 (part of the Tithe Map; Wigan Archive Service)

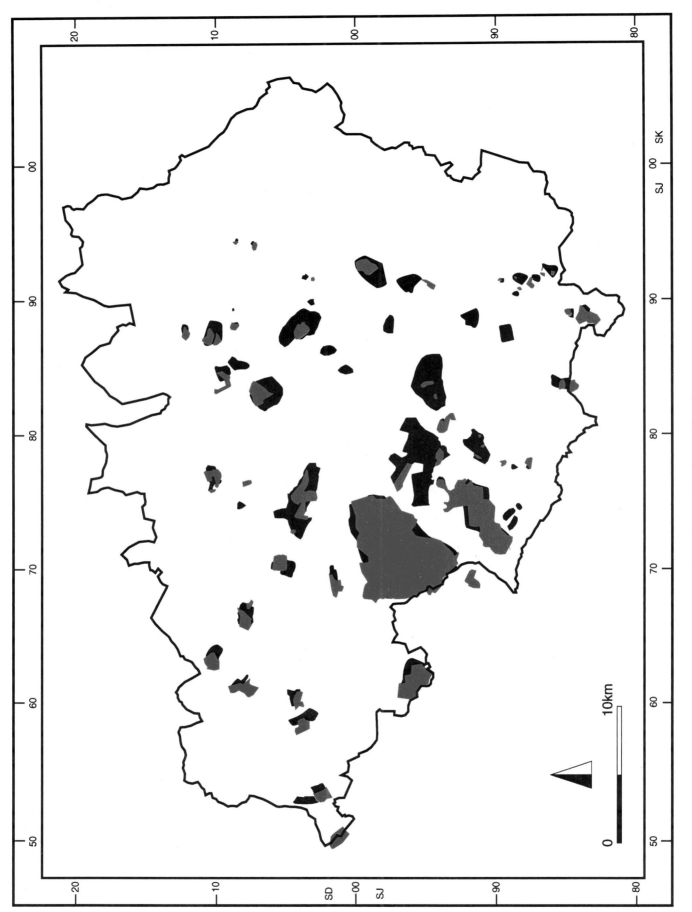

Fig 4 Greater Manchester: historic mosses
(For key see Appendix 1)

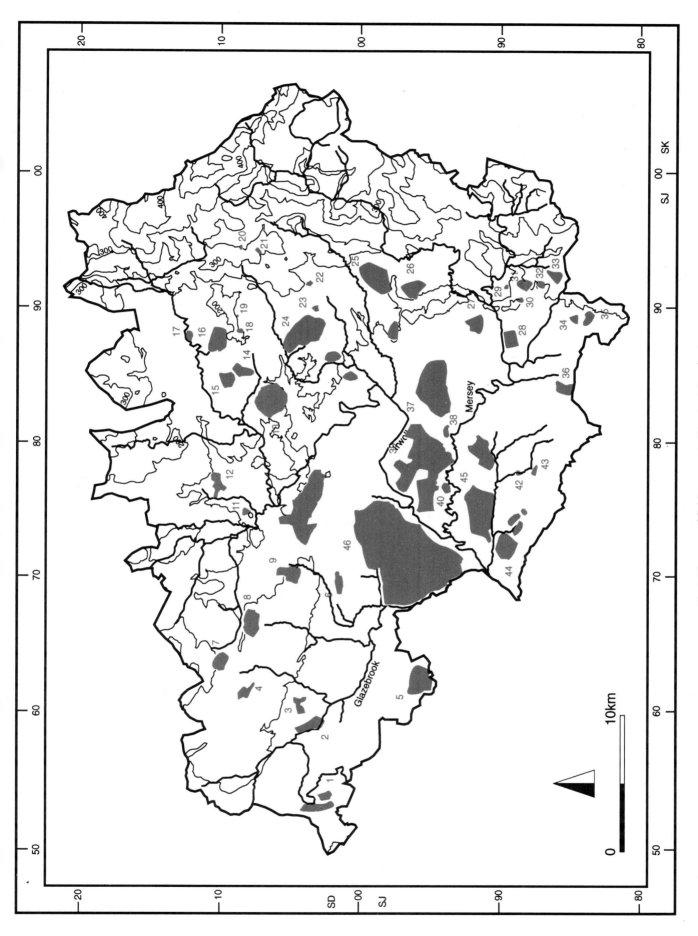

Fig 5 The area of field survey in Greater Manchester

Arable fields were walked, when suitably weathered, in 30 m strips to map soil boundaries and discover artefacts. In any area where artefacts were likely to be found (such as a sandhill located near water), walking lines were closed to 3 m to ensure that nothing was missed. This is essential in a region where artefact levels are extremely low. The crop and condition of every field was recorded to make a complete record of current land-use and to provide data for fieldwork-quality plans.

Within arable mosslands, where there were great extents of moss, each field was checked for high spots protruding through the moss, and for any buried old land surface exposed in dike sides. The few peat-cutting trenches available (at Little Woolden) were checked for exposed organic remains. Rapid survey techniques can be used in areas of deep peat, transects of 200 m being sufficient. Transects of 100 m were used on extensive areas of clay soil.

There are no trenches currently being exposed by commercial extraction. Old ones exist at Little Woolden and abandoned trenches at Red Moss are too overgrown to view. The large-scale extraction at Chat Moss is the only currently active site. No regular checking was made, but during the field visit all drainage dikes were monitored. The grading method of removing peat offers no opportunity to find archaeological material, but did expose high spots of the old land surface, on which one lithic site was found.

Pasture use of mosses prevents discovery of artefacts, and peat extents have to be measured in dike sections and from disturbances caused by molehills. Where there is arable land, care has to be taken in the interpretation of regions of dark soil. All that is dark was not necessarily caused by the former presence of peat (skirtland). Soil can be dark because of the extensive application of nightsoil that was prevalent throughout the region, or it may be where exposures of coal measures occur on the surface (as at Aspull). Lancashire ploughsoil is generally darker than arable land in southern England, because for much of its history it has been pasture, and there is a high content of dark humus, even where there has never been any peat.

Mosses lying in urban areas may be destroyed or masked by buildings. Thus the small moss at Mossgrove, Oldham, now a grassed play and recreation area, has been mined away and covered by a coal slag-heap. Slag, not moss, was visible in flower beds. No information could be added by the site visit to the historic record of the extent or type of moss. On the other hand, Kitts Moss, Stockport, which was completely built over with bungalows in 1958—9,

could easily be mapped from the street topography. The moss lay in a small basin that was easily picked out by a fall in road levels, and the centre of the basin, a small area of nearly a hectare lying below 75 m, was conveniently marked by a contour on the OS map. Garden soils at Kitts were peaty within the basin and had post-medieval debris from nightsoil deposits formerly disposed on the moss. A householder from the moss centre volunteered the information that there was no undisturbed peat left, only clay underneath the topsoil, which information completed the record.

Many urban mosses were seriously damaged and destroyed, such as the Trafford complex, now a large industrial estate, but information could often be obtained from sports fields, parks, cemeteries, or wasteland lying within former mosses. The edge of Barton Moss was plotted from molehills in an unused part of a cemetery; gardens in Wigan cemetery, at the edge of Ince Moss, confirmed that the moss had been destroyed by a coal mine and that the site was further buried by an enormous tell of modern rubbish.

An urban moss with surviving remains is Marland Moss, Rochdale, lying in a small deep basin, easily appreciated topographically, and forming part of an attractive municipal garden, unlike many northern moss sites which are derelict wastelands. Some Marland peat has been destroyed by an ornamental pond in the basin centre, but the surrounding peaty garden beds were said by a gardener to 'shake when a tractor was driven over', indicating that a substantial peat deposit remained, subsequently confirmed by augering (*see Chapter 5: Marland Moss*).

Each day, site locations, peat boundaries, and landscape data were plotted on an archive copy of the 1:10,000 scale map. The total area walked and surveyed is marked on Figure 5.

Figures and presentation of results

The regional figures (Figs 2—5, 61) show topography, locations of mosses, the areas surveyed etc, as explained in their keys. Significant individual mosses are illustrated by maps prepared from 1:10,000 scale plans and published at 1:25,000.

Each illustrated moss has an archaeological plan showing topographical features with the land-use and fieldwalking-condition of every field. The key to the field condition is given on a flap on the inside of the back cover.

Four categories of fieldwork quality have been recorded, according to the following scheme:

1   Field conditions good to ideal; walked in 30 m or narrower transects.

2   Fieldwork sufficiently detailed to record soil and peat. The coverage was considered adequate; this category was normally used for archaeologically sterile areas with deep peat (walked in 200 m transects), and was also used on extensive areas of till clays which have been walked in 100 m transects.

3   Potentially informative area walked in 30 m transects but with conditions of weathering or crop unsatisfactory. Mainly used for thick crop coverage of corn or rape.

4   Not visited.

The plan also shows the extents of surviving peat and the peaty soils where peat once extended (skirtland). Archaeological sites and monuments, and locations of boreholes, if any were made, are indicated. Some mosses are further illustrated by sections and contour maps produced from the palaeoecological studies.

Aerial photography

The aims and methods of aerial photographic survey in the North West Wetlands Survey were previously established and used in Lancashire (Cox 1991, Middleton *et al* 1995) and Merseyside (Cox 1992).

Specialist oblique-angled photographs were interpreted alongside the blanket coverage provided by the vertical surveys commissioned by county councils for administrative, archival and planning purposes. Vertical photographs are not taken for archaeological purposes, and their fixed viewpoint and often unsuitable timing (*eg* when crops are unripe, or earthworks are masked by vegetation) are problematic for archaeological interpretation. However, verticals are often the only data source in Greater Manchester, and require use of a stereoscope to provide a three-dimensional and enlarged view.

After interpretation, archaeological and environmental data are transferred to translucent overlays for 1:10,000 OS maps, either as defined areas of archaeological potential, or as accurate plans of the buried features, depending upon the nature of the recorded features and the validity of the photographic interpretation. An accurate cartographic representation of archaeological and environmental features is achieved by use of aerial photographic rectification software (Haigh 1989), which provides compatibility with GIS systems and allows integration with the main NWWS archive. Sketched additions to the maps can also be added to the GIS databank via digitisation. The graphical record is supported by a database which can be interrogated via location, site name or site type.

The details of the results with comment on the ground visits are described in Appendix 3.

Palaeoecological survey

Full details of the general rationale behind the palaeoecological survey and research strategies employed in the North West Wetlands Survey are described in the North West Wetlands Survey research design (Middleton and Wells 1990). In this section, the methods specific to survey in Greater Manchester are described and explained.

Biostratigraphical survey

Large-scale biostratigraphical studies across major hydromorphological units were undertaken with the following aims:

1   Elucidation of mire ontogeny;

2   Determination of the degree of preservation of the peat;

3   The provision of a contextual framework in which to place detailed (particularly palynological) studies.

A 30 mm bore Eijkelkamp gouge auger was used to obtain cores of peat for rapid field description. A Zeiss optical level was used for determination of relative surface levels of core positions. The exact location, number, and core intervals were varied according to the different conditions and depositional histories encountered in the various types of mire bodies occurring in the county. The more homogeneous large-scale peat deposits contained in many of the extensive raised mire deposits required relatively wide-spaced cores (*ie* >100 m), whereas the more complex seral developments recorded in small basin mires necessitated close sampling (<30 m) in order to achieve a coherent picture.

Field descriptions of stratigraphic elements were complemented by sampling at short depth-intervals from the cores collected and these were examined in the laboratory using a Leitz M3 Stereozoom microscope in order to check and add detail to field determinations of material and to establish the presence and frequency of any carbonised plant material.

A final complete stratigraphic description from each core was synthesised from amendments and additions derived from laboratory examination of peat samples and transcribed to a computer database. This information was then used to draw a series of stratigraphic sections through each major mire system surveyed. The sum total of all the data was then used to interpret mire ontogeny.

Humification values of *in situ* peats is measured crudely in the field using the same modified Von Post determination method as employed by the Soil Survey (Burton and Hodgson 1987). Table 1 reproduces the criteria on which the humification estimates are based.

Assessment of small mires

Because of the large number of small mosslands surviving in the county, minor sites (*ie* the small mosses, pp 101—14, and also Kearsley Moss, pp 82—86) received a palaeoecological assessment to determine research potential, rather than extensive survey. The aim was to establish condition of the surviving peats and to estimate their value as archives. This involved a less detailed and more selective exercise which could be undertaken rapidly and the assessments are therefore based on a more subjective appraisal of the field evidence.

Palynology

Pollen preparation followed standard procedures (HCl, NaOH followed by sieving, HF if necessary and Erdtman's acetolysis; Faegri and Iversen 1989). Any sand encountered in samples was resistant to hot or cold HF acid treatment and so was separated from the organic material by differential sedimentation.

A known volume of peat (using volumetric displacement; Bonney 1972) with the addition of *Lycopodium* tablets (Stockmarr 1972) was used to calculate pollen concentration values. Samples were mounted in silicone oil and examined microscopically with an Olympus BH2 at x 400 magnification, x 1000 for critical grains. Usually at least 500 pollen grains excluding mire types, aquatics, indeterminates and Pteropsida were counted except at some levels where pollen was less abundant. Identifications were carried out using the keys of Faegri and Iversen (1989) and Moore *et al* (1991) plus a very small reference collection, which restricted more specific identifications. Charcoal particles > 5 µm together with a known number of *Lycopodium* spores were counted

on the pollen slides (Peglar 1993). Mineral content was calculated by loss on ignition of the samples. These were dried at 105°C for at least 48 hours then heated at 550°C overnight.

Plant macrofossil analysis

Methods used for sub-sampling and description followed Barber (1981). The results display relative abundances of plant remains from constant volumes of peat (20 cm³) cut from a core. The material was systematically scanned using a low power (x8—40) Leitz M3 Stereozoom microscope and assessed on a semi-quantitative five-point scale:

1 rare	vegetative material occurring only once or one seed
2 occasional	vegetative material occurring only a few times or 2—5 seeds
3 frequent	vegetative material occurring regularly or 5—20 seeds
4 very frequent	vegetative material occurring in every portion of the sample examined or 20—40 seeds
5 abundant	vegetative material occurring in field of view all the time and dominating the sample, or 40+ seeds

Analysis and storage of palynological and macrofossil data was accomplished using computer-based tools to categorise data and aid its interpretation. Data were analysed to ascertain possible human influences on the natural environment and to relate this to known and newly-discovered archaeological remains.

All radiocarbon dating was carried out by the Scottish Universities' Research and Reactor Centre at East Kilbride. Full details of methods and procedures can be obtained from Dr G T Cook at the Centre.

Presentation of palaeoecological results

Peat stratigraphy
The results are presented as a series of stratigraphic diagrams. Each diagram displays stratigraphic information from an individual transect. Composition of the gross components of the peat and organic deposits are indicated using symbols based on Troels-Smith (1955) with some modifications (*see keys for details*).

The deposit elements recorded are based on a combination of field observation and rapid Stereozoom scans of sieved material from samples of *c* 25 cm³ of peat from selected segments of cores. The presence of charcoal was usually detectable only by the microscope examination except when present in exception-

Table 1 Von Post humification scale

Degree of composition	Nature of liquid expressed on squeezing	Proportion of peat extruded between fingers	Nature of plant residues	Description
H1	Clear, colourless	None	Plant structure unaltered, fibrous, elastic	Undecomposed
H2	Almost clear, yellow brown	None	Plant structure distinct almost unaltered	Almost undecomposed
H3	Slightly turbid, brown	None	Plant structure distinct most remains easily identifiable	Very weakly decomposed
H4	Strongly turbid, brown	None	Plant structure distinct, most remains identifiable	Weakly decomposed
H5	Strongly turbid, contains a little peat in suspension	Very little	Plant structure clear but becoming indistinct; most remains difficult to identify	Moderately decomposed
H6	Muddy, much peat in suspension	One-third	Plant structure indistinct but clearer in the squeezed residue than in undisturbed peat, most remains unidentifiable	Well decomposed
H7	Strongly muddy	One-half	Plant structure indistinct but recognizable; few remains identifiable	Strongly decomposed
H8	Thick mud, little free water	Two-thirds	Plant structure very indistinct; only resistant remains such as wood and root fibres identifiable	Very strongly decomposed
H9	No free water	Nearly all	Plant structure almost unrecognizable; practically no identifiable remains	Almost completely decomposed
H10	No free water	All	Plant structure unrecognizable completely amorphous	Completely decomposed

ally large quantity. In the charcoal diagrams the density of shading represents the relative abundance of carbonised particles based on a semi-quantitative scale identical to that used for recording macrofossil remains described earlier.

The core position in the diagrams has been drawn to various scales chosen for the best enhancement of presentation. The physical size of the cores displayed in the diagrams is exaggerated in order to enable clear presentation of the results. Heights relative to OD of the surface of the core positions are displayed where known, otherwise relative heights of cores are shown.

Palynology
The palynological results are presented graphically as pollen diagrams (Faegri and Iversen 1989; Moore *et al* 1991). The pollen taxon values shown on the diagrams are relative. These are calculated as a percentage of a pollen sum, which is defined

for each site; pollen concentration values are available from the authors at LUAU. The computer software used for the manipulation and presentation of the data is TILIA and TILIA*GRAPH (Grimm 1991).

Charcoal is expressed as a percentage of pollen sum plus the number of charcoal fragments. The vertical depth scale on the diagrams shows centimetres from the surface of the deposits (not OD). The pollen diagrams have been divided visually into phases or zones to aid interpretation.

Macrofossil analysis
The results are presented as a series of histograms indicating frequencies of vegetative remains and seeds from each level sampled. In the diagrams and descriptions, the word 'seed' is used to describe both fruits and seeds in order to simplify the presentation. The series of histograms display the frequency value for seeds, leaf fragments etc, found in 20 cm^3 from

each sample examined from stratigraphical elements down the core.

The appearance of a taxon name on the diagrams without additional qualifying information indicates the representation of totals of vegetative/reproductive remains found referable to that particular taxon. Watts and Winter (1966) are followed in the relative certainty of determination afforded to the macrofossils encountered.

Individual histograms are arranged such that taxa having similar stratigraphical occurrences are grouped together, as far as is possible.

Calibration of radiocarbon determinations

Where possible, radiocarbon dates have been calibrated to calendar years using the conventions established at the 1986 Trondheim Radiocarbon Conference (Switsur 1986; Pearson 1987). This uses a calibration curve based on dendrochronology and all dates thus expressed in this volume have been calibrated using the computer program CALIB (University of Washington Quaternary Isotope Lab 1987) and employing a curve based on a 20 year atmospheric record dating to 8100 radiocarbon years BP (Stuiver and Pearson 1986). Dates older than this have been calibrated using the as yet uncorroborated curve of Stuiver and Reimer (1993). Calibrated dates are presented at a 2 sigma (96.4% confidence) level. Such dates are denoted by the suffix cal BC or prefix cal AD.

The archive

The archaeological archive consists of the following:

1   Unedited notes made each day. The substance of them is published in Chapters 3—5 and in the Gazetteer (*Appendix 1*); they contain additional diary-type notes on the farms, farmers, crops, and weather;

2   The finds, bagged by site number;

3   Maps. Each day landscape data were plotted on dye-line copies of 1:10,000 maps. Land-use is fully indicated by a series of (mostly) self-evident letter codes (a full explanatory list is lodged with the archive). Geology and soils are mapped in green biro, and archaeological sites in red biro. Other archaeological information (such as ridge and furrow) is sketched in pencil;

4   The historical and SMR data consisting of a file for each moss.

The palaeoecological archive consists of the following:

1   Original pollen and macrofossil count sheets stored as raw data;

2   Electronic archiving of the same in both numerical and graphic form using the computer software TILIA/TILIA*GRAPH. The same information will be archived within GIS;

3   Pollen slides are archived at LUAU along with residues from each preparation;

4   Selected peat samples have been retained in cold storage. These will be stored for as long as is practicable at the Institute of Environmental and Biological Sciences, University of Lancaster.

The final text and digitised diagrams are stored in the GIS at LUAU and on disk.

3

CHAT MOSS

Chat Moss, the largest of the mosses in Manchester, lies north of the Mersey and west of Central Manchester, near the confluence of the Mersey and Irwell rivers at 17—27 m OD (Figs 4, 6; Plate 2). It is contained by the A580 in the north, the Glaze Brook to the west, the A57 in the south and the M62 near Worsley to the east.

Chat Moss extended to approximately 6000 acres of bog at the end of the nineteenth century (Baines 1836, 3, 132). The area of peat more than 0.5 m deep mapped in 1992—3 was 2450 hectares (6050 acres), being approximately 9 km long and 4.5 km wide at its extremities. The moss occupies alluvial sands and gravels and clayey till, which overlie Triassic Bunter Sandstone, Upper Mottled Sandstone and Manchester Marls.

Chat Moss was divided into parts relating to the surrounding townships, which, taken anticlockwise from the north, are Astley, Bedford, Little Woolden, Great Woolden, Cadishead, Irlam, Barton, and Worsley (Fig 6). In the past small sections of the Moss have been referred to under different names such as Pool Moss by Barton and Flow Moss by Astley. The major part of the moss now lies within Barton-upon-Irwell township of Eccles parish in the Hundred of Salford.

The townships were divided into (or made up of) small estates, that seem to have been established at a fairly early date, possibly taking advantage of enclosure of waste beyond the village nucleus. Most of the centres still survive as place names or modern farms, which help to identify the medieval landscape. Within Astley, at the north, were Morley's Hall, Peel Hall, and a grange belonging to Cockersand Abbey. To the west, in Bedford township, were Hopcarr Hall, Shuttleworth Hall and Bedford Hall. Along the western edge of Chat Moss, on the bank of the Glaze Brook, were Light Oaks Hall, and Great and Little Woolden Halls. South of the moss, on the narrow belt along the north bank of the River Irwell, Barton Hall lay close to the hamlet of Barton at a river crossing between Urmston and Davyhulme. Within Worsley township, Booths Hall and Worsley Hall were situated on the north-eastern parts of the moss.

Previous archaeological finds

Very little archaeological information was known from Chat Moss before the discovery of the Iron Age site at Great Woolden Hall on the south-western edge of the Moss (*Appendix 4* GM 1). A summary of the excavations at Great Woolden is given in the period discussion (*see Chapter 6*) taken from an interim report (Nevell 1989b). A perforated gritstone pebble-hammer (c 127 mm long, 110 mm broad, and 38 mm thick) was found when excavating foundations for the Partington Steel and Iron Company at Irlam (Boyd-Dawkins 1911, 101). It was called a net sinker at the time and presumed to be Roman. The fine working makes it unlikely to be a fish-net weight, and it probably dates from the Bronze Age (Plate 3). A bronze socketed spearhead, 150 mm long, was found 6—9 m deep in excavations for the Manchester Ship Canal near Irlam (Plate 4) (Harrison 1892, 250; McGrail 1978, 220—2). The Partington Iron Company excavations also revealed 'a rude boat', near Irlam, presumably a prehistoric dug-out (Bailey 1889a).

At Worsley, a human skull found in the moss in 1958, now held at Manchester Museum, was radiocarbon dated to cal AD 66—400 (1800±70 BP; Plate 5; Garland 1987). Close to the East Lancashire Road at Boothstown, in 1947, a hoard of more than 550 late third century Roman coins in two pottery jars was discovered buried beneath a flagstone. The findspot was close to the Wigan—Manchester Roman road, and c 0.5 km north of the historic boundary of Worsley Moss (Anon 1947, 224). Another Roman coin hoard was discovered c 0.5 km from the 1947 site, also at Boothstown.

A medieval cross, now at Eccles church, was rescued from Barton Hall when it was demolished in 1897 (SMR 1904, SJ 7540 9794). The remains of the pale of Barton Park, shown on Saxton's map of Lancashire

(1577), are alleged to have been found during the early nineteenth century. Steele describes 'ancient oak palings' found in a regular series, and marked by a mound enclosing about 400 acres (Steele 1826, 304).

These finds are few, there being only a skull from the moss itself, although this is interesting in view of other similar finds in the region that may be evidence of a severed head cult (*cf* Middleton *et al* 1995). In the south, the Irlam artefacts could be considered as deliberate deposits in the river rather than being directly concerned with the moss.

The name Chat Moss is first recorded in 1277 as Catemoss, apparently referring solely to the southern section, where an estate included 'salteye', half of 'Boysnope', and lands between the Irwell and Chat Moss, called 'copped greave, deep lache, derboche and the hay' (Farrer and Brownbill 1914, 4, 365). The name may derive from a personal name *Ceatta* (Ekwall 1922, 39), or from *ceat*, meaning a wet piece of ground (Mills 1976, 71). Local tradition, without satisfactory

evidence, connects Chat Moss with St Chad, the Bishop of Mercia in AD 669 (Baines 1836, 3, 135).

The earliest detailed description of the moss occurs in a 1322 survey of Manchester manor:

> Chat Moss is the soil of the Lords of Barton, Worsley, Astley, Workedley and Bedford. The wood being undivided is not measured because there is so small a goodness contained in so large an extent — in which all the tenants of the Lords have common of turbary; but the yearly profit of the said moor is not extended by reason that the little goodness of the said premised commons gradually grows less (Harland 1861, 392).

An unidentified Halmoss, probably a demesne Hall Moss in the vicinity of Barton Hall, is listed as:

> 12 acres of turbary in which the tenants of the lord of Barton have common of turbary; from which no advantage can accrue to the lord

20

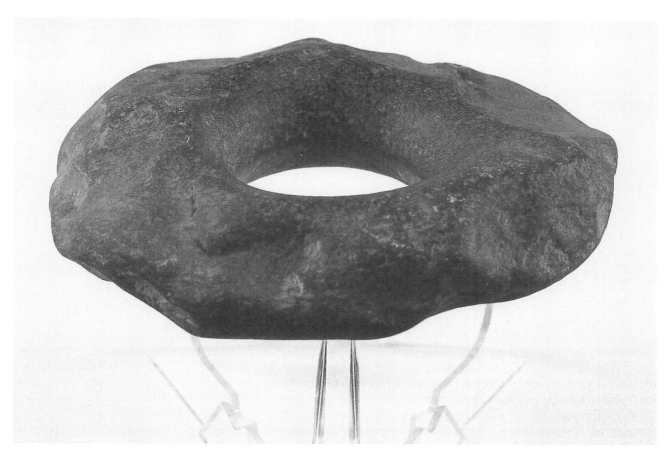

*Plate 3 Perforated hammer-stone from the Manchester Ship Canal at Irlam; c 120 mm diameter and 38 mm thick
(reproduced by permission of Warrington Museum)*

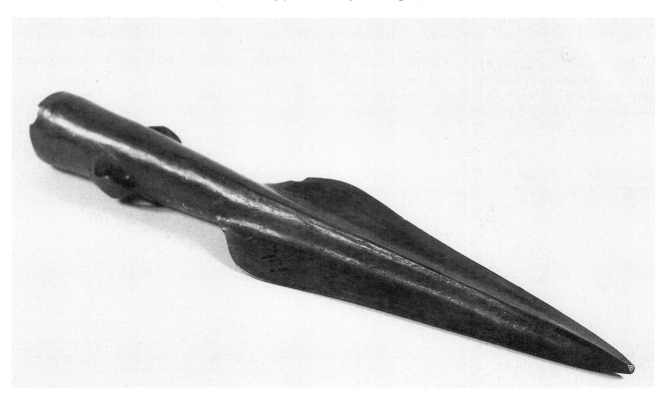

*Plate 4 Bronze spearhead from the Manchester Ship Canal at Irlam; length 150 mm
(reproduced by permission of Warrington Museum)*

beyond common, except in this, that the arable lands, by reason of these are rented higher.

Barton Hall had a garden with fruit and herbage in 1322, 11 acres of demesne worth 26s 8d, four fens partially enclosed for building upon, arable land let at 12s, a fishery worth 14d, two parts of a mill worth 30s, lands planted for 34s 8d, herbage and pannage worth 9s, and assize rents of free tenants worth £17 11s 5d. There was a covert of oaks in Boysnape (Harland 1861, 391).

The earliest references to other places surrounding the moss are as follows. On its western edge by the Glaze Brook, Great and Little Woolden are first recorded as Vuluedene ('wolf valley') in 1299 (Ekwall 1922, 39). On the south, at the confluence of the Glaze Brook with the River Irwell, lies Cadishead, a manorial site called *Cadewallisete* in 1212, probably meaning 'Cada's stream pasture'(Ekwall 1922, 39). The neighbouring hamlet of Irlam, lying between the Moss and the River Irwell, is referred to in 1190, as *Urwilham*, 'hamlet of the Irwell' (Ekwall 1922, 39). Between Irlam and Barton was another small hamlet, Boysnope, in 1277 called *Boylsnape*, 'bull pasture'

Plate 5 Roman skull fromWorsley Moss (Manchester Museum)

(Ekwall 1922, 39), the memory of which is retained only as a place-name. Barton (1196) is a common place-name, generally meaning a settlement connected with corn production, or a detached portion of a manor (Ekwall 1922, 38).

Parts of the moss and the surrounding townships have incidental references, some of them earlier than the general description of the whole moss. Between 1184 and 1190 land at Irlam was described as:

> ground between the crooked oak and the stub which is in the leach at the head of wolfpit-croft, and from that stub through the moss unto the hedge gate, thence following the hedge beyond the carr unto the arable land, along the carr so far as the said bounds, to wit, between Elmtree-pool and Elbrook, and the deep carr unto the bank of Irwell: with common right and easements of the said town (Farrer 1900, 719).

Irlam lands lying in the 'reudis' between the high road and marsh are mentioned in 1317, and in 1360 William del Ferry of Irlam is recorded, probably owning a ferry on the Mersey referred to in a grant of 1425 (Farrer and Brownbill 1914, 4, 371 n86).

The moated site of Morleys in Astley (SMR 4064) was described in *c* 1200 (Farrer 1903, 43) as:

> Dykefurlong and Moorlegh, and the moiety of Birches, and the ridding which is by the brook, and half the wood between the brook and blackleach, and the whole moiety of the Spen, between garthmoss and Blakemoor, and an acre of demesne in the croft which was Alexander's; with common right, easements and liberties and pannage of the pigs (Farrer 1900, 710).

A parcel of land called Mokenis, identified as the small manor of Booths in Worsley, had bounds, in 1299:

> beginning at Acornsyke, where it was met by the fall of Konksyker, between Worsley and Astley, along the fall of Blackbrook, thence by the bounds of Astley and Irlam, across the moss to Ringand Pits, and thence going down to the meadowyard (Farrer and Brownbill 1914, 4, 378).

Barton property, in 1343, was common of pasture on Pool Moss (unidentified), lying between Pool Brook and Sandyford under Harley Cliff in Boysnope, and between the fences of Poolfields and the bounds of

Worsley upon Chat Moss (Farrer and Brownbill 1914, 4, 366).

Sir William Leyland, owner of Morleys, received Leland the antiquary, who described the Hall in *c* 1535:

> Morle in Darbyshire [sic], Mr Lelandes place, is buildid — saving the fundation of stone squarid that riseth within a great moote a vi foote above the water — al of tymbre, after the commune sort of building of houses of all Gentilmen for most of Lancastreshire. Ther is as much Pleasur of Orchardes of great varite of Frute and fair made walkes and gardines as ther is in any Place of Lancastreshire. He brennith al turfes and petes for the commodite of Mosses and Mores at hand (Toulemin-Smith 1909, 6—7).

Medieval descriptions indicate the local dominance of Chat Moss, especially in the precarious threatened narrow strips of land by the Glaze Brook and Irwell, often likely to be overwelmed. A broadsheet described the eruption of Chat Moss that occurred in 1526:

> [the] sayd mosse doyth cou lx acres and more off sowying lond wt medowes on bothe ye syds off ye sayd ryv, wheche sowying land & medowes pt of they are vj scor rods & more ffrom the place were ye mosse brak owtt, and besydes yt fyllyd vpe dyvers valles & slacks wt grett marle patts, & ou & besydes yt ffyllyd uppe all ye holle Ryv to a place called Kelcheth melne & yt dame of the sayd mylne, ye wheche Ryv & dame is a hudreth Rods & more in lenght, & the mosse ley so hyght yt coved all the wallez to ye heyght off the hyghest place off the growne yt by wheche byextymacon is lx ffots off heght in some places, and aldo, by reason off the sayd mosse wheche copas [compass] was lxiiij howsez, were off xxvi bene. In grett jopde & stand in the watr some to ye eysing [eaves] and abowe yt, in so meche ye in habitors had myche payne & sorowe to save theyr & theyr cheldren undreowed, & yet, thanket be to god, no crystyn body was pisshed, butt yt they lost ther godes in ther howses, wt ther corne & hey in theyr bernes, & some syne & doggs drowyd. thys mosse & watr brake owtt ye same day in ye mornyg, abowt viijth of ye clock, in ye syght of Thoms Kelcheth, gent, Mathew lythegow, Robt Wattmoght, & Ryc fflytcroft, wheche all iiij bene takyn for honest men, & they say yt by yeyre trwyth, all they thoght yt had bene, domysday, by cause ye watr & mosse wt growyng trees yron dyd mete wt so grett vyloence & came tomlyng to ward yey, & they dyd cry apon theyre neghbors &

> demyd vupp wtin the sayd Ryv unto ye Settday in the mornying yen nect ffolowing, then pte off the sayd mosse, by Reason off the seyd watr & mudd so demyd uppe, brake owtt att on tyme & went so hyght yt went ov ye toppe of Kylcheth mylne & so ov medowes & ffyldes, & muddyd & slegged ye a yard depe & more in some places, unto Holcroft & to Glazebrook bredgh [bridge], and so to the watr off marcy [Mersey] where y is a ffeyre noyt [ferry boat] att y Holyn ffayre, wheche cold not goo for thycknes off mudd by the space off ctayne dayes, the wheche mudd colowred y watr lyck to yncke downe unto Warington & so to Livupole, qwych is owne arme off th esee, & is in lenght from ye sayd mosse ffollowyng y watr xxiiij mylez or more, & is so corruppyd that nother beasts nore Cattellez wyche dryncke y off (Crofton 1902, 142—4).

Court rolls from the sixteenth century and later list the names of moss reeves who ensured maintenance of drainage streams and ditches around the moss, and regulated turf cutting. The court at Astley, in 1564, recorded illegal encroachments on Chat Moss, and some references indicate apportionment of moss acres: Thomas Leyland, of Morleys, had *c* 25 acres of moss, rented out at 4d per rood. In 1572 every illegal load of turf taken by the tenants was valued at 2d (Lunn 1968, 114). In 1624 an agreement was made between George Leigh of Barton Manor and Cecil Trafford that certain tenants should be allowed leasehold on Barton Moor in proportion to their holdings in the township.

Much of Chat Moss, in the seventeenth century, became the property of the Egerton family, and descended as part of the Bridgwater estate, and its modern successor the Peel Estate. In *c* 1630 the Bedford estate consisted of 18 messuages, 10 cottages, one watermill, 640 acres of land and meadow and pasture, and 2560 acres of moss and turbary (Farrer and Brownbill 1914, 3, 432). Woolden and Cadishead became the property of the Bridgwater Trustees, as did Little Woolden Hall after 1868 (Baines 1867, 1, 597).

Recent land-use of Chat Moss

Drainage of the moss

Before 1762 there is little evidence for any concerted attempts at drainage and cultivation of the moss. Records suggest that, unlike other mosses around Manchester, Chat Moss was used solely for pasturage and turbary. Once peat had been exhausted, the turbary rights held by freeholders and tenants of Barton, Astley, and Worsley lapsed and the land was reclaimed by the lord of the manor and turned into cultivated plots (Lumb 1958, 32). Turbary rights would also lapse due to lack of interest, when a supply of cheap pit coal for fuel, much superior to peat, was available.

The first part of the moss to receive large scale drainage was near Worsley. The Duke of Bridgwater had underground and open canals at his Worsley coal mines in 1758, used to transport coal into Manchester. An extension of the canal was built on to Chat Moss *c* 1762, and used to carry waste from the mines on to the mossland. Young, in 1770, refers to the canal as a mossland drain (Mullineaux 1959, 21). By 1773 Worsley land was available on lease for cultivation. Trees were planted at the end of the canal, and the area was nicknamed Botany Bay Wood, after the Australian colony (Mullineaux 1959, 21; Hadfield and Biddle 1970, 22). The main canal, later called the Bridgewater Canal, was extended west into Cheshire in 1795, via Leigh and Runcorn, along the edge of Worsley Moss. It is probable that an intention to carry it across Chat Moss was abandoned because of the difficulties met when building the moss canal (Mullineaux 1959, 22).

The southern side of the moss, owned by the de Traffords and leased out to one Wakefield of Liverpool, had a small area of moss reclaimed in *c* 1788. In 1793 an Act of Parliament enabled Wakefield and his partner William Roscoe of Liverpool to lease 2500 acres of Chat, which was commuted solely to Roscoe in 1805 (Salford Archives, U84/T1). He began intensive drainage works on Barton and Irlam mosses. A main road was laid out, running east—west, 12 yards (11 m) broad with 6 feet (1.8 m) wide ditches on either side, fed by side open drains 50 yards (45 m) apart. By 1808 a farmhouse was erected and the surface moss levelled. Two innovations were introduced to assist cultivation, the horse patten (a broad overshoe for the hoof) and 10-inch wide wheels for the plough, to prevent sinking in the soft ground (Wheaton nd *c* 1987).

In *c* 1830 a movable rail track was laid next to the newly constructed main railway line joining up with the Mersey—Irwell navigation to the south, for the distribution of nightsoil and other manure on to the moss, and ultimately, for the return carriage of crops. In 1833 a report made to the Select Committee on Agriculture shows that an estate of 700 acres had 300 acres under crop. The Earl of Ellesmere (Worsley estate) and Colonel Ross of Astley Estate also began improvement of the moss. By 1849, the date of the First Edition of the Ordnance Survey maps, about a third of Chat Moss had been brought into cultivation. Subsequent maps show an increase in arable cultivation south of the railway line, which continues (Fig 6). Urban expansion had little effect on the moss, the great depth of peat being unsuitable for building.

Recent peat extraction

By 1905 peat cutting was quite extensive on Chat Moss, producing peat litter for animal bedding. Part of the Astley Estate was used in 1889 by the Astley Moss Litter Co Ltd, later taken over by the Griendstveen Moss Litter Company, which was still trading in 1908 (Plate 6). Payment was made on royalties of cut peat, expected to average 1333 tons annually, or a minimum of £100 *per annum* (Wigan Archives DDX/CAL/Box 13). Another company, Peatco Products Ltd of Red Moss, Horwich, was situated on the 12-yard road towards Cadishead and Woolden Hall.

At present, there are three methods of peat extraction taking place on the moss. North of the rail line, large-scale mechanical milling at Nook Farm removes all peat except the bottom metre, which is left for regeneration. On the southern side of the rail track (land formerly on lease to the Griendstveen Moss Litter Co), peat is extracted by machine in blocks and left to dry before collection by a powered locomotive and taken to an on-site milling factory. At Little Woolden Moss there is a continuation of the traditional method of hand peat cutting; dry blocks are gathered on a hand-pushed trolley guided by rail tracks, and milled and bagged by hand.

Communications

The Manchester to Liverpool railway line (London and North Western Railway), begun in 1828 and officially opened in 1830, had similar engineering problems to the Bridgewater Canal in 1768. The engineer George Stephenson used a technique of strapping barrel caskets together to act as a through drain. To counteract the sinking weight of the rails, the load was distributed by rafting sleepers on hurdles spread out on the moss surface. The Manchester end of the track was stabilised by an embankment of cut and dried turf bricks (Bailey 1889b 122—4; Smiles 1857).

The rail track might have been expected to improve transport of manure and goods, but at first there were too few stations to make it possible for agriculturists on the moss to take advantage of rail transport. Later, the Manchester Corporation installed two loading stations on the main rail line, at Barton and Irlam, and light rail tracks were built linking with the Mersey Navigation and the main rail line. The light rail network was mainly on the south side of the main line, Astley Moss was linked to the main railway line at

Plate 6 Hand peat-cutting at Chat Moss, 19th July 1916 (Manchester Central Library, Local Studies Library)

Astley station, and to the Irwell at Lower Irlam. The network of lines is illustrated by the Ordnance Survey Revised Edition in 1894.

Barton airport, situated on the edge of the moss, was first proposed by Sir Sefton Brancker, a Civil Aviation Director. Servicing the city, it was officially opened in 1930, after a two-year building programme (Kidd 1993, 195—6). The aerodrome is relatively small and is located on skirtland. There were no large-scale engineering works that affected the moss.

The M62 motorway cut across the southern part of Chat Moss in 1978, near to Irlam. The deep cutting has affected the drainage of neighbouring peat fields, according to reports from farmers (Plate 8).

Landfill

In 1893, after the success of the refuse disposal scheme on Carrington Moss, the Manchester Corporation bought the de Trafford estate, paying £139,000 for 2600 acres of land. In 1897 the Clate 8)orporation began to use the moss for nightsoil distribution (Phillips 1980, 230). After the end of the First World War, because of food shortages, there was demand to develop more of the raw bog and by 1930 waste dumping ceased.

The waste does not seem to have affected the moss very much. Modern debris lies profusely on the land surface, but stratigraphical survey shows little evidence of disturbance below the top 0.5 m.

Field survey and mapping in 1992 and 1993

The main mass of Chat Moss is a nearly flat plateau of peat lying a few metres above the surrounding land surface. The raised peat is most dramatically visible on the west, north of Little Woolden Hall (near site GM 2, Fig 6). Here, a modern ditch cuts slightly into the moss exposing a wall of peat some 2 m high. Even where not cut back, the peat rises about 2.5 m in 5 m horizontal. It is easy to appreciate how, in wet weather when 'untamed', the moss would swell, 'bursting' and falling into the deep, narrow channel of the Glaze Brook, as occurred in 1526. Peat is best preserved, as expected, where there is woodland or scrub, and no modern agriculture.

The uniform appearance of the central area belies its complexity, for near the centre-west, peat-cutting ditches, 2.5—3 m deep, strike sandy soil with intact tree roots in the old ground surface (SJ 698 955). An-

other buried ground surface undulation has been exposed by the Nook Farm peat quarry works (GM 5, *see below*), and to the east three 'islands' (SMR 2909) rise above the peat surface. One at Brighton Grange (SJ 733 676) consists of clay soils, mostly covered by grass. A cropmark to the south (SJ 731 974) is caused by another buried island nearing the surface. Within 300 m north-west and south of the island peat depths are 1.5 m and 1.2 m respectively. Two other islands nearby (SJ 736 982, SJ 741 983) could not be examined because of coverage by scrub and trees.

In the skirtland (areas with humose soils that were once covered with peat) the undulations of the old ground surface are plain to see. The borehole sections confirm and further illustrate the variable surface beneath the moss (*see below*). The amount of skirtland varies; it measures the shrinkage of the moss since the early Middle Ages, mostly during the last two centuries. It is greatest at the north-west, where peat has fallen back up to a kilometre from the Black Brook and the Glaze Brook which were the boundaries.

The land-use of the area and the extent of deep peat are marked on Figure 6. The skirtland and hinterland are described below, beginning at the north and working anti-clockwise, after the description of archaeological sites.

The earliest site found at Chat Moss, dated to the Mesolithic period, lies in the area of peat, on a small sandy knoll of the buried old ground surface. It was exposed by the large-scale peat quarry excavations at Nook Farm (*c* 80 ha). A series of parallel ditches about 15 m apart, made to drain the peat-working area, cut through the base of the quarry which has *c* 1 m peat left on the old ground surface to allow regeneration of wetland when quarrying ceases. The dikes penetrate the old surface which is boulder clay on the lower ground to the north, but in the west-centre the base rises to form a sandy knoll on which occur worked flints, chert, and burnt stone. The site (GM 5, Fig 6) lies at SJ 7907 9797, and the area over which flints were discovered is 150 x 100 m (Plate 7). At the time of discovery in October 1992, several flints were found. Others have been collected since. The ditches were deepened in March 1993, yielding a few more burnt stones; the upcast masked most of the find spots until weathering occurred. Several peat cores were taken for palaeoecological analysis and the results are detailed below. The flints are mainly Mesolithic with some early Neolithic types, dating from about the fourth millennium BC. Further work has taken place on the site with sondage sampling (pers comm J Walker 1995).

This is to date the most significant early site in the area, even though the level of finds is low compared

Plate 7 Chat Moss: Nook Farm quarry and exposed sand knoll

to other parts of the country. Since the site is still largely buried and undisturbed, it has great potential for determining if there were any structures and waterlogged artefacts.

A low sandy patch of skirtland, 240 x 60 m, near Moss Side (SJ 6998) on the northern skirtland of Chat Moss, produced a little burnt stone, one poor-quality worked flint and one other worked piece of probable Mesolithic date (site GM 4, Fig 6). Although the level of finds is extremely low, this is likely to represent some form of activity, possibly a short-lived settlement or camping site, perhaps related to the larger site at Nook Farm. The location is typical of a prehistoric site: on a sandhill, next to water, and near a wetland.

On the east of the moss, near Grange Farm and the M62, are two other sites that produced burnt stones, and a third lies to the south-east near Boysnope. No flint or other artefacts were discovered, but from their location and nature the burnt stones are likely to represent further prehistoric activity in the region (GM 7, GM 9, GM 10).

The Iron Age enclosure at Great Woolden Hall (GM 1) (*described in Chapter 6: The Iron Age*), lay under pasture with nothing visible. A modern ditch opposite to

Little Woolden Hall revealed an 'ancient' ditch emerging from underneath the moss, buried by 0.3 m peat. The ditch (GM 2, Fig 6) is 22.5 m south from the north-west corner of Hollin Wood spinney (SJ 6838 9496). In section, from the top, there is 0.3 m peat, 0.3 m of old soil, and 0.6 m loamy fills within the ditch, which was 2.4 m wide; there were no finds. Probably the ditch is prehistoric, possibly connected with the Iron Age site (GM 1) lying about 1 km distant at Great Woolden Hall.

The new archaeological discoveries at Chat Moss are limited to the prehistoric period, and are sparse, in keeping with the number of sites and finds known previously. The palaeoecological studies (*see below*) show that the base of Chat Moss is variable with several deep hollows that became wet in the Mesolithic period. Ridges of ground remained 'dry' into the Neolithic period, and there was clearance of the land and probably some small-scale agriculture around the Nook Farm site. By the late Bronze Age the whole area became wetter and the moss developed into the large area now familiar, preventing settlement except around the edges.

The Iron Age site at Great Woolden sat precariously between the Moss and the Glaze Brook when it was occupied, peat probably having already reached to nearly its medieval extent. The land adjacent to Chat Moss seems to have been of no interest to the Romans, and was not utilised until the expansion of settlement and population in the late Saxon era. The following notes describe the archaeological topography of land around Chat Moss.

Astley

North of the Black Brook, the upland west of Astley Lower Green has boulder-clay soil. Most of the area was probably meadow until recently; there are no medieval furlong boundaries or ridge and furrow. Immediately south of the Black Brook, the skirtland has pools of peat and sand lies underneath the clay surface. A low sandy patch of skirtland contained the prehistoric site (GM 4) discussed above. The site location shows that the moss did not extend this far north at the time.

To the south, peat rises up sharply by 2 m. A farm borehole revealed 3.6 m of peat at SJ 6964 9792, only 200 m from the moss edge. Most of the south part of Astley township is formed of deep peat covered by scrub and birch woodland.

Bedford

Morley's Hall, Crompton House, and Grange Farm lie on the hinterland. Immediately west of Morley's Hall is the site of a late medieval or early post-medieval building, consisting of a platform made of stone with a little brick (site GM 3, SJ 6888 9927). Dark soil and a small amount of post-medieval pottery lie around. Morley's Hall is referred to in c 1200 and the moat, still in existence, was created by 1533. Site GM 3 is likely to be part of the manorial complex.

The whole area has clay soils; some ploughed fields were in good weathered condition, but there were no sites. Iron Age and Roman settlement would be expected on this kind of terrain in other parts of England. The natural drainage has complex gullies and the Black Brook does not lie in a well-defined valley.

The arable fields around Crompton House yielded no finds. Most of the land lies low and dark, possibly having a colluvium content, and may have been meadow and pasture until the nineteenth century. The higher boulder clay ground next to the Bridgewater Canal had medieval cultivation with wide ridge and furrow, visible as cropmarks with reverse-S shapes in young corn near Bedford Hall (SJ 676 991).

Grange Farm has a few grass fields around its buildings. One on the west was ploughed 25 years ago, with old furrows just visible. The unploughed fields are very uneven, possibly as left by glacial melt-waters. At SJ 6775 9860, bisected by the A580, are the earthworks of narrow rig of probable nineteenth-century date, not ploughed since c 1925. There are many ponds, some of which may have been dug to obtain sand that lies 9 m deep, underneath c 1.2 m of boulder clay.

South of the Black Brook, Windy Bank Farm has clay skirtland, and peat rises sharply at the east under woodland. To the south-west, Light Oaks Hall (SMR 4138) has much skirtland with soils consisting of clay and sandy clay. East of the Glaze Brook, at SJ 679 966, is a sandy and pebble bank with other sandy 'hills' near peat and water. Examination in excellent weathered condition failed to produce any finds. Peat at the east of the farm caught fire in the dry summer of 1921.

Little Woolden

The northern part of Woolden continues in a wide skirtland belt at Moss House Farm and White Gate Farm, east of Glaze Brook. Peat at the south-east lies 3.6 m deep. Several areas of the skirtland next to Glaze Brook were in good condition but without finds. A spur of skirtland at Moss Lodge Farm yields large logs in the peat. In the mossland sandy clay is exposed at a depth of 1—1.5 m in ditches, with roots into the old land surface.

Towards Little Woolden Hall the moss crowds to the Glaze Brook, leaving a narrow belt of skirt next to the steep and narrow valley. The 'ancient' undated ditch revealed in a modern ditch section (GM 2) has been described above.

Farther south near the Keeper's Cottage, a weathered field was in good condition, showing a disturbed belt along the Glaze Brook with nightsoil, then a band of undisturbed sandy soil, next to skirtland and peat. No finds occurred on the sand, even though it is located between a water supply and peat. A 'rectangular earthwork' (SMR 1867) on the surface of the peat is not likely to be of archaeological significance; no finds were discovered.

Most of the moss was cleared of surface vegetation and drained for peat cutting. As well as the high spot in the old ground surface already mentioned, other ditches on the west, 2.5—3 m deep, strike sandy soil with *in situ* tree roots. Nothing archaeological was exposed. To the south-east the pre-Flandrian surface falls away and there are old cuttings with rail lines for small trolleys. Cleanly-cut sections 1.5 m deep show undulating peat profiles; heather and cotton-grass grow on the surface.

Great Woolden Hall has skirtland near to Glaze Brook, with a clay and sandy-clay base. The Iron Age site (GM 1), described fully in Chapter 6, lay under pasture with nothing visible. Small patches of sand on a ploughed field to the east of the site were carefully searched but yielded no finds. A possible mill dam lies in a spinney south-east of the Hall, at SJ 6960 9343. On the south-west of Great Woolden Hall is a band of clay and sandy-clay skirtland about 100 m wide, leaving 150 m of mineral soil falling to Glaze Brook. The skirtland is arable but yielded no finds, and nothing was discovered at the spot of a proposed cropmark (SMR 1873).

Cadishead and Irlam

The surviving parts of these mosses are deep peat, which lies high and level. The ground begins to slope down towards Irlam, but skirtland and mineral soils are built over. One small area of skirt is exposed on a playing field at SJ 727 956, with no finds.

Barton

All the ground to the north-west of the M62 is deep peat, except for the mineral 'island' already referred to. Between the M62 and the old Irlam main road (A57) soils are variable, with peat at the north, sandhills by the road, and some clayey skirtland. Exposed mineral land is skirt, so any upland not covered by peat must have been a narrow belt under Irlam, next to the Irwell. All the ground south of the main road appears to have been built on or quarried. To the west is a high mound of boulder clay subsoil and rubbish with 0.3 m of peat over it (SJ 731 960), probably dumped from the Manchester Ship Canal or M62. This mound is likely to give rise to the cropmark recorded as number 430 (see Appendix 4).

The fields near the old A57 road are all skirtland, even though there is a ridge that would be a natural moss limit, suggesting that the moss once spilled over into the Irwell. Although the setting is likely to be of high archaeological potential, the only finds on the sandhills were a few burnt pebbles (GM 10, SJ 7376 9659). There were no other burnt stones anywhere in the field and they are likely to represent archaeological activity and are not derived from the general light spread of nightsoil. A farmer who had worked the fields said he had found nothing in 35 years.

Barton Aerodrome is now pasture land, with a natural gully at its south-west. A cemetery to the east is sandy in the middle and peaty at the west.

Worsley

Grange Farm lies at the north-east of Chat Moss and has skirtland of various types, mostly clay. A little sand occurs near the farm, and at the south, cut by the motorway, where it extends to more than a hectare (SJ 7495 9875). Two of the sand patches yielded burnt stones, representing probable prehistoric activity (GM 7, GM 9). Nearby, site GM 8 is a dark patch of post-medieval rubbish.

To the west lies deep peat, rising high. A canal runs north—south near to the SJ 74 northing and continues into deep peat next to a farm track, where it is marked by a deposit of nineteenth-century rubbish (GM 6, SMR 1903), finishing at SJ 7390 9905. This is a branch of the Bridgewater Canal, made to drain the moss in 1762. Farther south, two fields have a raised linear mound made of post-medieval rubbish crossing them, which seems to be an old track; it was expected to be a continuation of the canal, but is not in line with the lower part.

Botany Bay Wood lies on peat and north of Shaw Brook are two areas of peat that are probably separate basins from the main moss. Malkins Wood Farm (SJ 718 989) stands on a clay 'island' that has skirtland below it. Most of the farmland lies on peat, visible in dikes and mole upcasts, probably of varying depths with skirt patches. All of it is pasture ground with mining subsidence at the north; the whole farm has Astley mine workings underneath.

East of Botany Bay Wood, centred on SJ 745 999, is a large field containing several blocks of nineteenth century narrow-rig ploughing.

The SMR has cropmarks and soilmarks noted for various parts of Chat Moss. They are all likely to be of recent drainage and agricultural date. No finds were made at any of the sites examined during the course of the present fieldwork. The SMR numbers are 857, 1470, 1628, 1783, 1860, 1863, 1864, 1866, 1867, 1869, 1870, 1873, 1875, 1876, 1895, 1010, 3014, 3019, 3020, and 3209.

Palaeoecological evidence

The area conveniently defined here as the Chat Moss complex forms the largest mire in the Greater Manchester region, measuring approximately 9 km long and 4.5 km wide.

For simplicity, 'Chat Moss' is used in the following section to define the land between the boundaries formed by the Glaze Brook to the west, the M63 to the east, the A580 in the north and the A57 in the south. As such it incorporates those areas of peat known traditionally as Bedford, Little Woolden, and Barton Mosses. The complex interaction between elements of landownership, traditional place-names, and stratigraphy makes it difficult to be certain what relation current geographical nomenclature has to former hydromorphological units. Development and disturbance have further blurred distinctions between former mesotopes so that accurate identification of the subtle boundaries of such units is impossible without much more fieldwork. Because of this it is convenient for descriptive purposes to treat the peats as a whole, although a crude attempt is made to estimate former mesotope (*sensu* Ivanov 1981) extents in the general palaeoecological reconstruction outlined below.

Previous research

Despite the size and importance of Chat Moss, the only palaeoecological investigations which had been undertaken prior to the current survey were those of Erdtman (1928) and Birks (1964—5). Erdtman's study was limited in scope and concentrated on a small section of the Chat Moss peats. He nevertheless established the existence of sub-Boreal and post-Boreal peats. Birks produced a detailed pollen diagram documenting regional vegetation change from the late-glacial period to later prehistoric times. The palynological work was supplemented by a long (5 km) peat depth profile transect. Some retrospective radiocarbon dating of important stratigraphic features was subsequently undertaken (Godwin and Switsur 1966). The results from this research are summarised below in order to produce a context for the current study's palynological work.

Sampling strategy

The present-day mire expanse was surveyed by means of a series of five transects of 68 cores (Figs 7, 9—18). Stratigraphic survey procedures followed those outlined in Chapter 2.

Palynological study was restricted to an archaeologically important area in the northern section of the moss known as Nook Farm. Methods followed those already outlined and detailed sampling procedure is described in a later section devoted to this specific site. Limited

Plate 8 The M62 motorway which cuts across the southern part of Chat Moss

Fig 7 Chat Moss: transect locations

time precluded the detailed analysis of representative peat profiles from the whole complex. Instead, plant macrofossil analysis was confined to two areas which were chosen in order to compare mire ontogeny between widely spaced locations and test the synchroneity of similar stratigraphical changes.

Condition and survival of the peat archive

In some parts of the complex peat has been stripped to the underlying mineral ground (*eg* Nook Farm, SJ 710 985) or has been lost due to development (*eg* the construction of the M62 in the south of the site (Plate 8) and landfill operations at Astley Green), but elsewhere peat depth varies from 0.15 m to over 7 m.

Although the survey did not discover any part of the mire which retained 'topmoss' characteristics (*ie* remained completely untruncated), it is possible that isolated relict profiles of this nature remain undetected in the less intensively farmed areas of the complex. The amount of truncation varies across the site, but perhaps averages some 2 m in most areas. However, in a few places relatively light truncation (*ie* probably <1 m) appears to have occurred, particularly in the Botany Bay Wood (SJ 725 980) area; this supposition is based on the relatively large amount of *Sphagnum imbricatum* surviving in the upper part of the stratigraphy on this part of the moss. Birks recorded that *c* 2 m of *S imbricatum* peats formed the upper stratigraphy in many parts of the moss in the early 1960s. Compared with this, the findings of the current survey, some 30 years later, show that this has been reduced to *c* 0.5—1 m on average, allowing a rough estimate for rate of loss of perhaps 30—50 mm a year during that period.

Below the upper half metre or so of the stratigraphy, the peats of Chat Moss generally remain in a good state of preservation for palaeoecological purposes with humification values varying from 2 to 9 on the Von Post Scale, but mostly occurring around 5—7 (Table 1).

Topography and mire classification
The altitude of the current peat surface varies widely, the highest point being recorded at 27.21 m OD and the lowest at 16.68 m OD.

The mire complex is underlain by an irregular mineral ground surface of clays and sands probably dating from the late and immediate post-glacial periods (Fig 8). In broad terms the deepest peats appear to occupy at least two prominent basins, one situated in the south-west portion of the moss and the other in a more central location. To the north and south of these lie higher ridges overlain by thinner peats displaying less complicated stratigraphies. Along its longer (south-west to north-east) axis, the mire can be seen to be occupying a raised section of mineral ground in its south-central portion (*ie* between SJ 6940 9530 and SJ 7165 9625) with lower ground at either end. This in turn affects the surface topography of the peats creating the impression of a 'raised mire', whereas in fact the highest central portion is responding to changes in underlying mineral ground topography.

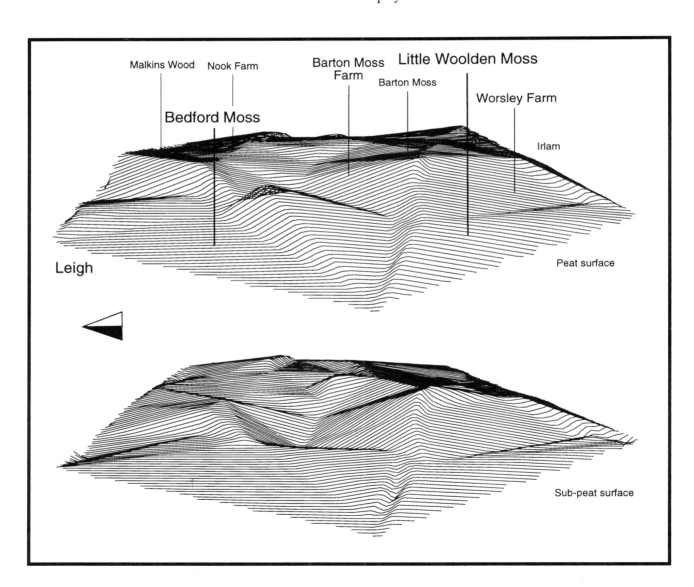

Fig 8 Terrain model of the Chat Moss peats (x 71 vertical exaggeration, lines spaced 50 m apart)

The complex can therefore be seen to have developed by progressive overgrowth of peat onto intervening mineral ridges, as is clearly demonstrated by the series of radiocarbon dates from the Nook Farm area of the moss (*detailed further below*). Here, a series of basal peat deposits exhibit gradually decreasing ages as the stratigraphy is followed upslope, indicating the progressive paludification of higher, 'hard' ground.

In this respect the Chat Moss peats seem to emulate the stratigraphic behaviour of blanket mire systems, albeit at gentler amplitudes of slope, rather than the classic model of a raised mire with its domed cross-section — an observation also noted by Tonks *et al* (1931) and Taylor (1980). It may be best, therefore, to regard the whole Chat Moss complex, or macrotope, as an *intermediate* or *ridge-raised mire* (*sensu* Moore and Bellamy 1974; Lindsay 1989) with paludification of mineral ground extending from several originally

Plate 9 Block cutting of peat in the central deep peat basin of Chat Moss. To the north of this area the peat is exploited using the more destructive milling process by a different company

discrete sources, resulting in the eventual coalescence of the peat bodies. The implications of this sequence of events are discussed later in Chapter 6.

General peat stratigraphy

Deep isolated basins containing basal fen and open water deposits occur within the general mire expanse (*eg* SJ 7115 9725, SJ 7200 9640; Figs 9—18) marking the sites of earliest organic accumulation, possibly dating from the immediate post-glacial (Birks 1964—5). In particular, two obvious basins occur, one in a central location at SJ 7150 9700, and possibly another to the south-west centred on SJ 7125 9400 (Plate 9). Over much of what was to become Chat Moss, however, peat accumulation was evidently slow until the Neolithic and a relatively dry surface was maintained, as shown by the widespread occurrence of wood peats, particularly of *Betula*. These are often followed by or are associated with *Polytrichum/ Aulocomnium palustre/Calluna*-dominated communities, also indicating relatively dry peat-forming vegetation with slow accumulation rates. Birks' inference that most of the wood peats at Chat are referable to Godwin's Zone VIIb (approximately the beginning of Flandrian III) (Birks 1964—5) is partly confirmed by radiocarbon dates from wood and associated peats of around 3500—3000 cal BC from the Nook Farm region of the moss (GU-5271 and GU-5273).

Eriophorum-dominated vegetation usually succeeds the latter communities, sometimes forming long sequences interrupted by occasional bands of *Sphagnum sect Cuspidata* peats. This suggests periods of stable, relatively dry hydrological conditions punctuated by rapid, but relatively short-lived shifts, to very wet conditions. At least two *S sect Cuspidata* bands are a sufficiently consistent stratigraphical feature in parts of the Chat Moss complex as to indicate major hydrological changes, and there are hints that others may occur. It is commonly assumed that in ombrotrophic stratigraphy these features represent climatic shifts towards wetter conditions (*eg* Barber 1981) although Birks (1964—5) records a 'flooding horizon' which he tentatively ascribes to anthropogenic forest clearance in the vicinity.

The next, and final, widespread stratigraphical element present in the Chat Moss peats is the switch to dominance of *Sphagnum imbricatum*, interpreted as an indication of increasing surface wetness, originally ascribed by Birks to a change to wetter conditions in the late Bronze Age or early Iron Age. Two radiocarbon dates are available from analogous stratigraphy (Godwin and Switsur 1966, 391) which date the boundary between the *Sphagnum* peat and underlying *Eriophorum*-dominated peats to 1010—530 cal BC, and the end of the drier phase at 1680—910 cal BC

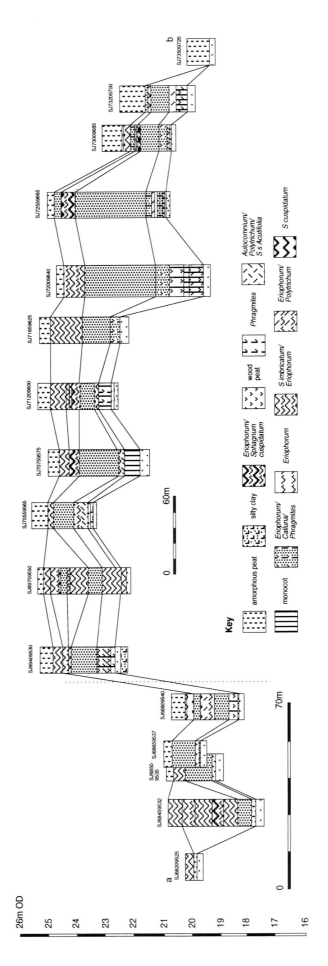

Key

amorphous peat	silty clay	*Aulocomnium/ Polytrichum/ S s Acutifolia*
monocot	*Eriophorum/ Calluna/ Phragmites*	*Phragmites*
	Eriophorum/ Sphagnum cuspidatum	*Eriophorum/ Polytrichum*
	wood peat	*S cuspidatum*
	Eriophorum	*S imbricatum/ Eriophorum*

Fig 9 Chat Moss: east-west transect

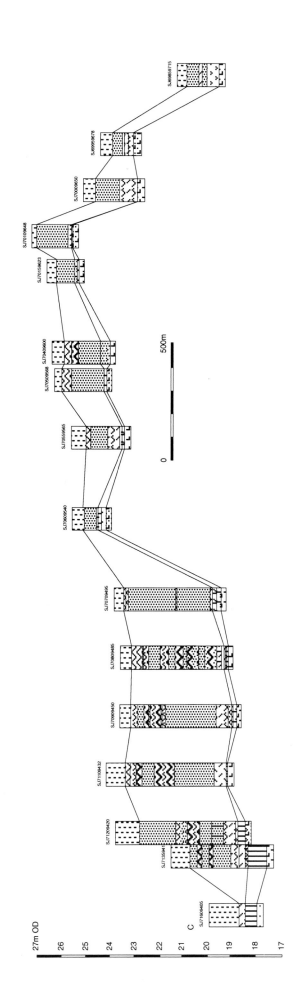
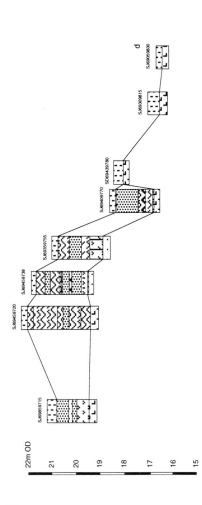

Fig 10 Chat Moss: south-north transect 1

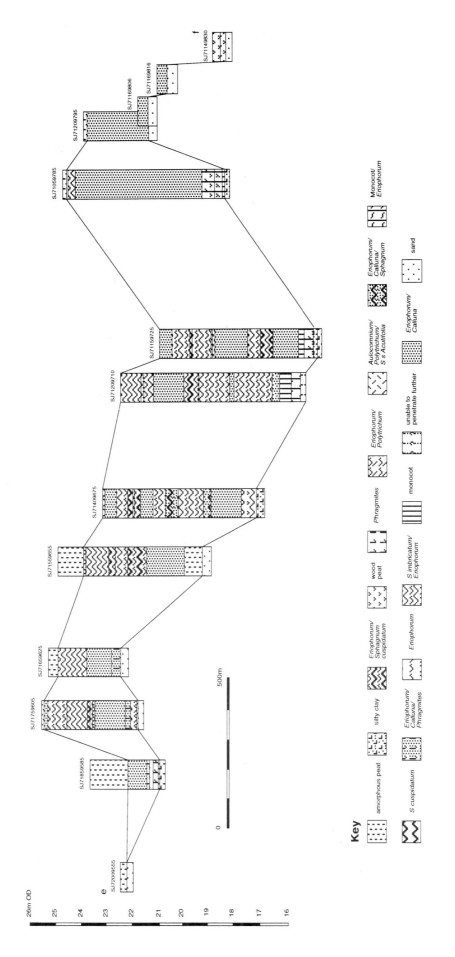

Fig 11 *Chat Moss: south-north transect 2*

Key

	amorphous peat
	silty clay
	Eriophorum/ Sphagnum cuspidatum
	Eriophorum/ Calluna/ Sphagnum
	Aulocomnium/ Polytrichum/ S s Acutifolia
	Monocot/ *Eriophorum*
	Eriophorum/ Polytrichum
	Phragmites
	Eriophorum/ Calluna
	sand
	Eriophorum/ Calluna/ Phragmites
	wood peat		*Eriophorum*
	S imbricatum/ Eriophorum
	unable to penetrate further
	S cuspidatum
	Eriophorum/ Sphagnum cuspidatum
	S imbricatum/ Eriophorum
	monocot

500m
0

SJ72009555
SJ71859585
SJ71759605
SJ71659625
SJ71559655
SJ71409675
SJ71209710
SJ71159725
SJ71059785
SJ71209795
SJ71169806
SJ71169816
SJ71149830

26m OD
25
24
23
22
21
20
19
18
17
16

e
f

36

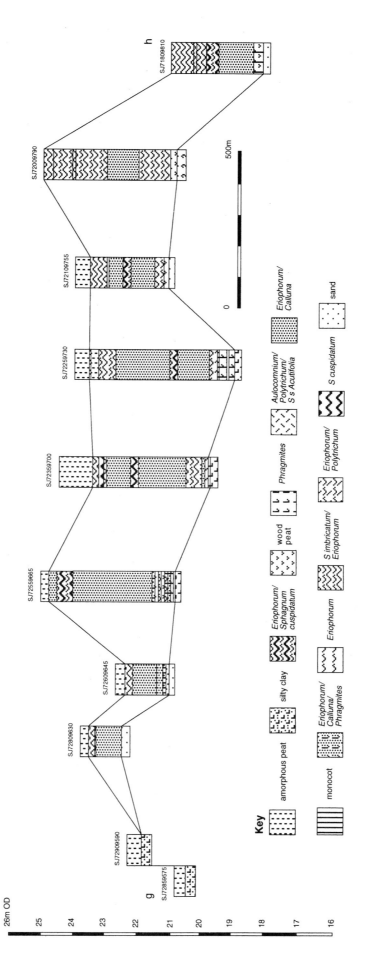

Fig 12 Chat Moss: south–north transect 3

Key

amorphous peat	silty clay	Eriophorum/ Sphagnum cuspidatum	wood peat	Aulocomnium/ Polytrichum/ S's Acutifolia	Eriophorum/ Calluna
monocot	Eriophorum/ Calluna/ Phragmites	Eriophorum	Phragmites	S imbricatum/ Eriophorum	S cuspidatum
				Eriophorum/ Polytrichum	sand

500m

0

26m OD
25
24
23
22
21
20
19
18
17
16

SJ72859575 g
SJ72909590
SJ72809630
SJ72609645
SJ72559665
SJ72359700
SJ72259730
SJ72109755
SJ72009790
SJ71809810 h

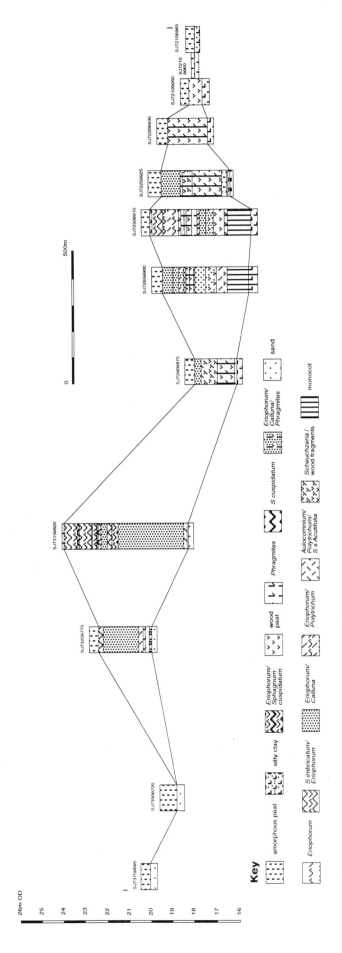

Key

| amorphous peat | silty clay | *Eriophorum/ Sphagnum cuspidatum* | *S cuspidatum* | *Eriophorum/ Calluna/ Phragmites* | sand |
| *Eriophorum* | *S imbricatum/ Eriophorum* | *Eriophorum/ Calluna* | *Phragmites* | wood peat | *Eriophorum/ Polytrichum* | *Eriophorum/ Polytrichum/ S s Acutifolia* | *Aulocomnium/ Polytrichum/ S s Acutifolia* | *Scheuchzeria/ wood fragments* | monocot |

Fig 13 Chat Moss: south-north transect 4

38

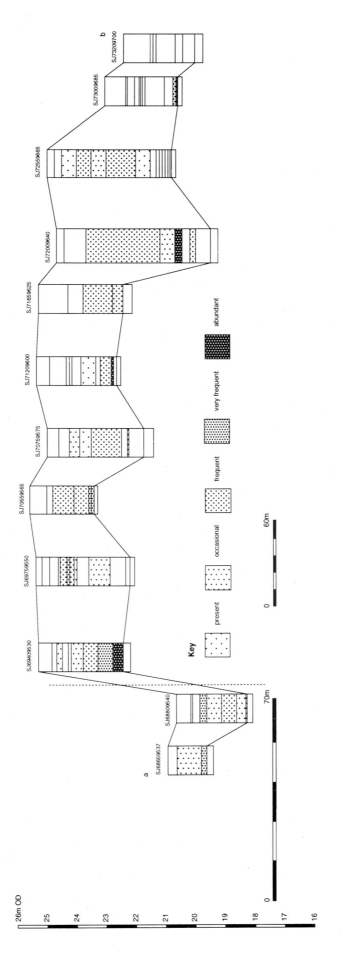

Fig 14 Chat Moss: west-east charcoal distribution

Key

present

occasional

frequent

very frequent

abundant

0   60m

70m

0

39

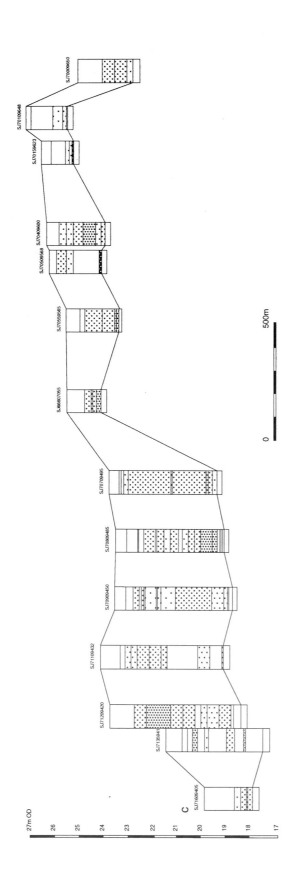

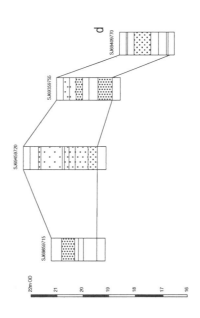

Fig 15 Chat Moss: south-north charcoal distribution 1

40

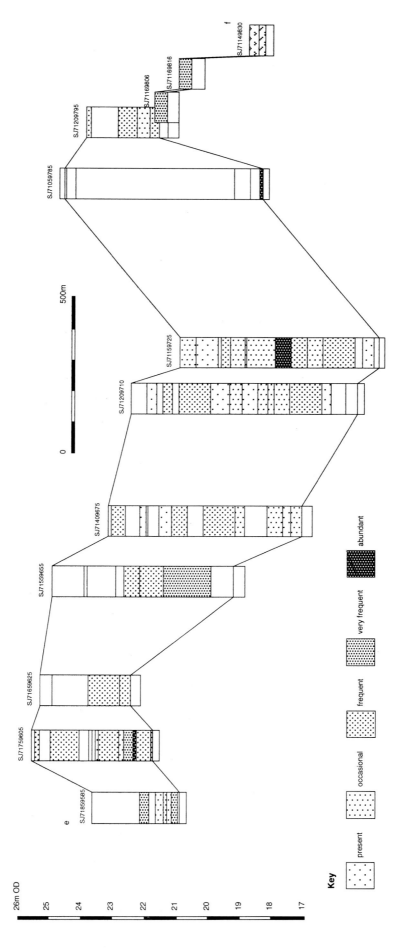

Fig 16 Chat Moss: south-north charcoal distribution 2

Key

present

occasional

frequent

very frequent

abundant

500m

0

26m OD
25
24
23
22
21
20
19
18
17

SJ71859585
e
SJ71759605
SJ71659625
SJ71559655
SJ71409675
SJ71209710
SJ71159725
SJ71059785
SJ71209795
SJ71169806
SJ71169816
SJ71149830
f

41

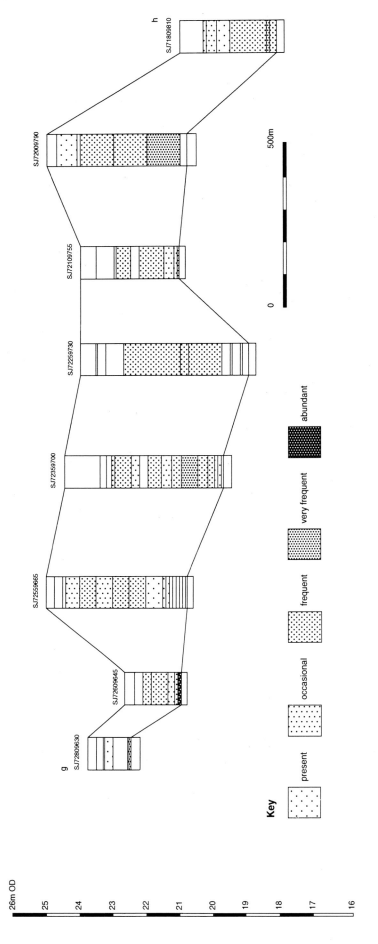

Fig 17 Chat Moss: south-north charcoal distribution 3

42

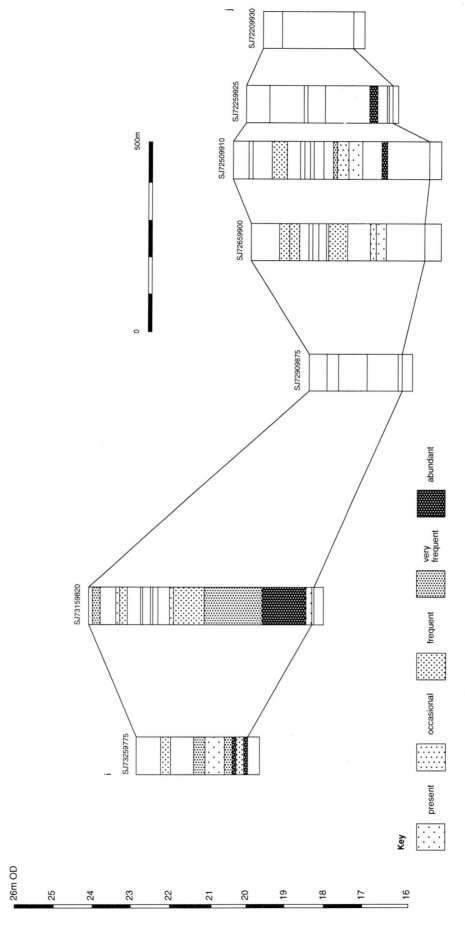

Fig 18 Chat Moss: south-north charcoal distribution 4

43

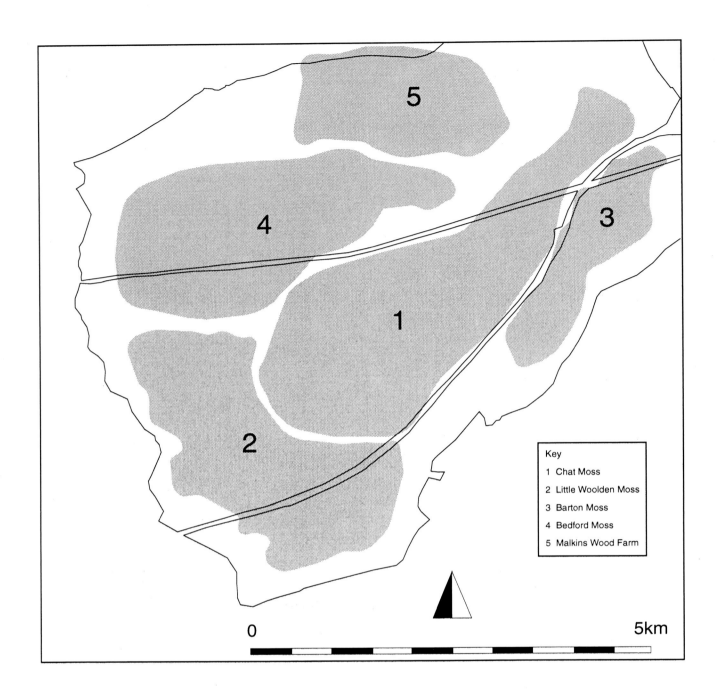

Fig 19 Chat Moss: probable mesotopes

Key

1 Chat Moss

2 Little Woolden Moss

3 Barton Moss

4 Bedford Moss

5 Malkins Wood Farm

0   5km

(3070±150 BP, Q-682, 2645±100 BP, Q-683). In the current study, radiocarbon determinations for the end of the *Eriophorum*-dominated phase range from 1690 to 1440 cal BC (3280±50 BP; GU-5366) and 1735—1430 cal BC (3280±60 BP; GU-5359). Such a switch to dominance by this species has been noted throughout the North West's ombrotrophic mires during their later development, and an attempt is currently being made by the authors to establish whether the change is regionally synchronous.

This general sequence is broken to the north-east of the site near Malkins Wood Farm where a strati-graphically distinct basin mire is separated from the rest of the system by the Shaw Brook. Here a sequence of fen and carr peats reaching up to *c* 6 m deep characterises the deposits. This separate mesotope was initiated in a central hollow containing a *Phragmites* reedswamp fringed by peripheral fen-carr communities of *Betula*, *Salix*, and monocotyledons. Later a stable carr community persisted over the whole mesotope for some time, eventually being succeeded by ombrotrophic *Eriophorum* and *S imbricatum* communities as the stratigraphy became contiguous with the peats across the Moss Brook. Although no radiocarbon dating was possible in this section of the moss, it seems probable that the fen peats span at least the Mesolithic period with the change to ombrotrophic stratigraphy being contemporaneous with the rise to *S imbricatum* dominance observed in the rest of the system.

Identification of possible mesotope boundaries

The Malkins Farm basin mire forms an obvious distinct mesotope within the general Chat Moss complex, but it is difficult to define the boundaries, or definite existence of others. Figure 19, however, shows a first attempt to identify the possible original mesotope components from which the complex is formed, with five possible sites of original mire inception.

A central site contains the deepest peats (*c* 6—7 m) in the whole complex enclosed within an elongated basin, the lowest section of which reaches *c* 15 m OD. The basin boundaries may approximate to an original independent 'Chat Moss' (1). This is flanked by a putative south-western mesotope (2), which might be associated with the historic Little Woolden Moss. The peats contained within the approximate boundaries delineated for this unit are *c* 3—5 m deep and occupy

a mineral surface *c* 18 m OD. They display prominent *S sect Cuspidata* pool layers set within long *Eriophorum*-dominated sequences.

To the south-east another possible mesotope, perhaps associated with the place name Barton Moss (3), is characterised by shallower peats occupying higher ground at *c* 20—22 m OD.

To the north-west an original Bedford Moss (4) may have stretched from Moss Brook and Glaze Brook to the edge of the present-day Botany Bay Wood. The peats are relatively shallow (*c* 2 m) and are dominated by facies of mire communities indicating relatively dry conditions, with species such as *Eriophorum* and *Polytrichum* (Plate 10). However, it is likely that much of the upper *Sphagnum* layers have been truncated. The peats occupy a raised ridge at around 24—25 m OD, dropping to *c* 18—20 m at the northern fringe. To the east, towards Nook Farm, the peats thicken to 6 m but are still composed primarily of *Eriophorum* in contrast to the peats of similar depth immediately adjacent to the south.

A north-eastern mesotope (5) is formed by the basin mire described earlier from the Malkins Wood Farm district.

Plate 10 The Bedford Moss component of the Chat Moss complex. Although the current vegetation appears more 'natural' than the ploughed areas, the palaeoecological record is severely truncated here compared to many of the cultivated zones

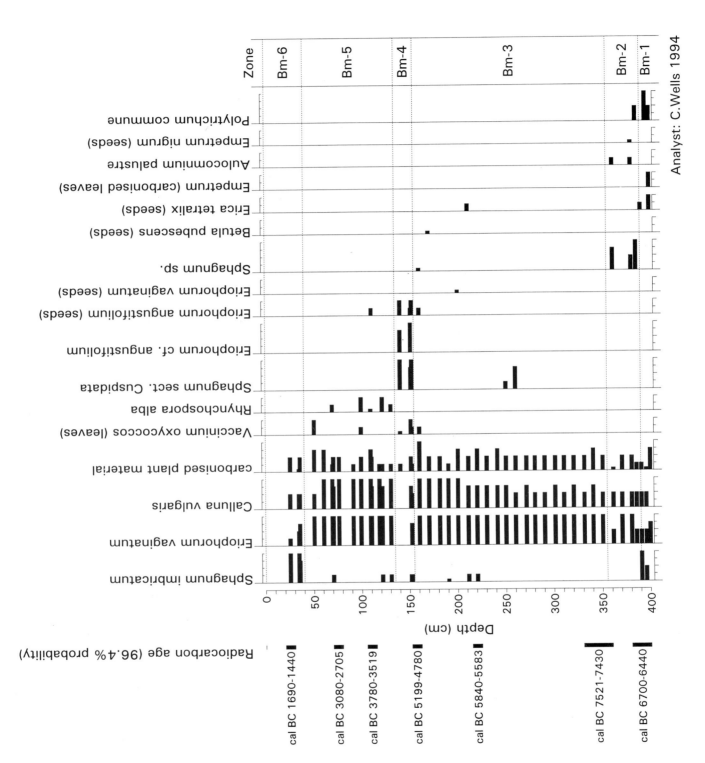

Fig 20 Chat Moss: Barton Moss macrofossil diagram (for details of radiocarbon dates see Appendix 2)
Key to relative abundance of macrofossils ▪ rare ▪ occasional ▪ frequent ▪ very frequent ▪ abundant

Analyst: C.Wells 1994

46

It must be stressed that the crude divisions outlined above remain extremely tentative. The stratigraphy is complicated by the numerous sand ridges running through the site and there are almost certainly extra subdivisions to be made to this picture. Further stratigraphical survey, backed by systematic radiocarbon dating, is necessary to characterise fully the ontogeny of the Chat Moss complex.

Macrofossil analysis

Detailed plant macrofossil analysis was limited to two locations 3 km apart: Barton Moss (SJ 7235 9700) close to the heart of the mire complex, and Worsley Farm (SJ 7110 9432) towards its south-western extremity (Figs 20 and 21). Similar stratigraphical development had been identified from both sampling sites previously during the initial gross stratigraphy survey, and this had included a number of 'pool peats' (ie stratigraphic elements characterised by the occurrence of abundant remains of *S sect Cuspidata*) and obvious macroscopic charcoal inclusions. The main aims of the analyses were to add more detail to the gross stratigraphical information, particularly to ascertain a more precise knowledge of the distribution of carbonised particles through the peat profiles, to compare the dating of various stratigraphical changes, and to test whether the supposed 'wet phases' represented by the pool peats were synchronous. The diagrams have been zoned by eye into macrofossil assemblage zones in order to aid description.

Barton Moss (BM) (SJ 7235 9700)

Although the depth of the peat at this location was known to be 4.75 m, it was impossible to penetrate the lowest 0.75 m, which included a wood layer overlaying silty clays. Consequently the analysis was limited to the upper 4 m of stratigraphy.

BM-1 (4.04—3.90 m)
The lowest 0.14 m of the core display an assemblage characteristic of relatively dry, acid, mire communities with abundant remains of *Polytrichum commune* along with seeds of *Betula pubescens* and leaves of *Empetrum nigrum* (some of which are carbonised) and frequent remains of *Calluna vulgaris* and *Eriophorum vaginatum*. The zone is also marked by a peak of carbonised material at the base of the core. *Sphagnum imbricatum* remains are also abundant, but it is thought that this is likely to be more recent material which contaminated the bottom samples during the difficult extraction experienced at this depth at the time of sampling. This would explain the unlikely radiocarbon date for this level of 6700—6440 cal BC (7750±60 BP; GU-5372) which does not conform with the rest in

the series from the site and appears too recent. Instead it is more likely that the age of the peat from these levels lies before *c* 7500 cal BC.

BM-2 (3.90—3.55 m)
A zone marked by the abundant remains of *Eriophorum vaginatum* and *Sphagnum* sp, along with frequent remains of *Calluna vulgaris* and *Aulocomnium palustre*. Carbonised plant material remains consistently present at frequent intervals.

BM-3 (3.55—1.55 m)
A uniform peat dominated by remains of *Eriophorum vaginatum* and *Calluna vulgaris* occurs in this zone. There is a slight interruption to this at 2.45—2.65 m when a minor pool peat is indicated by the occurrence of *S sect Cuspidata* leaves. No absolute date is available for this episode, but, judging from the available radiocarbon determination from 2.15 m to 2.25 m (5840—5583 cal BC; 6850±60 BP; GU-5370), it possibly occurred before 6000 cal BC. Carbonised material remains frequent throughout the phase, particularly after the 'pool' phase above 2.45 m, but reaches a noticeable peak at its end at *c* 1.60—1.62 m. A date of 5199—4780 cal BC (6020±60 BP; GU-5369) approximately refers to this point (1.52—1.62 m).

BM-4 (1.55—1.35 m)
A very marked pool peat occurs at this stage in the profile, dominated by *S sect Cuspidata*, and *Eriophorum cf angustifolium*. The age of this episode appears to lie shortly after 5199—4780 cal BC (1.52—1.62 m; 6020±60 BP; GU-5369).

BM-5 (1.35—0.40 m)
Eriophorum vaginatum and *Calluna vulgaris* regain dominance and there are sporadic occurrences of remains of *Vaccinium oxycoccus*, *Rhynchospora alba*, and *S imbricatum*. Carbonised material remains frequent, with peaks at *c* 1.10 m (dated to *c* 3780—3519 cal BC (1.05—1.15 m; 4870±60 BP; GU-5368) and 0.50—0.60 m, occurring after 3080—2705 cal BC (0.70—0.80 m; 4300±60 BP; GU-5367).

BM-6 (0.40—0 m)
S imbricatum takes over dominance of the assemblage. This appears to have occurred around 1690—1440 cal BC (0.20—0.30 m; 3280±50 BP; GU-5366).

Worsley Farm (WM) (SJ 7110 9432)

Once again it was found impossible to sample the lowest 0.75 m of stratigraphy at this point and the core covers only the upper 4.30 m of stratigraphy.

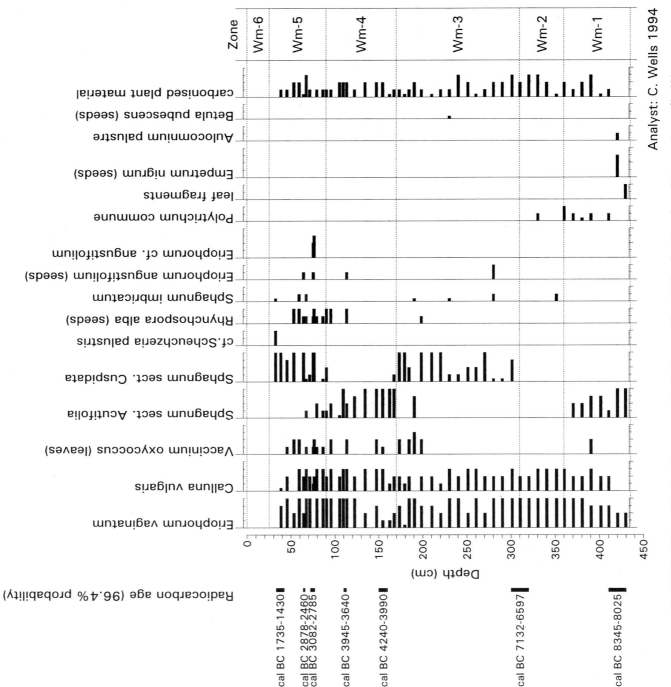

Fig 21 Chat Moss: Worsley Farm macrofossil diagram (for details of radiocarbon dates see Appendix 2)
Key to relative abundance of macrofossils ▪ rare ▪ occasional ▬ frequent ▬ very frequent ▬ abundant

Analyst: C. Wells 1994

WF-1 (4.30—3.60 m)
From 4.30—3.60 m the peats are dominated by *Sphagnum sect Acutifolia* with frequent remains of *Eriophorum vaginatum,* and *Calluna vulgaris* along with *Polytrichum commune, Empetrum nigrum,* and *Aulocomnium palustre.* Carbonised material makes an early appearance, starting at *c* 4.04 m and forming a continuous curve thereafter. There is a noticeable peak of charcoal at *c* 3.90 m. The beginning of this phase (4.10—4.30 m) is dated to 8345—8025 cal BC (9140±70 BP; GU-5365).

WF-2 (3.60—3.10 m)
This zone is characterised by the dominance of remains of *Eriophorum vaginatum* and *Calluna vulgaris.* Carbonised material reaches consistently high levels of representation towards the latter part of this phase (3.00—3.20 m), which is radiocarbon dated to 7132—6597 cal BC (7980±80 BP; GU-5364).

WF-3 (3.10—1.70 m)
A zone characterised overall by the varying abundance of *S sect Cuspidata* interacting with *Eriophorum vaginatum* and *Calluna vulgaris,* but can be further subdivided into four obvious peaks of *S sect Cuspidata* influence (3.10—2.95 m, 2.80—2.45 m, 2.25—1.95 m, 1.85—1.70 m). This composite zone ends emphatically with a sudden switch to assemblages dominated by *Sphagnum sect Acutifolia* at 1.70 m.

WF-4 (1.70—0.90 m)
The zone, beginning before 4240—3990 cal BC (5270±50 BP; GU-5363; 1.50—1.60 m) is characterised by the marked dominance of *Sphagnum sect Acutifolia* along with less frequent amounts of material referable to *Eriophorum vaginatum* and *Calluna vulgaris.* It ends around 0.90 m (3945—3640 cal BC; 4950±60 BP; GU-5362) and 3082—2785 cal BC (4320±50 BP; GU-5361), when *S sect Cuspidata* reasserts dominance of the macrofossil assemblage.

WF-5 (0.90—0.25 m)
This zone, primarily marked by the dominance of *S sect Cuspidata* remains, also includes much *Eriophorum vaginatum, Calluna vulgaris, Vaccinium oxycoccus,* and *Rhynchospora alba,* as well as a short-lived peak of *Eriophorum cf angustifolium* near the beginning of the phase at 0.70—0.77 m (radiocarbon dated to 3082—2785 cal BC (4320±50 BP; GU-5361). A peak of charcoal occurs shortly after this at *c* 0.62—0.64 m and is radiocarbon dated to *c* 2878—2460 cal BC (4050±70 BP; GU-5360).

WF-6 (0.25—0 m)
The final zone shows a switch to *S imbricatum* dominance at 0.25 m, preceded by a short-lived occurrence of *cf Scheuchzeria palustris,* and radiocarbon dated to 1735—1430 cal BC (3280±60 BP; GU-5359).

Interpretation
Both diagrams display a broad succession from relatively dry, acid mire communities with remains of *Polytrichum commune* and *Empetrum nigrum* through to *Eriophorum vaginatum / Calluna vulgaris*-dominated mire, punctuated by wetter episodes marked by the occurrence of *S sect Cuspidata* remains, and finally *Sphagnum imbricatum* dominance.

Disregarding the putative pool peats for the moment, it would seem that the timing of the other major stratigraphic changes is broadly similar between the sites — assuming the basal radiocarbon assay from Barton Moss (GU-5372) to be anomalous — with the early mire communities beginning to develop around 8500—8000 cal BC in both places. *Eriophorum vaginatum / Calluna vulgaris*-dominated communities persisted at both locations until *c* 1700—1400 cal BC, when *Sphagnum imbricatum* took over as the chief peat former. However, the diagrams appear to show that pool development occurred independently at the two sites, with the possible exception of zones BM-4 and the latter part of WF-3 which may date to *c* 5000—4000 cal BC.

Carbonised material
The presence of carbonised plant material was an almost ubiquitous feature of the Chat Moss peats (Figs 14—18). Charcoal occurred in most of the stratigraphic elements present on the moss, but perhaps the most interesting was its widespread occurrence in the lower reaches of the stratigraphy, sometimes at a high level of abundance (*eg* Fig 18).

Burnt wood and associated burnt peats near the base of the moss were particularly noticeable at some core positions (*eg* SJ 7110 9432, SJ 7050 9568; Fig 15), and this was also seen to be the case in the detailed macrofossil diagram from Barton Moss. The macrofossil studies also provide more detail about the distribution of macroscopic carbonised material in the Chat Moss peats than was possible with the relatively crude samples taken as part of the gross stratigraphical survey. They show that burning was virtually a constant feature during the mire development, and noticeable peaks of charcoal are discernible from time to time.

On the basis of the extant data it is impossible to be certain of the cause or archaeological significance of the peat charcoals although it is interesting to note that carbonised wood (dated to the Neolithic/Bronze Age) was also present at a putative late Mesolithic lithic scatter site on the moss (detailed below), and apparently associated with clearance activity recorded in the palynological record.

Birks (1964—5) has also speculated that increases in *Pteridium* spores associated with clearance indicators in his pollen study suggest the occurrence of fire.

Palynology and regional vegetational change
A comprehensive series of pollen analyses from deep peat cores was undertaken at Chat Moss in the 1960s by Birks (1964—5). This approach was designed to gain an understanding of vegetation change on a regional basis, whereas the present study concentrated on a series of closely spaced peat monoliths from the northern part of the moss (Nook Farm) in order to gain insight into localised changes in the vegetation which might be associated with archaeological material discovered here.

In the following section a brief summary of the regional vegetation change at Chat Moss is presented (based on Birks 1964—5), which provides a context for the present palynological study from Nook Farm and the ontogeny of the mire.

Pre-Flandrian vegetation was represented at Chat Moss by a typical late-glacial succession dominated by herb communities together with grasses and sedges (Birks 1964—5). Open conditions with high pollen values of *Rumex* and *Artemisia* persisted until juniper began to make inroads on the open landscape. The first appearance of trees followed on from this situation, with sporadic colonisation from birch and pine until a short interlude of colder conditions. The final clay bands (the so-called 'blue-grey' clays) have been interpreted as representing open water solifluction deposits from unstable and relatively unvegetated margins. After this episode, the surrounding vegetation retained much the same composition, but probably increased in cover due to more equable temperatures, which also allowed the sporadic appearance of trees. Eventually birch and hazel scrub became dominant as temperatures rose in the immediate post-glacial. Oak and pine began to increase in representation until pine declined allowing a mixture of alder, birch, hazel, oak, and elm to dominate the arboreal vegetation. The elm decline is recorded shortly after this phase. Later, the first obvious signs of forest clearance occur which led to the tentative identification of four periods of disturbance in the

later prehistoric period. These possible clearance phases vary in their characteristics, but generally involve falls in tree pollen and increases in pollen of herbs, grasses, *Plantago*, and occasionally cereal.

Nook Farm — a small Neolithic clearance?
Whilst Birks' regional study supplies a broad context of vegetation history at Chat Moss, the work presented in the following section provided an opportunity to examine detailed vegetation change in the immediate vicinity of an archaeological feature revealed during fieldwork: the Nook Farm lithic scatter. The discovery of this site, situated on a sand island and characterised by flint, chert, and charcoal remains, prompted an attempt to establish whether the artefactual evidence could be related to local palaeoecological changes.

The Nook Farm site is located on the northern edge of Chat Moss, approximately 700 m south of Moss Brook (SJ 7107 9797). A broad sinuous ridge of sand, comprising material probably related to late-glacial fluvioglacial deposition, was exposed by commercial peat removal and revealed lithic material and carbonised wood on the surface in places. Peat shelved into the ridge to the north and south of the artefact-yielding mineral ground, providing sampling points for palaeoecological study (Plate 11).

Sampling strategy
The peat/sand interface was sampled using monolith tins at three positions in a north—south transect across the sand island on which the lithic scatter lay.

Nook Farm 1 (NF1) was situated immediately adjacent to the lithic scatter. Nook Farm 2 (NF2) was positioned 100 m to the south of NF1. Nook Farm 3 (NF3) was positioned 100 m to the north of NF1. Figure 29 shows the position of the three sites together with the height OD and the radiocarbon dates for the basal peats.

The samples were analysed for pollen to investigate the possibility of a small-scale local clearance on the sand island. The monoliths were subsequently subsampled in the laboratory for pollen, mineral content, and charcoal. Samples were prepared routinely for pollen at 40-mm intervals and 10—20 mm from critical levels. Macrofossil sub-samples of 30 cm³ of peat were taken at various intervals down the monoliths depending on the location of visually-identified stratigraphic elements.

Preparation methods and data analysis followed the procedures outlined in Chapter 2.

Pollen

Figures 22—24 are simplified percentage diagrams with a pollen sum including trees, shrubs, dryland herbs, and *Calluna*. The more detailed results of the pollen analysis along with the pollen concentration data are in the archive and can be obtained from the authors at LUAU. The diagrams are divided by eye into phases to aid interpretation and discussion. Charcoal values on the diagrams were calculated as a percentage of the pollen sum plus charcoal.

The percentage mineral content was measured at 40-mm or 20-mm intervals. In the pollen diagrams, however, a continuous curve is shown in order to aid clarity of presentation. This has been calculated by entering values for gaps in the record calculated from the average of the two adjacent samples.

Macrofossils

The macrofossil results are presented as a series of bar histograms representing the frequencies of plant remains with depth (Figs 25—27). The semi-quantita-tive scale is recorded by four different bar lengths related to the frequency categories. For a description of the lithology of the sediments see the stratigraphic diagram (Fig 13).

Radiocarbon dating

Bulk samples immediately above the sand at the three sites, a *Betula* stump *in situ* adjacent to Nook Farm 3, and a piece of charcoal associated with lithic finds, were collected for radiocarbon dating. The results are shown in Table 2.

The radiocarbon dates from the bases of Nook Farm 1 and 3, suggesting a Neolithic date, provide results not statistically significant at the 95% level of confidence (Ward and Wilson 1978, 21), indicating that the pollen record may represent contemporaneous events at both sampling positions. An examination of the difference between the calibrated probability distributions of these results (Bronk-Ramsey 1994, 24) suggests that these events are unlikely to be more than 500 years apart. The date from the *Betula* stump excavated from alongside the Nook Farm 3 sample

Plate 11 The lithic scatter site at Nook Farm, with its discoverer, Mike Peace. The peat can be seen to shelve into the sand 'island' as the latter emerges from the wasting milled peats. A transect of peat monoliths was taken across this island to enable a three-dimensional palynological study to be undertaken

gives a similar date. The *in situ* carbonised wood associated with lithic material gave a date range indicative of the early Bronze Age—late Neolithic period.

Subsequently, radiocarbon samples were taken at 0.05—0.10 m from the Nook Farm 1 and 3 monoliths to try and ascertain whether they were contemporaneous. However, the disturbed nature of the upper stratigraphy, due to recent peat milling operations and the transit of heavy tracked machinery, militated against the chances of successful radiocarbon dating of the uppermost stratigraphy because of the likelihood of the introduction of more recent material into the dating sample. A relatively small quantity of younger carbon can significantly alter the date of an older deposit (Bowman 1990) and this probably explains the inconclusive nature of the results. The radiocarbon date of 380—100 cal BC (2170±50 BP; GU-5356) at 0.05—0.10 m at Nook Farm 1 is probably erroneous, as it seems unlikely that 0.15 m of peat would have taken *c* 3000 years to accumulate and for this reason it has been disregarded.

A Bronze Age date is indicated from the material at the base of Nook Farm 2. The latter pollen diagram therefore provides evidence for the vegetation during the early Bronze Age, whilst Nook Farm 1 and Nook Farm 3 represent the Neolithic development.

In the following sections, the pollen record from each individual monolith is presented in order of decreasing age of the basal peats (*ie* NF3, NF1, NF2).

Macrofossil diagrams

Nook Farm 3
Betula wood peat forms the base of the sequence, some of which is burnt. The middle section of the diagram is dominated by *Polytrichum commune* remains, along with wood fragments. Charcoal appears to be absent from this phase. Finally the top of the profile shows a change to ombrotrophic or poor fen conditions with the appearance of *Eriophorum* and *Calluna* (Fig 25).

Nook Farm 1
The base of NK1 is very badly humified and is dominated by indeterminate monocotyledon remains, probably of *Eriophorum/Calluna* and *Eriophorum*, dominate the profile. Charcoal frequencies are high and remain so up to 0.15 m below the surface.

Nook Farm 2
Eriophorum and *Calluna* dominate the profile throughout, although *Betula* wood fragments are present in small quantities in the lower 0.10 m. Charcoal is ubiquitous through the profile, although at reduced frequencies compared to Nook Farm 1 (Fig 27).

Pollen results

Nook Farm 3
NF3-a (0.55—0.41 m)
Percentages for tree and shrub pollen are 95—100%, with *Alnus* as the major component reaching a maximum of 40%. Values for Pteropsida spores are quite high, as are those for indeterminate pollen grains. There is a continuous curve for dryland herbs with pollen from a variety of types recorded. The pollen concentration values for herbaceous pollen are higher than at subsequent levels, as they are for all pollen types. Percentages of charcoal range from 5% at 0.40 m to 43.5% at 0.44 m (Fig 22).

NF3-b (0.41—0.30 m)
Tree and shrub pollen is >95% and is dominated by *Betula*, with a maximum of 50—60% when herbaceous pollen is not recorded. Concentration of pollen is high and the mineral content falls steadily. Charcoal values are low. The upper boundary is marked by the abrupt transition from a wood peat to a less humified *Polytrichum*-dominated one (Fig 22).

Table 2 Radiocarbon dates from Nook Farm, Chat Moss

Sample	Radiocarbon age (BP)	Date (cal BC 2 σ)	Lab code	Depths
Nook Farm 1	4590±70	3599—3047	GU-5271	0.25—0.40 m
Nook Farm 2	3710±60	2300—1940	GU-5272	0.38—0.48 m
Nook Farm 3	4670±60	3629—3207	GU-5273	0.30—0.40 m
Betula	4570±50	3497—3100	GU-5280	
Carbonised wood	3930±80	2853—2149	GU-5325	
Nook Farm 3	4020±50	2861—2460	GU-5354	0.05—0.10 m
Nook Farm 1	2170±50	380—100	GU-5356	0.05—0.10 m

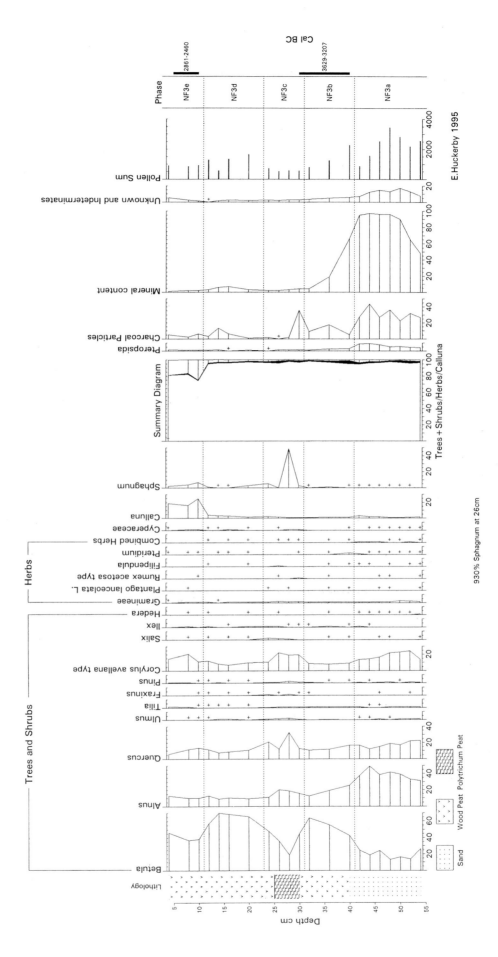

Fig 22 *Chat Moss: Nook Farm 3 percentage pollen diagram—selected taxa*
(for information about radiocarbon dates see Appendix 2)

53

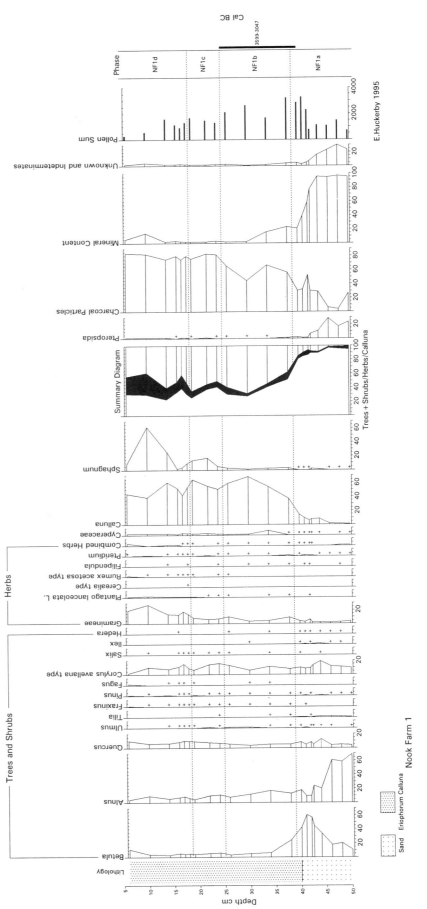

Fig 23 Chat Moss: Nook Farm 1 percentage pollen diagram—selected taxa
(for information about radiocarbon dates see Appendix 2)

54

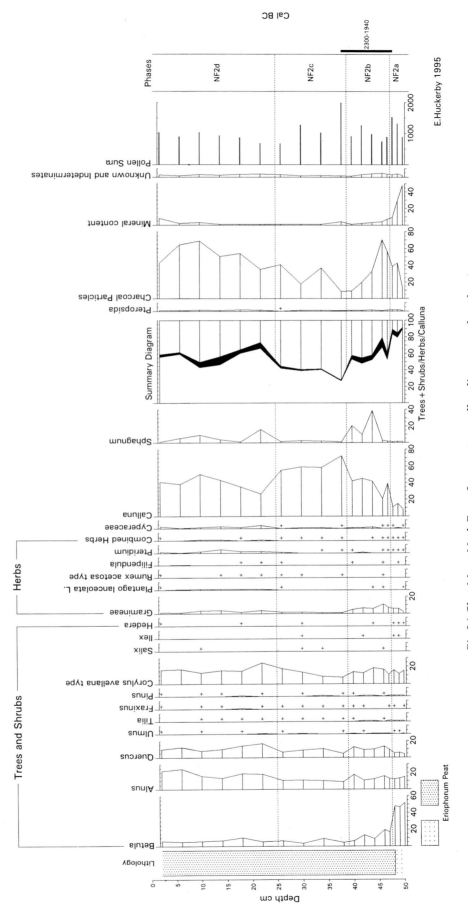

Fig 24 Chat Moss: Nook Farm 2 percentage pollen diagram—selected taxa
(for information about radiocarbon dates see Appendix 2)

55

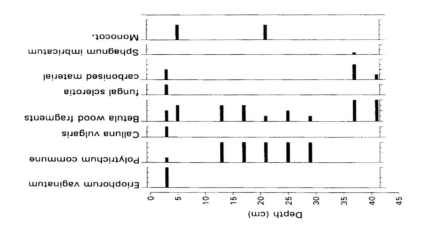

Analyst: C E Wells 1995

Fig 27 Nook Farm 2

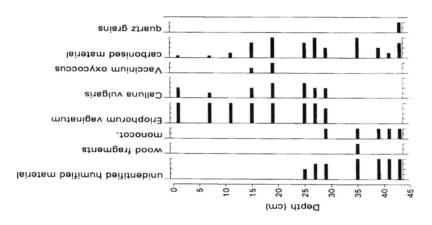

Fig 26 Nook Farm 1

Chat Moss: Nook Farm macrofossil diagrams (for information about radiocarbon dates see Appendix 2)

Key to relative abundance of macrofossils ▪ rare ▪ occasional ▪ frequent ▪ very frequent ▪ abundant

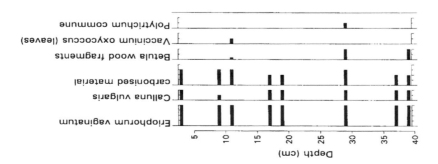

Fig 25 Nook Farm 3

NF3-c (0.31—0.23 m)
Tree and shrub pollen is >95%, but *Betula* values fall briefly to 20%, with other trees rising as a consequence. Pollen from dryland herbs becomes more consistent and *Sphagnum* spores reach a peak of 930% of the pollen sum. Mineral content and charcoal values are generally low, but the latter increase to 35.7% at the start of the *Betula* fall. The pollen concentrations are lower during this phase than in others.

NF3-d (0.23—0.11 m)
Values of *Betula* pollen recover to 60—70%, with resulting falls of other types whose concentrations do not alter a great deal. Pollen concentrations recover to earlier values. Charcoal percentages are very low. Mineral content is low, but rises to 6.58% at 0.16 m.

NF3-e (0.11—0.06 m)
The lower boundary is marked by the rise of *Calluna* pollen to 20%. Tree and shrub pollen falls to <70%, but recovers to 80%. *Betula* pollen falls sharply, that of other trees falls, but *Corylus avellana*-type pollen increases. Mineral content is low as is charcoal. Pollen concentrations remain high.

Throughout the pollen record from Nook Farm 3 trees dominate the pollen sum. The early part of the monolith suggests a damp alder carr which was replaced by birch woodland as a result of burning. Pollen concentrations of herbaceous plants are higher than at subsequent levels. A number of pollen types representative of wet meadows, heaths, and pastures within non-regenerating woodland (Behre 1981) were identified (*eg Succisa pratensis, Plantago lanceolata, Filipendula*, Rubiaceae, and others). It is possible that alder was being cleared on the site and was replaced by birch. This was dominant throughout the period represented by the monolith except for a brief period of wetter conditions in NF3-c (0.28 m). At this stage there are some indications of more open conditions when birch declines sharply. This is preceded by higher charcoal values and a corresponding increase of *Corylus avellana*-type, *Quercus*, *Alnus*, some herbaceous pollen, and a massive peak of *Sphagnum* spores. Two possible interpretations may be advanced to account for these changes:

1   A brief period of birch clearance from the mire surface associated with burning, causing a temporary rise in the water-table due to increased run-off;

2   A natural increase in surface wetness caused by climatic deterioration and the resulting failure of tree regeneration.

The increase in charcoal percentages prior to the decline of *Betula* pollen perhaps adds weight to the first interpretation. *Betula* recovers until 0.12 m when there is evidence of possible further clearance. This takes the form of an expansion of *Corylus avellana* type and *Calluna* pollen preceded by higher peaks of mineral matter.

Nook Farm 1
NF1-a (0.50—0.39 m)
A maximum of *Alnus* pollen (60—70%) is replaced by one of *Betula* at 60%. Pteropsida spores and indeterminates are high. Throughout the phase, tree and shrub pollen has values of 85—90% of the sum. Charcoal particles range from 3% to 53%. The pollen concentration ranges from 65,302 to 1,249,723 grains/cm^3. Mineral content decreases at 0.44 m when concentrations of herbaceous pollen rise.

NF1-b (0.39—0.25 m)
Tree and shrub pollen is at <60%, falling to 35%. The lower boundary is marked by rapidly rising *Calluna* pollen to high percentage values. Gramineae pollen is 5—10%. Pollen concentration values are high, charcoal percentages range from 45% to 67%, and the mineral content falls to 1.97%.

NF1-c (0.25—0.19 m)
Tree and shrub pollen increases slightly to 40%, with *Corylus avellana* type pollen as the major component. Gramineae pollen maintains values of 10% and *Calluna* is >50%. *Sphagnum* spores rise. Pollen concentration values are reduced, mineral percentages are low and charcoal values reach a maximum of 83%.

NF1-d (0.19—0.06 m)
Gramineae pollen rises to 20—30% at the lower boundary of this phase with a cereal-type grain recorded at 0.18 m. Pollen from plants indicating open ground were present in this phase. Total tree and shrub pollen is <40%. *Calluna* pollen is recorded in lower concentrations than the previous phase. *Sphagnum* spores increase to a maximum of 60%. Charcoal percentages are very high, > 74%, and mineral content reaches a peak of 13.64% at 0.1 m. The pollen concentrations are low, falling to 3,397 grains/cm^3 at 0.06 m.

Vegetation immediately prior to true peat formation consisted of a fairly open damp *Alnus* community which was replaced by a *Betula*-dominated one. At the transition from the sand to the pure organic deposits there occurs an abrupt change in the vegetation, marked by rapid expansion of *Calluna* pollen preceded by greater values for the pollen of herbaceous plants indicative of damp meadows or non-regenerating woodland. Major fire disturbance is indicated by the exceptionally high number of microscopic charcoal particles in the organic sediments, reaching values of >84% of the pollen sum. Macroscopic carbonised wood is also recorded, suggesting that the origin of the burnt material is not limited to the influx of carbonised particles from regional vegetation. At a depth of 0.17 m the stratigraphy changes from humified *Eriophorum/Calluna* peat to a less humified fibrous *Eriophorum*-dominated one, and this is preceded by an increase in *Sphagnum* spores indicating the onset of wetter conditions. At this transition there is greater evidence of grassland, with a continuous curve for *Plantago lanceolata* pollen and that of the combined herbs, together with slightly lower values of *Calluna*. The record of a cereal-type grain at 0.17 m suggests that there may have been a little arable farming associated with clearance activity. The total tree and shrub pollen curve above 0.38 m is < 50% of the pollen sum, in contrast to Nook Farm 3 where it is > 95% up to a depth of 0.12 m, when it falls to 80%.

Nook Farm 2

NF2-a (0.50—0.475 m)
Tree and shrub pollen, predominantly *Betula*, comprises 85—90%. The upper boundary is marked by a rise in *Calluna* pollen and a fall of *Betula*. There is 13—43% charcoal and the total pollen concentration is high.

NF2-b (0.475—0.39 m)
Tree and shrub pollen falls steadily from 80% to 30%. Pollen from dryland herbs reaches a maximum of 10%. *Calluna* values and *Sphagnum* spores are rising. Percentages of charcoal range from 9% to 71% and mineral content from 4% to 0.99%. Pollen concentration values are high but falling.

NF2-c (0.39—0.25 m)
Calluna pollen dominates the diagram with that of trees and shrubs at < 45%. Percentages of charcoal are low and the mineral content is very low. Pollen concentrations are generally high.

NF2-d (0.25—0.02 m)
Dryland herb pollen and that from trees and shrubs rise, with continuous curves for *Plantago lanceolata* and *Pteridium* pollen. Herbaceous pollen decreases at 0.06 m. *Calluna* values fall, but *Sphagnum* spores and Cyperaceae pollen increase. Charcoal values fluctuate from 35% to 69%. Mineral content is low, but rises to 7.9% at 0.02 m. Pollen concentration varies from 22,019 to 180,357 grains/cm³.

The pollen diagram from Nook Farm 2 represents the vegetation in the early Bronze Age. Here the *Betula* woodland was cleared. This clearance was followed by a period of wetter conditions, indicated by the peaks of *Sphagnum* spores, and there is some evidence of wet meadows and/or pastures in non-regenerating woodland. Percentages of char-

coal particles declined after the fall in *Betula* pollen. The diagram suggests that a community dominated by *Calluna* developed with little suggestion of grassland until NF2-c when conditions again became wetter, charcoal increased, and dryland herbs were more abundant. *Calluna* recovered as conditions became drier but did not reach their earlier values. Pollen from trees increased, except for *Betula*, suggesting that the mire surface was not recolonised.

Summary of results
The three monoliths from Nook Farm have produced a record of the changing vegetation during the Neolithic and early Bronze Age. An attempt has been made in Figure 28 to illustrate the major differences in the pollen diagrams by showing the sum of the trees and shrubs, dryland herbs, and *Calluna*.

The pollen record from Nook Farm 3 represents a damp *Betula* woodland which developed as the *Alnus* was destroyed by burning with very little evidence of clearance. At Nook Farm 1, however, which is closer to the lithic scatter, the vegetation differs considerably. It illustrates the destruction, possibly by fire, of *Alnus* then *Betula* being replaced by a *Calluna*-dominated vegetation with some grassland. Macroscopic charcoal is also present in greater quantities at this site than at the flanking monoliths. Where the peat changes to a less humified *Eriophorum* peat there is greater evidence of agriculture, and it is suggested that it is of local origin from the sand island during the Neolithic. Nook Farm 2, which is dated to the Bronze Age, has less evidence of pastoral farming, but percentages of *Calluna* pollen and microscopic charcoal particles are high. Macroscopic charcoal is also present in significant amounts.

Discussion of Nook Farm results
The basal peats from Nook Farm 1 and 3 provide dates that are not significantly different statistically (Ward and Wilson 1978), therefore the vegetation change indicated by these results suggests there is possibly represented at Nook Farm a small local (ie < 100 m diameter) clearance of the vegetation on the sand island during the Neolithic.

Natural or artificial burning?
Throughout the following discussion an assumption is made that the burning was artificial and not natural. There are several reasons for this. Evidence from Norway and Japan suggests that a uniform and continuous curve of charcoal percentages indicates lengthy periods of habitation and cultivation, whereas isolated peaks probably characterise natural burning (Faegri and Iversen 1989; Kaaland 1986). At Nook

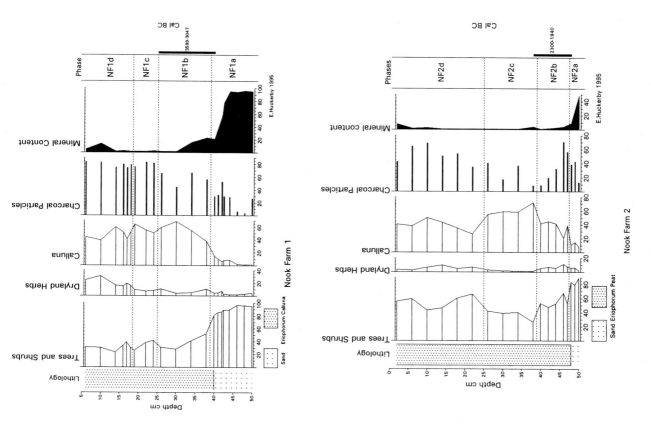

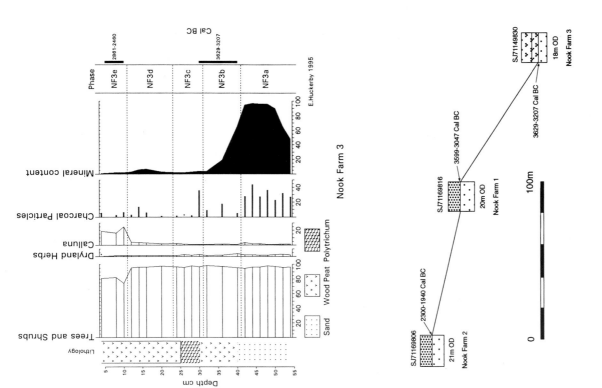

*Fig 28 Chat Moss: Nook Farm summary percentage pollen diagrams together with their relative positions
(for details of radiocarbon dates see Appendix 2)*

Farm 1 a continuous curve is present. In addition, a lack of woodland regeneration following burning may indicate human interference, although climatic deterioration could also produce a similar effect (Kaaland 1986; Odgaard 1988). At Nook Farm 1 there is little evidence for regeneration.

It is also uncertain how likely the occurrence of completely natural fires might be in British mires. Radley (1965), in his historical survey of fires on Derbyshire peatlands, came to the conclusion that there were few major fires before 1900 and that the case for natural (ie lightning-initiated) peatland fire was not proven.

Another possibility that must be considered is whether conformity of the charcoal particles with the stratigraphy is reliable. There has been speculation that fire may spread laterally beneath the surface of a mire to the mineral substrate (Chistjakov et al 1983). At Nook Farm, however, the evidence for changes in the pollen diagram contemporary with the charcoal would seem to make this possibility less likely.

Another feature that occurred in the three monoliths is the presence of swollen and distorted pollen grains, generally those of *Corylus avellana* type, *Betula, Alnus,* and *Calluna*. Andersen (1988, 1993) suggests that this is evidence of increased temperatures on the surface or in the soil, and therefore of *in situ* burning rather than the indications of distant fires in the catchment. In samples exhibiting this feature other plant material is modified, causing the preparation of slides for pollen analysis to be problematic.

Effects on mire development
There is an increasing body of evidence from the British Isles and Europe that suggests forests on poorer soils were cleared by human populations from the Neolithic onwards, with the subsequent development of heathland and/or peat deposits in areas of oceanic climate. Work in Denmark and Norway suggests that heathland was managed by fire to provide winter grazing from the Neolithic until recent historic times. *Calluna* is an evergreen low shrub whose growth is stimulated by fire and grazing. Kaaland (1986) puts forward the suggestion that repeated burning and grazing in conjunction with a deterioration in the climate prevented the development of *Empetrum* heath and tree regeneration supporting a species-poor community dominated by *Calluna* in Western Norway. This is a view also advanced by Odgaard (1988) for heathland areas in Jutland. The values of *Calluna* pollen in a pollen diagram are thought to be directly related to the abundance of the plant in the community (Moore et al 1986; Evans and Moore 1985). Pollen diagrams from Scandanavia record a greater variety of herbs after the initial burning, suggesting tempo-

rary clearances, a feature seen at Nook Farm after the decline of *Betula* and *Alnus*. The presence of charcoal in conjunction with the expansion of *Calluna* pollen points towards clearance of the woodland in the area of Nook Farm 1 in the Neolithic, continuing to a lesser extent in the Bronze Age. Birks (1964—5) considered there was little or no evidence of Neolithic or Bronze Age agriculture at Chat Moss. *Calluna* pollen, however, was not regarded as indicative of clearance when his paper was published. The sampling interval in this study was also closer than that used by Birks, and this would have a greater chance of recording brief periods of clearance. In view of these results it is possible that a Neolithic population caused local changes to the surface vegetation of the mire.

Scale of clearance
The effects of small clearances in pollen diagrams are known to be limited to relatively short distances. In studies of medieval sites from northern Germany it has been shown that frequencies of cultural indicators decline rapidly at a distance of 400 m from known fields (Behre and Kucan 1986). A similar effect has been recorded in the Somerset Levels (Coles et al 1973).

At Nook Farm it appears probable that the interference was on a small scale. On the basis of the available data, it is provisionally estimated to have been <80 ha in area (calculated using the distance between Nook Farm 1 and 3). However, time constraints and the removal of peat from the sand island precluded the mapping of its full extent and so the figure must remain tentative.

Despite this, the scale of disturbance is similar in range to estimates for Bronze Age clearances in Ayrshire recorded by Turner (1975) who speculated that they were possibly only a few hundred metres across, and therefore may be no more than 10 ha in extent.

General reconstruction and discussion of environmental change at Chat Moss
The earliest recorded (mineral) deposits from the Chat Moss basins appear to date from the Late Devensian 1 stage, ie the immediate late-glacial period (>12,000 cal BC). The upper boulder clays can probably be ascribed to this period when glacial conditions obtained, with little opportunity for organic accumulation to occur due to the periglacial environment. The sparse surrounding vegetation might typically have consisted of low-growing herbs, sedges and grasses with some dwarf trees (eg *Betula nana*).

Glacial outwash sands and gravels were deposited over wide areas of the boulder clay terraces characterising the upper Mersey valley, including the site of what was to become Chat Moss, probably during the subtle climatic amelioration of the Late Devensian 2 stage (the Windermere Interstadial). This period, dating from c 12,000—10,500 cal BC, is thought to have left a signature at Chat Moss in the form of subtle stratigraphic changes in underlying inorganic deposits (the current study did not examine these) manifested by the occurrence of a 'clay mud'. Birks (1964—5) established the presence of this feature in his deepest profile and found it to be slightly richer in organic remains than the clays above and below. The vegetation of the surrounding area at this time might be characterised as 'park tundra', ie much open ground with copses of pine and birch woodland.

A return to colder conditions marking Late Devensian 3, sometime after 11,000 cal BC, triggered solifluction deposits in the Chat Moss basins, while further afield the vegetation remained broadly similar to that found previously, but became more open in response to the deterioration in temperatures.

It was not until c 10,500—10,000 cal BC that periglacial conditions finally ended with the opening to the Flandrian I stage, and this was marked by the initiation of organic deposition in the deepest hollows. The landscape at this time was dominated by birch and pine, with birch predominating on the sands, gravels, and clays of what was to become Chat Moss. Sporadic wetter hollows dominated by monocotyledons (particularly reeds and sedges) formed rare exceptions in this lightly-wooded situation.

By c 8,500—7,500 cal BC a slow peat formation was widespread, generated by communities dominated by brushwoods, *Polytrichum*, *Aulocomnium palustre*, and *Eriophorum*-species indicating a gradual change to wetter conditions. Damp conditions encouraged the spread of poor-fen communities, deriving at least some of their nutrients from mineral ground run-off.

In the broader landscape of the time, the surrounding 'uplands' were predominantly wooded, and trees such as pine and oak were beginning to make their presence felt among the dense birch and hazel stands.

Burning of the mire vegetation began to occur over extensive areas of the moss at this time. The characterisation of artefacts from the north of Chat Moss as Mesolithic points to some level of human activity in the complex contemporary with the burning episodes, and a link between them and Mesolithic groups cannot be ruled out. In this connection it is perhaps worth noting that there appears to be evidence, from mires further west in the Mersey Basin, for Mesolithic-age burning associated with palynological evidence for clearance in the catchment (Cowell and Innes 1994). More work is clearly required before resolution of this interesting possibility one way or the other can be achieved. This would be particularly valuable to establish as it is conceivable that the effects of burning may have included the encouragement of a trend towards acidification and degradation of the system.

Eventually true ombrotrophic mire communities dominated by cotton-grasses (*Eriophorum* sp) succeeded this phase at c 6000 cal BC, forming more open conditions over wide areas of Chat Moss. Drier conditions still obtained on higher sand and gravel ridges where brushwood and related communities persisted. Burning of the vegetation continued sporadically while the surrounding vegetation was dominated by alder, birch, hazel, oak, and elm along with much lime. As time progressed, peat accumulation probably started to build up to the extent that the mire began expanding laterally, paludifying intervening mineral ridges and encouraging coalescence of the peats in the complex. As true ombrotrophic conditions became established over the bulk of the moss, the effects of any climatic perturbations will have become more pronounced. A series of possible wetter episodes may be reflected in *Sphagnum sect Cuspidata*-dominated pool peats, which are detectable at various points in the stratigraphy. There is a substantial body of information available which suggests that pool development across mires was often synchronous in North West England during prehistory and that it was fundamentally determined by climatic factors (eg Barber 1981; Barber *et al* 1994; Smith 1985; Wimble 1986). The detailed macrofossil studies from the Barton Moss and Worsley Farm sections of the complex, however, appear to suggest that pool development was largely independent at these two sites. More data is required before this can conclusively be shown to be a widespread characteristic of Chat Moss as a whole, although it points to the possibility that, unlike the studies cited above, climatic factors were not sufficiently strong to force widespread phase shifts over the whole mire complex before the Bronze Age.

The elm decline occurred around the same period as the later of these stratigraphical changes and, during Flandrian III (after c 4000—3700 cal BC), the effects of human activity became more obvious, with deforestation episodes affecting the surrounding woodlands. Until the current survey, the nature and scale of these clearances could only be guessed at due to the lack of resolution afforded in the sole regional pollen study. The use of multi-core pollen diagrams in the recent intensive study of the Nook Farm site, however, allows an insight into the nature of at least some of the clearance episodes. An apparently localised and short-lived clearance episode, involving predominantly the removal of birch, was here associated with burning of vegetation; the influence of the clearance seems to have extended no further than c 50—100 m. With the present level of data it is impossible to be certain how widespread such small-scale clearances were in the region. It is perhaps interesting in this context to note that a peak in charcoal recorded from the Worsley Farm macrofossil diagram is dated to 2878—2460 cal BC (4050±70 BP; GU-5360), not dissimilar to the determination achieved from charcoal at the Nook Farm site (2853—2149 cal BC; 3930±80 BP; GU-5325).

If the record from Nook Farm is typical it might be possible to envisage Chat Moss during the Neolithic as a shallow basin complex, with developing valley mires in wetter depressions and lightly wooded higher sand ridges subject to sporadic human disturbance in the form of limited clearances and burning. This activity might possibly have encouraged the further spread and development of the encroaching ombrotrophic communities.

Following on from the *Eriophorum*-dominated peats (probably sometime around 1700—1400 cal BC) *Sphagnum imbricatum* became the dominant peat-forming plant on the moss. This may have occurred as a result of a change to wetter climatic conditions, although Birks suggested that a 'flooding horizon' referable to this stage in the mire development may be connected to forest clearance activities implied in the pollen record. Whatever the cause, the subsequently accelerated rate of peat accumulation allowed further lateral expansion of the mire and paludification of any remaining drier ridges. This period, from the Iron Age onwards, saw Chat Moss reach its biggest and wettest state. It would have dominated the landscape around this section of the Mersey Basin, forming an inconvenient obstacle of agriculturally poor terrain. This may be reflected in the apparent reduction in intensity of burning recorded from the upper *Sphagnum*-dominated parts of the profiles referable to this period. However, from the early Middle Ages onwards it is likely that human exploitation of the mire would have increasingly affected the hydrology of the system, initially by peat cutting at the fringes, but gradually increasing in intensity through the medieval period until the reclamation schemes of the eighteenth and nineteenth centuries began to make severe inroads into the hydrological integrity of the system.

Due to the truncation of the peat profile over most, if not all, of the system it is impossible to discern whether any further change in stratigraphy occurred after this period, but by comparison with other sites in the region it seems likely that *Sphagnum*-dominated communities persisted until historically recent times when drainage and development commenced.

Recommendations for further research

Further stratigraphic survey is required to test ideas about mesotope development at this very large and complicated site. A series of macrofossil analyses from each putative mesotope would elucidate mire ontogeny in greater detail and would help to identify causes of stratigraphic changes (*eg* climatic/anthropogenic).

The identification of a possible small-scale prehistoric clearance at Nook Farm suggests that a series of pollen studies across the complex may be worthwhile in order to attempt to isolate anthropogenically-caused small-scale changes in the pollen record undetectable by other means.

More radiocarbon dating of stratigraphical changes is necessary to test uniformity of development across the complex. Wider dating of charcoal bands identified from the survey is desirable in order to identify any chronological trends.

Summary
The survey revealed that, despite many years of disturbance, extensive areas of well-preserved prehistoric peats still remain in the Chat complex.

Deep hollows within the complex began to accumulate organic deposits in the late-glacial period, and fen-carr and brushwood communities began to develop over many parts of the present site in the Mesolithic/early Neolithic periods. These communities became restricted to higher and better-drained ridges during the late Neolithic and Bronze Age as valley mire communities began to develop on wetter terrain. True ombrotrophic communities dominated by *Eriophorum* and *Calluna* became established over wide areas and caused the coalescence of formerly hydrologically discrete mires into a larger complex. Wetter episodes, possibly caused by climatic changes during the Neolithic and Bronze Age periods, caused the *Eriophorum* sequences to be punctuated by a series of widespread pool peats characterised by *S sect Cuspidata*, although it is not clear how widespread these were. A permanent change to wetter conditions, probably at some time in the mid-to-late Bronze Age, is indicated by the widespread development of *S imbricatum*-dominated peats.

There is extensive evidence for fire disturbance throughout the prehistoric period. The site contains large areas of peat with the potential to produce pollen records covering much of the prehistoric period. The peat archive in its current condition represents an extremely important palaeoecological resource, but it is threatened in the medium-to-long term if current rates of wastage continue.

4

THE LARGER MOSSES

This chapter describes the larger mosses, other than Chat Moss, where peat survives. The mosses of the group, which have large amounts of peat covering many hectares, are Ashton Moss, Carrington Moss, the Kearsley Moss complex, and Red Moss.

Ashton Moss
(SJ 920 985)

Ashton Moss (Figs 4, 29, Plate 12) is located at the foot of the western Pennines, close to the 100 m contour, lying on boulder clay that blankets faulted Bunter Sandstone and the shales and mudstones of the Middle Coal Measures, and covering an area of c 125 ha. It incorporates Little Moss on the north side, and Droylsden Moss on the west, and lies on the western side of Ashton-under-Lyne parish, between the River Medlock to the north, and the River Tame, to the south.

A stone axe is said to have been found on Droylsden Moss (Higson 1859, 29—30; SMR 613, SJ 9098). Recent fieldwalking over the moss recovered a number of worked flints, 50% of which were found on organic soils and the remainder on peaty-loam soils. The material includes one possible microlith and nine other flints, possibly Neolithic (Nevell 1992a, 24, and pers comm). Three Roman coins of first or second century AD date were found during excavation of a ditch on Droylsden Moss, together with pottery sherds and a 'tomahawk-type' bronze implement (Higson 1859, 30; SMR 612; SJ 9119 9903). An undated, possibly early, human skull is preserved from a find spot on Ashton Moss (Stead 1986 et al, 184; Nevell 1992b, 6, plate 1.3).

The low level of finds suggests that, as with other Manchester mosses, there was little human activity in the immediate region of the wetlands after the prehistoric period until the Middle Ages.

Nico ditch, a linear ditch and bank (discussed in the Gazetteer, under Hough Moss), begins at Ashton Moss

and circles South Manchester as far as Stretford. It probably dates from the pre-Conquest period and may be a territorial boundary. Cinderland Hall, on the north side of Little Moss, is a sixteenth—seventeenth century hall (SMR 769, SJ 9135 9976).

Domesday Book records a church of St Michael's, identified as the church at Ashton. This, together with the possible etymology of the place name from British lyne referring to the 'elm' (lemo), suggests that Ashton originated from an early settlement core, although an alternative interpretation derives lyne from the boundary of the Honour of Lancashire (Ekwall 1922, 23; Nevell 1992b, 101).

Ashton was certainly a village during the twelfth and thirteenth centuries, and was granted a market and two fairs by 1414 (Morris 1983, 27). Early agricultural references indicate that by 1190—1212, woodland, field, meadow, and pasture in Audenshaw, on the south side of Ashton Moss, were being exploited. At Ashton, in 1346, the demesne included 40 acres of arable worth 12d per acre, 12 acres of meadow at 2s per acre, and 220 acres of woodland at 12d per acre; an early custom roll for Ashton refers to pig breeding and pasturage for cattle (Nevell 1991a, 52; Farrer and Brownbill 1914, 4, 341 n38).

Enclosure of Ashton Moss was clearly taking place during the medieval period, as illustrated by a dispute between 1400 and c 1425 relating to the meres (bounds) of Ashton Moss. The moss was of importance to the lord and tenants of Ashton manor; a custom roll for 1422 records the lease of turbary rights for £5, and the 16s paid by Sir John Byron was nearly as much as the 16s 4d he paid for his milling rent (Nevell 1991a, 59). Rentals list field-names such as 'moor hey' and 'hayscroft' at Ryecroft Farm on the southern fringe of Ashton Moss, suggesting waste land had been reclaimed.

Inventories of wills show that mixed farming was practised during the sixteenth and seventeenth centuries; the 12 acres of corn, oats, and barley listed in an inventory of 1617 for Alexander Newton of Newton township south of the River Tame were typical of this

Plate 12 Ashton Moss showing the peats overlying the deepest basin

area (Nevell 1991a, 83). Ashton Moss was divided into strips, known as 'rooms', and was managed by moss reeves, bailiffs and 'moss lookers'. Court rolls show the effective control made by these officers. Separate areas of the moss had different names, such as 'shadow moss' in 1613, and 'donome moss' of 1617 (which may be the same as 'donean moss' referred to in a 1422 rental from which ten loads of turves were to be conveyed to the manor house (Nevell 1991, 88)). By the early seventeenth century most of Little Moss had been reclaimed, and in 1831 Ashton Moss was drained and cultivated for market gardening. In the nineteenth century local farms supplied the growing urban industrial population of the town with market garden produce from drained land on Ashton Moss.

No new fieldwork was undertaken in 1992—3, as an assessment was made and a report published in 1991 (Nevell 1991b). The palaeoecological work undertaken at Ashton Moss in 1990 is reported fully below, and shows that the moss may have begun to develop in the Mesolithic period, surrounded by deciduous woodland.

The moss will be substantially destroyed by the planned route of the M66, and the metrolink, which will cut through its eastern and southern side (Plate 13). A full record of the archaeology and palaeoecology with radiocarbon dates should be made before any destruction of the moss occurs.

Palaeoecological evidence
Ashton Moss is the best preserved of a small series of former mires, now largely destroyed, situated at altitudes of 90—120 m OD and occupying broad shallow hollows in boulder clays in the upper Mersey catchment within the Irwell—Irk—Medlock hinterland (Shimwell 1985).

Previous research
Historical studies of the land-use of Ashton Moss have been undertaken by Fletcher (1982) and Nevell (1991b, 1992b), and the Soil Survey included the site in its inventory of English and Welsh lowland peat deposits (Burton and Hodgson 1987). The latter recorded a broad tripartite division into *Sphagnum* peats, *Sphagnum / Eriophorum* peats, and fen wood peats over boulder clay. Turbary Moor series soils were recognised on the site. Shimwell and Robinson (1993) undertook a small study of a cutting exposed during construction work as part of an archaeological evaluation in advance of proposed commercial development, and produced

pollen analyses from four spot samples down the peat profile.

Sampling strategy

A series of four transects of cores (Fig 29) was employed in the present survey in order to obtain a three-dimensional view of the stratigraphy of the peat deposits (Figs 30—3). A supplementary core (SD 9210 9860) was taken near to the approximate centre of the mire in order to obtain more information on the extent of the deepest deposits of peat.

Levelling of the surface height of the coring positions was obtained relative to a temporary bench mark situated on the gate entrance to the cricket ground at the eastern end of the site. Relative heights of the surface of the core positions are displayed.

Condition and survival of the peat archive

The current study proved the widespread survival of peats deeper than 4 m, with several substantial areas carrying organic deposits reaching >6 m. This contrasts with the Soil Survey investigation which asserted an average peat depth of 3.7 m (Burton and Hodgson 1987). The peats thin towards the margins of the mire, but still retain healthy depths (*ie* >1.5 m) close to the boundaries of the site.

The state of preservation of the organic deposits is also good, with upper *Sphagnum imbricatum* peats often displaying fresh, relatively unhumified characteristics ranging from H2 to H5. The lower parts of the stratigraphy are more humified, averaging H7—9, but are sufficiently well preserved to enable detailed palaeoecological analysis.

Topography and mire classification

The peat surface/depth profiles indicate that the moss occupies a shallow elongate basin with an uneven base. The longitudinal transect AMa (Fig 30) indicates that the depth of the clay/peat interface increases to the south-west, with a large basin occupying the south-western part of the site. The lateral transects show that a raised sub-surface ridge appears to exist in the topography, running parallel to the long axis of the moss and situated approximately to the north-west of Rayner Lane. Adjacent to this, a deeper channel appears to run along the north-west margin of the mire. The south-eastern margin of the moss appears by contrast to be rather shallower along much of its length. As with the other large mires in the

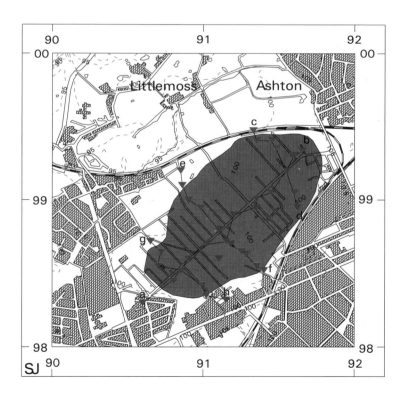

Fig 29 Ashton Moss: transect location and peat extent

Based upon the 1991 Ordnance Survey 1:10,000 map with permission of The Controller of Her Majesty's Stationery Office © Crown Copyright

65

Greater Manchester region, the morphology of the peat body appears to conform best to the characteristics of an intermediate mire.

Immediate pre-peat stratigraphy
The sub-peat surface consists primarily of silty clays forming an irregular topography of hollows and ridges. The substrate appears to be primarily formed of boulder clays.

Peat stratigraphy
Over much of the site the deposits display a series of basal wood peats overlain by a long sequence of *Eriophorum/Calluna*-dominated peat with some *Sphagnum*. Shimwell and Robinson (1993) noted the presence of a silt band interrupting the *Eriophorum/Calluna* peats from their sampling site in the north-east periphery of the moss. This feature was not recorded from any other part of the site and most probably represents a localised feature.

The final stratum is formed by fresh peats almost totally dominated by *Sphagnum imbricatum*. The degree to which this sequence is preserved varies over the moss, being heavily truncated particularly around its present margins and best represented towards the middle of the site south of Rayner Lane where the organic infill is deepest (cores at SD 9215 9872, SD 9218 9870, SD 9225 9865, SD 9228 9863, SD 9210 9860). In the deepest cores, however, the wood peat layer is absent and there are indications of fen and/or bryophyte communities having developed over the basal silty clays.

Carbonised material
Time constraints meant that the Ashton Moss peats were not studied as thoroughly as those from other sites in the county and hence relatively few field samples were taken for analysis in the laboratory. Because of this, the distribution and frequency of charcoal in the peats from the site is poorly known. However, from the few sequences that were examined in detail it seems that carbonised material is relatively scarce, the main occurrence being manifested at the south-west part of the mire in basal wood peats (SD 9180 9845; Fig 30). This limited evidence conforms to the general trend for concentration of carbonised material in the lower stratigraphy noted from other mires in the county.

General overview and discussion of environmental change
Considering that Ashton Moss is known to have been reclaimed and under cultivation as early as the 1830s

Plate 13 Ashton Moss showing the urban development encroaching its margins and which seems fated to destroy this exceptional environmental archive completely

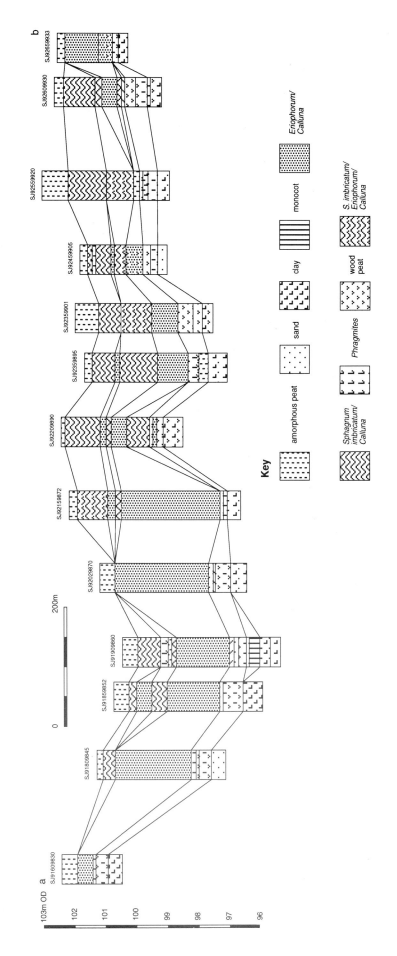

Key

Eriophorum/ Calluna

monocot

S. imbricatum/ Eriophorum/ Calluna

clay

wood peat

sand

Phragmites

amorphous peat

Sphagnum imbricatum/ Calluna

Fig 30 Ashton Moss: transect A

67

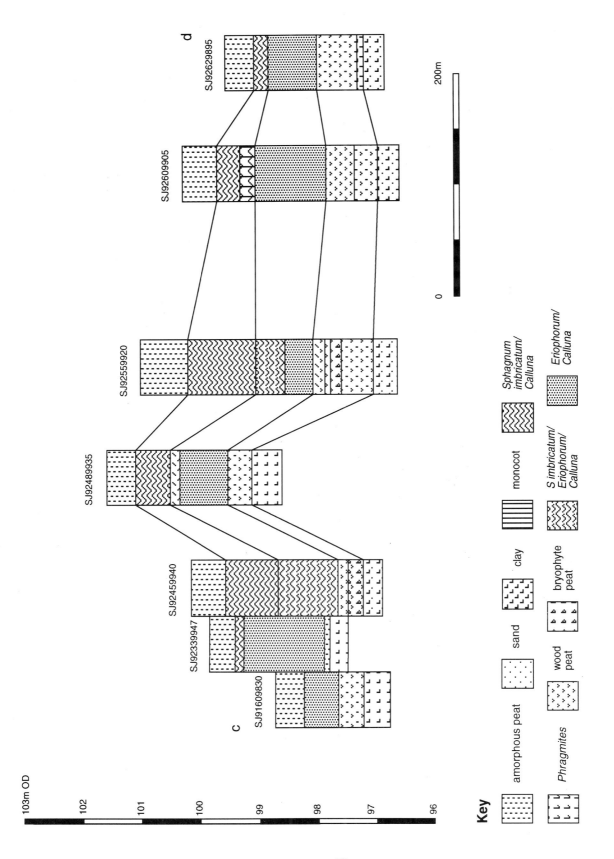

Fig 31 Ashton Moss: transect B

103m OD
102
101
100
99
98
97
96

d

SJ92629895

SJ92609905

SJ92559920

SJ92489935

SJ92459940

SJ92339947

c

SJ91609830

0   200m

Key

amorphous peat

sand

clay

Phragmites

wood peat

bryophyte peat

monocot

Sphagnum imbricatum/Calluna

S imbricatum/Eriophorum/Calluna

Eriophorum/Calluna

68

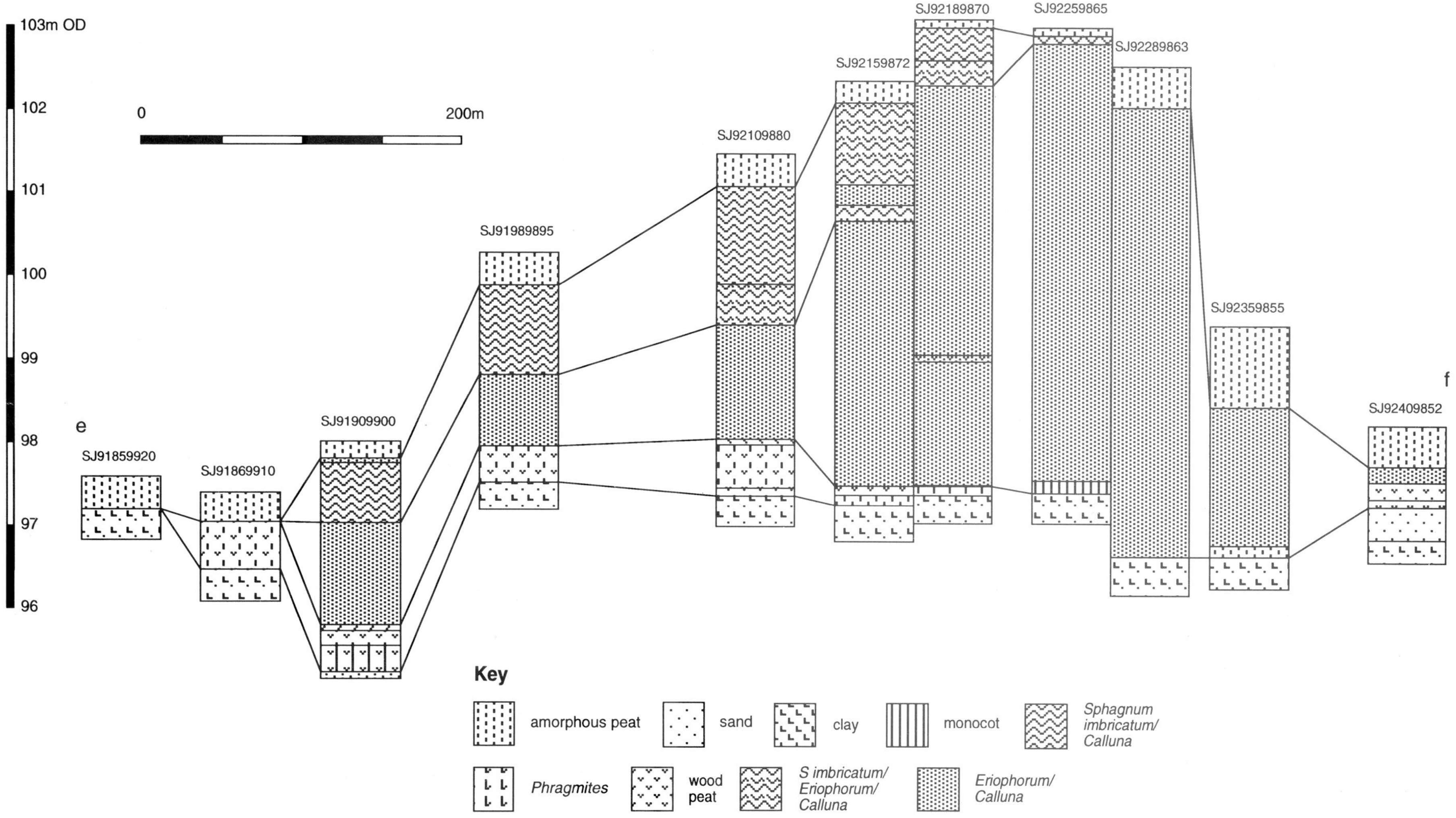

Fig 32 Ashton Moss: transect C

Key

- amorphous peat
- sand
- clay
- monocot
- *Sphagnum imbricatum/ Calluna*
- *S imbricatum/ Eriophorum/ Calluna*
- *Eriophorum/ Calluna*
- wood peat
- *Phragmites*

SJ92409852
SJ92359855
SJ92289863
SJ92259865
SJ92189870
SJ92159872
SJ92109880
SJ91989895
SJ91909900
SJ91869910
SJ91859920

e
f

103m OD
102
101
100
99
98
97
96

0
200m

69

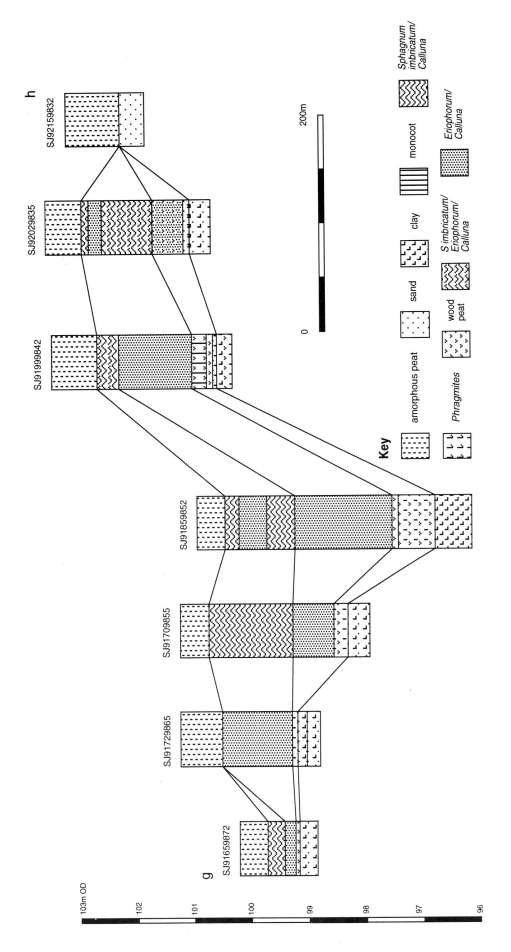

Key

Sphagnum imbricatum/Calluna

Eriophorum/Calluna

monocot

S imbricatum/Eriophorum/Calluna

clay

sand

wood peat

amorphous peat

Phragmites

h

SJ92159832

SJ92029835

SJ91999842

SJ91859852

SJ91709855

SJ91729865

SJ91659872

g

103m OD
102
101
100
99
98
97
96

0   200m

Fig 33 Ashton Moss: *transect D*

(Nevell 1991b), the investigation proved the survival of surprisingly well preserved and often deep peat deposits. It is probable that the relatively low intensity of agricultural activity (most of the moss is subject to market gardening cultivation) has enabled good sections of the stratigraphic record to remain intact.

The sequence of deposits recorded suggests that aquatic and fen communities were the earliest peat-forming plant communities to develop on the site, occupying a deep hollow in the clay surface slightly offset to the south-east of the approximate centre of the site. By comparison with similar topographical situations at Chat Moss, it would seem possible that organic deposits of at least immediate post-glacial age occur in this area. Mire vegetation in this wetter area probably persisted while the rest of the site became dominated by woodland and scrub. No radiocarbon dates are available for an independent chronology of the stratigraphy at the site, but by inferring dates based on their spot pollen preparations, Shimwell and Robinson (1993) suggest tentatively that fen-carr formation may have begun around the late Mesolithic/early Neolithic period, perhaps some 5000—4500 years BC. The site at this time would have formed a wet scrubby habitat set amongst the denser deciduous woodland occupying the higher boulder ridges surrounding the site. Edaphic conditions subsequently became wetter over the whole site allowing *Eriophorum/Calluna*-dominated peat-forming vegetation to develop over all areas. This would have produced a damp, open ecosystem passable on foot and which probably persisted for several hundred, or even thousand years. The presence of a silt band at the edge of the system recorded by Shimwell and Robinson hints at the possibility of catchment disturbance inducing temporarily increased run-off into the mire basin during this period.

The *Eriophorum*-dominated phase ended abruptly with a change to much wetter surface conditions, enabling an almost totally *Sphagnum imbricatum*-dominated mire to develop. A soft, spongy surface with pools of open water would have resulted making for difficult terrain in terms of access. Truncation of the uppermost profile obscures the later development of the mire, but indicates that the value of the peat accumulation as a domestic energy resource was recognised and exploited from at least the medieval period onwards (Nevell 1991b).

Recommendations for further research
The existence of extensive, deep, well-preserved peat deposits at Ashton Moss holds great potential for detailed palaeoecological work. A more detailed survey to establish the extent of carbonised plant material distribution is recommended and further palynological elucidation of the nature of the silt band recorded from the north-east of the site is necessary to verify its archaeological significance.

Due to the outstanding degree of preservation and survival of the peats at Ashton Moss, and also because this part of north-east Manchester (and south-east Lancashire as a whole) enjoys only poor palynological coverage, it is strongly recommended that a detailed and complete pollen investigation of the site be undertaken prior to any major disturbance of the stratigraphy, along with plant macrofossil studies of the peats.

Summary
Ashton Moss is a former active ombrotrophic intermediate mire, now reclaimed for agriculture. Stratigraphic survey shows that peat formation began in its irregularly-contoured shallow basin with aquatic and fen communities and after a period of scrub woodland domination over much of the site, ombrotrophic peat-forming vegetation became established and later expanded to dominate the whole mire. Later on still wetter conditions allowed the domination of the vegetation by *Sphagnum imbricatum*. Despite truncation in recent times, it is extremely likely that a pollen record covering most, if not all, of the prehistoric period is preserved in the peats at the site. It is also possible that a substantial portion of the post-Roman pollen record survives. The site in its present form has considerable value as a palaeoecological archive.

Carrington Moss
(SJ 740 915)

Carrington Moss is situated on the southern side of the River Mersey and the confluent River Irwell, at *c* 29 m OD (Front cover, Fig 34). The moss formed over fluvial sand and gravels deposited on top of clays; 325 ha of peat currently survive. The underlying solid geology is a band of Permo-Triassic sandstones and shales.

The township of Carrington is bounded on the west and north by the River Mersey, and on the south by Sinderland Brook leading into Red Brook, a tributary of the Mersey. The eastern side is formed by the eastern edge of Carrington Moss, adjacent to Ashton Upon Mersey township. The two villages in Carrington township, Carrington and Partington, are hemmed in between the river and the moss; the main road (A6144) linking the two villages, shown on the OS First Series 6" map (Ordnance Survey 1882, Cheshire Sheets 8, 9), is likely to be an ancient route

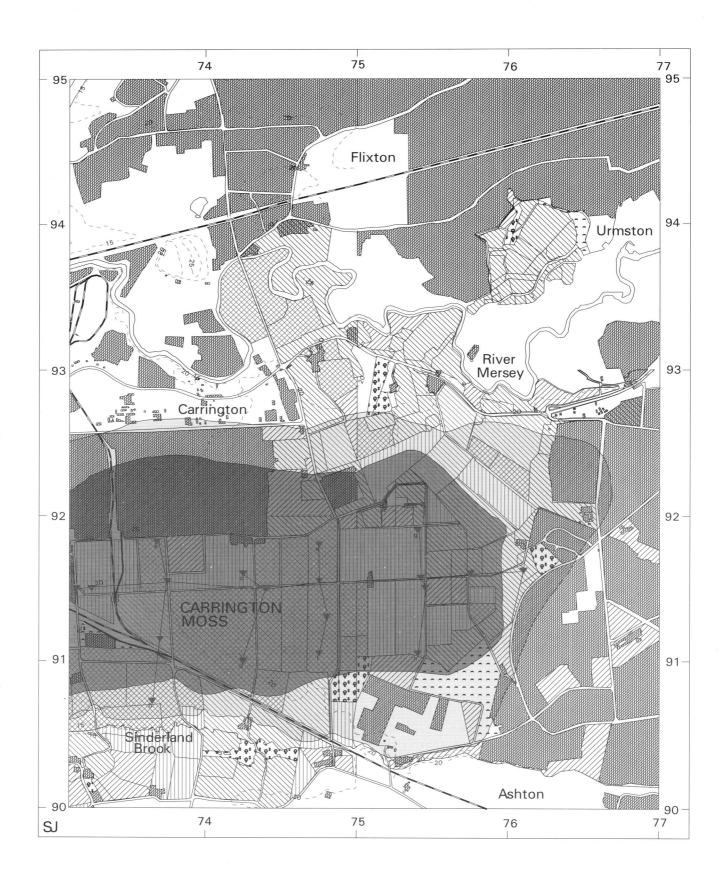

Fig 34 Carrington Moss and Urmston E'es: land-use, fieldwork, and peat extent with boreholes
Based upon the 1981/1991 Ordnance Survey 1:10,000 map with permission of The Controller of Her Majesty's Stationery Office © Crown Copyright

between the river flood-terrace and the edge of the moss. Access across the Mersey was at Carrington Bridge to Flixton, and at Partington a ferry crossed to Cadishead. The Old Hall and village core of Carrington are sited on the southern side of a bend in the river, with flood meadows to the north. Partington is located on a slight elevation, opposite to Cadishead, leaving a narrow channel for the river to pass. The ancient village cores have been largely rebuilt.

No archaeological sites or artefacts are known from the moss. Neither village is listed in the Domesday Survey, but a family of Carrington is recorded from 1191 until c 1600 when the estates passed by marriage to the Booths of Dunham. John de Carington had 26 acres of land, one acre of meadow, about two acres of pasture, and 100 acres of turbary in 1336—7, and George Carrington occurs in a 1402 rental of Dunham Massey (Ormerod 1882, 542, 543). A survey of 1553 lists for John Carrington, the last male heir of the Carringtons, 200 acres of cultivated land, 40 acres of meadow, 60 acres of pasture, three acres of wood, and 500 acres of moor, moss and turf (Trafford Archives: Local History Packs: Altrincham Section 2, appendix 3; Carrington and Partington). Partington is similarly recorded through the family name in c 1216 (Henry de Partington, in Ormerod 1882, 545).

The Elizabethan Carrington Hall, reputedly on the site of an earlier building, was demolished in 1849 and rebuilt as a farm in 1858—9. A beam from the hall, bearing the arms of the Carrington family, was reused in a nearby barn (Ormerod 1882, 542; White 1860, 909; SMR 1228, SJ 7430 9292). Another site, Danes-at-the-Well Farm (Dainewell), is said to have existed since the fourteenth century, although there is no evidence for this (White 1860, 909). A rental for 1851 records 536 villagers and 130 houses at Carrington, with an annual value of £2500 (White 1860, 909). Part of the estate was sold to Samuel Brooks in 1858, including the whole of Carrington Moss; he revived the court leet and ordered the leaseholders to maintain hedges, drains, and ditches.

The Cheshire Lines Railway, constructed between 1865 and 1873, crossed the south-west part of the moss. At this time the Earl of Stamford attempted the drainage of c 200 acres of mossland on the western edge, although the Ordnance Survey First Series 6" map (Ordnance Survey 1882) shows that 600 acres of raw bog still remained.

A report to the Health Committee for Manchester by Henry Whiley in 1883 stressed the need for a new waste and nightsoil disposal site for Manchester City (Phillips 1980a, 228). Carrington Moss was used for nightsoil in 1885; a new central east—west road with crossroads was laid. Waste was also transported by boat along the Rivers Irwell and Mersey until the opening of the Manchester Ship Canal in 1894, when a new large wharf was built at Carrington.

To distribute nightsoil over the moss, a two-foot gauge track was constructed along the roads, extending to six miles, with its own locomotives and trucks. Fields between the roads were rectangular in shape and eight acres in area. They were subdivided into four two-acre parcels by open drains four feet deep. Each two-acre plot was to be drained by spit drains four yards apart. In fields adjoining the roads, additional open drains, six feet by nine feet deep, were dug. The surface was delved and nightsoil added at a rate initially of 60 tons per acre. Potatoes would then be planted for two years. In the third year a crop of oats was to be taken, while clover and rye grass were to be sown in the fourth and fifth years (MCA Report to the Health Committee, Phillips 1980b, 99).

As reclamation developed, the site was let to farmers, the first tenancies on 570 acres being made in 1887, and by 1899 all the land was on lease for cultivation. By 1910, with the advance in sanitation in the town, very little nightsoil was sent for dumping on the moss. The rail tracks were taken up for smelting c 1940.

The area to the north of the moss is now extremely industrialised, particularly when compared with Dunham and Warburton. One of the first mills was erected in 1755 on land bought from the Old Quay Company. It was a paper mill, and was originally supplied with water through a tunnel from the Mersey. In 1793 it was converted to steam power. The nearby Millbank slitting and rolling mill was converted to a corn mill in 1796, and in 1855 was bought by the owner of the paper mill and formed into a second paper mill (Trafford Archives: Local History Packs: Altrincham Section 2, appendix 3; Carrington and Partington). In the 1920s the Manchester Corporation used some of the land they had bought at Carrington to build a gasworks at a cost of a million pounds. It was opened in 1929 (Trafford Archives: Local History Packs: Altrincham Section 2, appendix 3; Carrington and Partington).

Carrington Moss is currently used mainly for arable agriculture with encroaching industry on the north which has destroyed some of the peat (Fig 34). Future threats are likely to be further industrial encroachment by large companies which own the whole site, causing complete piecemeal destruction.

Field mapping in 1992 and 1993

The north part of Carrington Moss is much built over by industry (Fig 34). The centre is a plateau of slightly

Plate 14 Carrington Moss showing the surrounding industrial development and the nightsoil-enriched peat surface

raised, deep peat with a nightsoil covering (Plate 14). Throughout this volume the term 'nightsoil' is applied to areas with late nineteenth century rubbish consisting mainly of broken sherds, glass fragments and cinders, being domestic refuse that was added to nightsoil and became distributed with it. The rubbish is fairly ubiquitous throughout the region and forms a noisy background for archaeological finds. The soils and topography of Carrington Moss are described below, beginning from the north-west, working anticlockwise (Fig 34). The total area surveyed of peat more than 0.3 m deep is 325 ha (800 acres). On Figure 34 peat extent has been accurately mapped where the moss is arable; elsewhere the extent is estimated.

Near Heath Farm Lane (SJ 727 781) deep peat falls to 0.3 m depth in Broad Oak Wood down to sandy skirtland running south to the Sinderland Brook. Dark-stained skirt soils go much farther south-west than has been previously recorded, almost joining up with Warburton Moss. At Broad Oak Farm two fields of skirtland next to the brook, with peat on the north side, were in good condition for archaeological finds,

but nothing was found. Beyond Sinderland Brook sandy undulating soils had no dark stain, showing that the brook bounded the moss on the south.

Brook Heys Farm is mostly pasture, but one field of sandy skirt was arable and had peat in the north-west corner. No finds were made in a freshly-dug service trench. Birchhouse Farm, at the south-east of the moss, consists mainly of deep peat rising to a slightly higher plateau. Skirt fields lie near the farm; at the north-east of Birchmoss Cover (SJ 7490 9095) is a 'bog oak' tree trunk about 0.9 m in diameter and 12.3 m long.

All the southern skirtland was carefully examined when planted, rolled smooth, and weathered after rain and wind scouring. A few sandy islands showed through the peat; the tiniest piece of pottery, flint, or chert would have been visible; but the sand was clean and barren. This should be a prime area for settlement with the moss and Sinderland Brook in close proximity.

Peat was mapped satisfactorily from dike-side sections and molehill upcasts at Ackers Farm and Ash Farm. North of Ash Farm there is dark-stained sandy skirt 0.5 km wide. A skirt band lies at the north-east and beyond that is the skirt/mineral soil boundary. A ridge just south of the A6144, between the Mersey and the moss, seems to be the northern limit of peat, leaving only a very narrow strip along the river bank for settlement of later periods.

In the meadows by the Mersey, opposite to Flixton Bridge, sections revealed by slips of the river bank show there is 0.25 m of sandy alluvium, over 3.5 m of uniform sand lying on brown clay. There is no river gravel usable as a source of flint, and the slight alluviation shows there was little medieval arable land. The river floodplain is edged by a small scarp on the south where occupation sites would be, but all the fields are pasture. No finds were made on the ploughed floodplain. Carrington vill was sited on this limited dry bank similar to Irlam on the north side.

Above Danewell Farm, on the north of the moss, peat lies against a high ridge, although skirtland extended slightly over it towards the Mersey. The farm lies on a low spur between two small tributaries of the Mersey, in an ideal location for settlement. One field was ploughed, but no sites were observed.

The fieldwork repeated the historical record; no archaeological sites could be discovered. In view of the sandy soils and terrain it is surprising that prehistoric lithics were not present. The absence of a flint source in the Mersey floodplain highlights the difficulties of finding sites; flint was not readily available and would not be thrown away in quantity on a domestic site. Some of the prehistoric finds from the Manchester Ship Canal, described under Chat Moss, may relate to Carrington Moss. Mesolithic sites may well lie buried underneath the present moss.

Palaeoecological evidence

Carrington Moss comprises a roughly rectangular area of shallow peats which formerly covered an area approximately 3 x 2 km, situated in a very shallow basin on a terrace of the south bank of the Mersey. Most of the site is under arable cultivation.

Sampling strategy

The present study employed 22 cores in five transects spread across the moss (Figs 34—44).

Condition and survival of the peat archive

All the peat profiles were truncated to some degree. Peat depths range from 2.70 m to 0.3 m, where still present, while thin organic soils occur around the periphery of the site marking the former extent of the mire. Over much of the moss, however, the peat cover ranges between *c* 1 m and 2 m. On the basis of evidence derived from Ordnance Survey First Series maps, it would appear that perhaps *c* 100—150 ha of peat has been destroyed by industrial development along the northern and western edges of the moss, and a small amount has probably been removed from the eastern part.

Because of the extensive disturbance to the peat surface it is difficult to generalise about the former shape and profile of the mire system. However, based on comparison of field observations with similar systems it seems possible that the site originally carried only a relatively shallow peat cover (*ie* <4 m) before reclamation, and that subsequent agricultural activity has removed perhaps 1.5—2 m of peats on average. The remaining peats are quite badly humified, frequently in the range H7—9 (*see Table 1*), although the degree of preservation is currently adequate for detailed palaeoecological analysis to be carried out.

Topography and mire classification

The highest point recorded on the peat surface reached 22.63 m OD while the lowest was 18.75 m OD.

The underlying mineral surface occurs mostly at an altitude of 18—20 m OD, forming an extremely shallow and broad basin in the fluvioglacial terrace of the Mersey, within which peats have developed. Close to the approximate centre of the site, at SJ 7440 9150 (Fig 35), the silty clays form an 'island' of slightly higher ground which slopes away on all sides. The highest peats occur on this raised portion of the site, giving the superficial impression of a raised mire with a characteristic domed profile. It is clear, however, that the peat topography is largely dictated by the underlying stratigraphy, witnessed by the highest mineral surface being coincident with this point. The site could probably best be regarded as resembling an oblong pie crust with the highest part near the centre and the pastry contained within slightly crimped upraised edges. The question of mire type definition of the site is difficult, given the limited amount of stratigraphic data obtained, but on the basis of the available evidence it would seem likely that the site is a variant of an intermediate mire (*sensu* Moore and Bellamy 1974).

The immediate pre-peat stratigraphy

Although the sub-peat mineral ground was not analysed, field observation indicated that a predominantly sandy substratum was widespread over much of the moss area. This is likely to be a facies of the fluvioglacial deposits characteristic of the Mersey terrace; these are described in more detail in the Chat Moss section (*see Chapter 3*).

Peat stratigraphy

A relatively simple wood peat / *Eriophorum* / *Sphagnum imbricatum* peat succession is characteristic of most of the site. Much of the *Sphagnum imbricatum* layer has been truncated by reclamation activities. In a few locations on higher, better drained, sandy ridges, a *Polytrichum*-rich peat precedes the *Eriophorum*-dominated phase (*eg* SJ 7375 9155; Fig 35; SJ 7540 9190; Fig 38), and in a few places there is evidence for a *S sect Cuspidata* pool formation prior to *Sphagnum imbricatum* succession (*eg* SJ 7325 9150; Fig 39), although this seems to be a localised feature.

Carbonised material

Burnt plant fragments (particularly of Ericaceous species) are a common feature of the Carrington Moss peats. Carbonised material is particularly noticeable in the lower sections of the stratigraphy in *Eriophorum* / *Calluna* and wood peats (Figs 40—44).

General overview and discussion of environmental change

Because no palynological work was possible during the current survey the detailed discussion of environmental change will be restricted to the local landscape changes within the mire basin. However, the general sequence of regional vegetation change is likely to be similar to that recorded by Birks (1964—5) for Chat Moss (*see Chapter 3*).

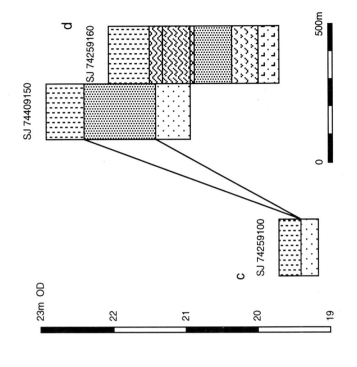

Fig 35 Carrington Moss: transect A

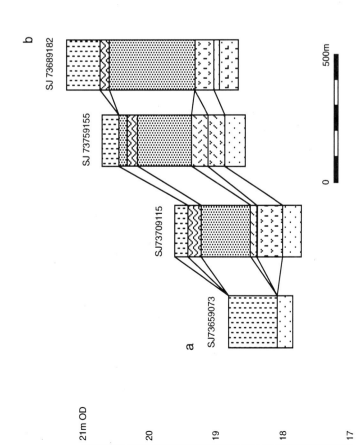

Fig 36 Carrington Moss: transect B

Key

amorphous peat

Eriophorum/
Calluna

wood/
Eriophorum

Sphagnum/
Eriophorum/
Calluna

wood/
monocot

Sphagnum

Scheuchzeria

S. cuspidatum

Eriophorum/
Polytricum

silty clay

sand

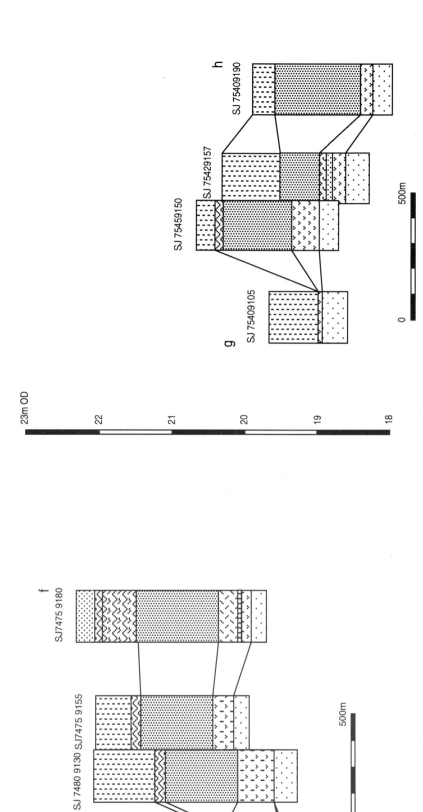

23m OD

22

21

20

19

18

SJ 7459150

SJ 7529157

h
SJ 7409190

g
SJ 75409105

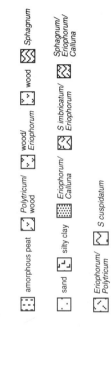

500m

0

Fig 38 Carrington Moss: transect D

23m OD

22

21

20

19

f
SJ7475 9180

SJ 7480 9130 SJ7475 9155

e
SJ 7475 9105

500m

0

| amorphous peat | *Polytricum/* wood | *wood/* *Eriophorum* | wood | *Sphagnum* |

| sand | silty clay | *Eriophorum/* *Calluna* | *S imbricatum/* *Eriophorum* | *Sphagnum/* *Eriophorum/* *Calluna* |

| *Eriophorum/* *Polytricum* | *S cuspidatum* |

Fig 37 Carrington Moss: transect C

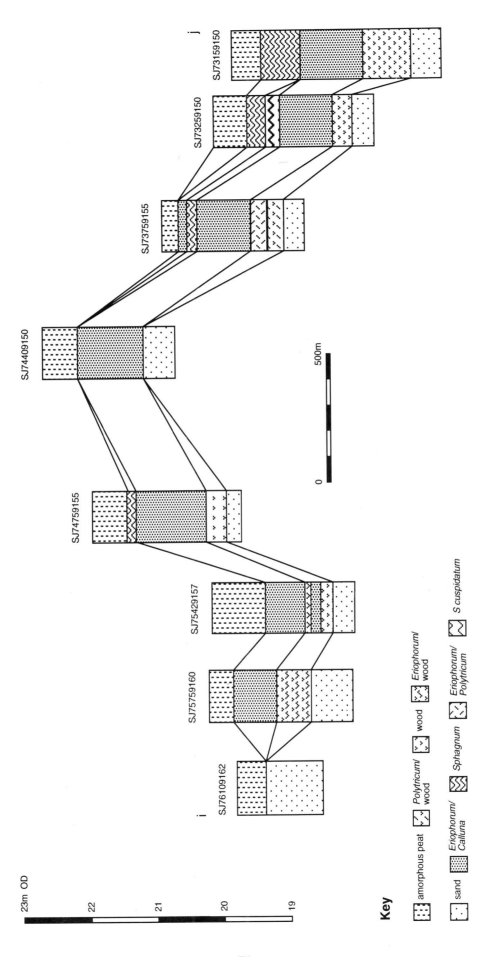

Fig 39 Carrington Moss: transect E

Key

amorphous peat [Polytricum/wood] — Polytricum/wood — [wood] — Eriophorum/wood

sand — Eriophorum/Calluna — Sphagnum — Eriophorum/Polytricum — S cuspidatum

23m OD
22
21
20
19

500m
0

SJ73159150
SJ73259150
SJ73759155
SJ74409150
SJ74759155
SJ75429157
SJ75759160
SJ76109162

i
j

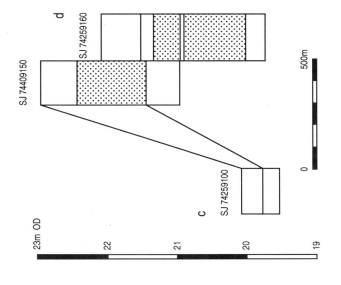

Fig 41 Carrington Moss: transect B charcoal distribution

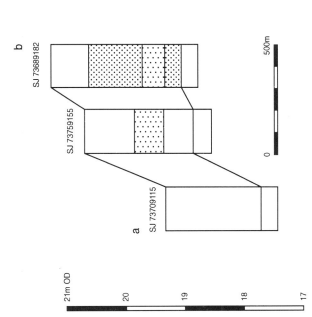

Key

present   occasional   frequent

Fig 40 Carrington Moss: transect A charcoal distribution

79

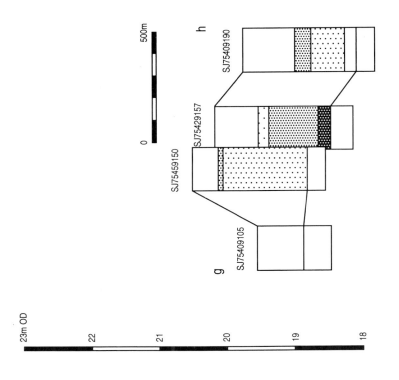

Fig 43 Carrington Moss: transect D charcoal distribution

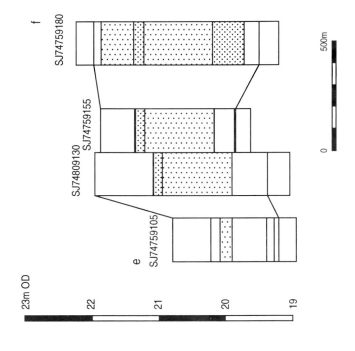

Fig 42 Carrington Moss: transect C charcoal distribution

80

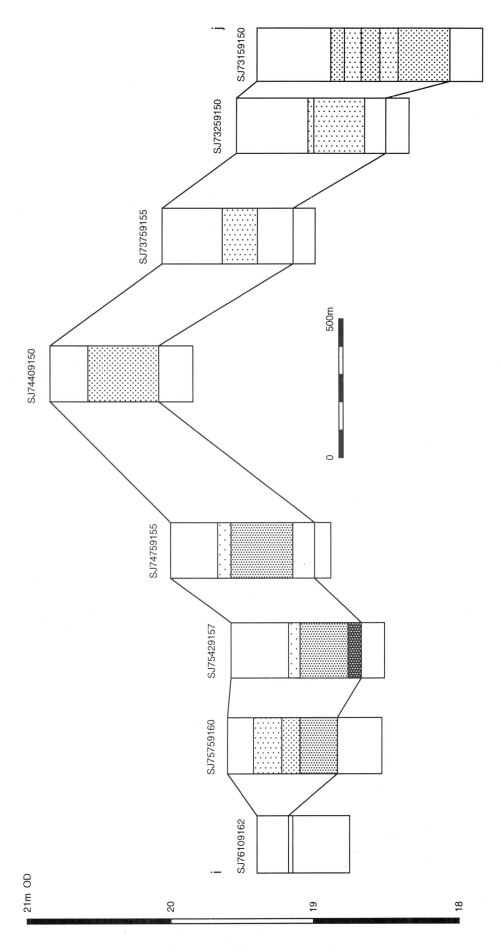

Fig 44 Carrington Moss: transect E charcoal distribution

21m OD

20

19

18

500m

0

SJ73159150

SJ73259150

SJ73759155

SJ74409150

SJ74759155

SJ75429157

SJ75759160

SJ76109162

j

i

No radiocarbon dates are available for the initiation of peat accumulation in the Carrington system, but on the basis of the relatively shallow peat profiles and the lack of deep hollows within the mineral substrate it seems probable that mire initiation occurred here later than in the deepest parts of the Chat Moss complex immediately to the north. One possibility is that this was at some point during the Neolithic period, c 4000—3000 cal BC, when a number of similar mires in North West England are thought to have begun peat formation (Shimwell 1985), and may be contemporaneous with a great expansion of peat accumulation over the Chat Moss system.

The very shallow basin profile seems to have produced relatively dry conditions initially, and ensured only slow peat accumulation. The basal organic deposits, uniformly composed of wood peats with variable amounts of monocotyledon material, indicate the widespread development of fen-carr woodland over the site. The presence of small amounts of burnt fragments in the peat from this stage, except at SJ 7542 9157 where burning appears to have been intense, points to a low level of fire disturbance throughout the site in the early stages of its development. A change from minerotrophic fen-carr to ombrotrophic or poor-fen communities next occurred. This switch appears to have been quite abrupt, and there is a hint of a flooding horizon having preceded it in one location, with a sequence of *Scheuchzeria palustris* peats evident in the stratigraphy (Figs 40—44; SJ 7370 9115). The site at this time would have presented a picture of stagnating carr woodland, with a small area of higher, better drained ridge carrying a mossy *Polytrichum commune*-dominated vegetation.

Eriophorum/Calluna peats became dominant over the site following this phase, representing the establishment of true ombrotrophic conditions with slow peat accumulation rates. Burning of the vegetation was a consistent feature of the mire ecology during this period, and the frequent carbonised plant fragments found in the deposits referable to this stage may represent the effects of numerous light fires, clipping the leggier heather from a surface that would have been moist but not difficult to traverse on foot.

The *Eriophorum/Calluna* vegetation was succeeded abruptly by *Sphagnum imbricatum*-dominated communities, marking a change to wetter hydrological conditions at the site. Again, there is limited evidence for some localised flooding of the surface prior to the establishment of *Sphagnum*-dominated vegetation, indicated by the presence of short-lived pool peats characterised by an abundance of *S sect Cuspidata* remains (Figs 40—44; SJ 7335 9150). The surface of the mire would have become extremely wet and spongy and would have proved difficult to cross.

Burning seems largely to have ceased after the transition to these wetter conditions.

The upper sections of the stratigraphy are now lost, but by comparison with other sites in the Manchester region it would seem likely that *Sphagnum imbricatum* dominance continued up to historically recent times.

Recommendations for further research

A substantial body of peat remains at Carrington Moss, perhaps spanning the period from the Neolithic to the Iron Age. Reclamation and the relatively thin peat deposits have conspired to create a situation where it is perhaps one of the more threatened sizeable mire sites in the county, and agricultural activity will continue to erode the palaeoecological value of the peats rapidly. Because of this, any palynological work at the site should probably be undertaken in the near future in order to maximise the information available from such study. If wasting of peats continues as at present and if it is occurring at a similar rate to that calculated for Chat Moss, the sandy clays, probably dating to a pre-Neolithic context, may become exposed on a wide scale within the next 25—50 years.

Summary

Carrington Moss is a former active ombrotrophic intermediate mire now reclaimed for agriculture. Stratigraphic survey shows that only moderately deep peats survive over much of the site, probably due to modern truncation of naturally shallow deposits. Peat formation began at an unknown date, but on comparative stratigraphic grounds is likely to have been later than other mosses in Greater Manchester. The development of the moss began with widespread fen-carr wood peats which were succeeded by *Eriophorum*-dominated vegetation and finally by *Sphagnum imbricatum* communities. There is extensive evidence for fires in the prehistoric period affecting the vegetation on the moss. A pollen record covering a significant part of the prehistoric period is likely to survive in the peat archive.

Due to the shallowness of the peats and the intensity of agriculture affecting the moss, the quality of the site as a peat archive is seriously threatened.

Kearsley Moss
(SD 750 040)

Kearsley Moss (Figs 4, 45) lies in Bolton and Salford Districts between the A6 and A666 roads, 3 km north-east of Chat Moss. In this report the name 'Kearsley' is used for the whole complex that comprises several parts, each associated with a neighbouring township or hamlet: Clifton Moss, Kearsley Moss, Linnyshaw Moss, Morton Moss, Red Carr Moss, Swinton Moss, Walkden Moss, and Wardley Moss. Details of each moss are to be found in the Gazetteer (*Appendix 1*).

The Kearsley Mosses formed in an area of complex geology and topography at the southern end of a glacial drainage channel. To the north, the River Irwell flows east and south around the ridge on which Kearsley is located. The solid geology consists of Coal Measures and muds and shales, with overlying clays and sands, and some gravel. The Coal Measures incline northwards, outcropping at the Irwell Cut, and provide a source of water springing from the rock, acting as a collection point for slow-draining water from the uplands.

The place name Kearsley probably means 'watercress clearing or hill' (*Cherselawe* 1187; Ekwall 1960, 269). It was a small settlement in the eastern part of the hamlet of Farnworth in the parish of Deane. Deane parish was split from the northern part of Eccles parish in 1541, although a church and parson are recorded from the thirteenth century (Farrer and Brownbill 1914, 5, 1, 3).

There are no known archaeological sites or finds from the region. Historic sites around Kearsley Moss include the moated site of Wardley Hall, on land held by the Knights Hospitallers, tenanted in the fourteenth century by Jordan de Wardley (Farrer and Brownbill 1914, 4, 384). In 1301 Richard de Worsley demanded common of pasture in 300 acres of wood and 100 acres of moor which had been enclosed from the waste, where Jordan de Wardley was tenant (Farrer and Brownbill 1914, 4, 385). In 1760 the Hall was sold to the Duke of Bridgwater.

An Enclosure Act for Kearsley was made in 1796 (LRO EDA/14/1), Bridgwater Estate maps of the eighteenth and nineteenth centuries show leasehold parcels, including a block of land held with Wardley Hall. This land had taken in ground close to the edge of Wardley Moss, exposing an underlying sand deposit, indicated by 'sand pits'. Walkden Moor is mapped as a large area of common land. Land farther east around Swinton Moor had field parcels with names indicating wet and boggy conditions, such as 'turf croft', 'little hey', 'water fall', 'meadow'. Irregular field parcels suggest earlier encroachments and enclosures of wasteland (Bridgwater Estate Plans in Irlam Archives M10/14).

The boundaries of piecemeal reclamation of moss are shown on the Bridgwater Estate plans and on the Ordnance Survey First Series and Revised Edition 6" maps (Ordnance Survey 1850, 1891). On the Ordnance Survey First Series map, Kearsley and Linnyshaw Mosses are still raw bog, with square fields of reclaimed land. The northern edge of Kearsley Moss has a strip of fields between Moss Lane, on which Moss House and Taskers Farm are located, and the edge of bog that had been reclaimed since the date of Yates' map of Lancashire (1786). At the centre of Kearsley Moss is a smallholding called the Bent Spur, a partly reclaimed plot of land with a small tree plantation.

Around the mosses, to the north, lay a string of collieries associated with the historically famous first water-powered colliery of Wet Earth, at Clifton (Ashmore 1969, 106). On the southern side is a small scattering of coalpits and a sandpit at the site of Wardley Moss.

During the 47-year period between the two Ordnance Survey map editions, the whole of the moss was drained. The Manchester and Bolton railway line (London and Yorkshire Railway, opened 1838; Farrer and Brownbill 1914, 5, 34), running between the River Irwell and Bolton Road, had been linked with the London and North-Western Railway, Worsley and Bolton line, south of the Chorley Road. A series of branch lines was constructed for the Bridgewater collieries, to supply the mines and the Kearsley Chemical Works as well as other sites. On the far eastern side of the moss complex there had also been urbanisation around Swinton and Pendlebury.

By *c* 1929—30, urbanisation around the eastern side spread out into the open landscape. Walkden became a place with substantial housing and industry, having collieries, mills, brickworks, a smallpox hospital, a cricket and football ground, churches, schools, and a town hall. Kearsley Moss around Bent Spur still retained most of the original drainage pattern in 1926, but the rest of the mossland had either been reclaimed or abandoned. In 1960—70 the construction of the M62 and M61 destroyed a large part of the mossland; borehole records show that excavated peat was dumped over nearby undisturbed peat (Department of Transport records). The current land-use is mapped on Figure 45; most land south of the M61 and M62 is derelict, with pasture to the north. The moss is threatened with further motorway work which will destroy most of it.

Field mapping in 1992 and 1993
The Kearsley Moss complex is described below in anticlockwise order, beginning at the north-west. There has been much damage by industry and building, and more recently by motorway construction.

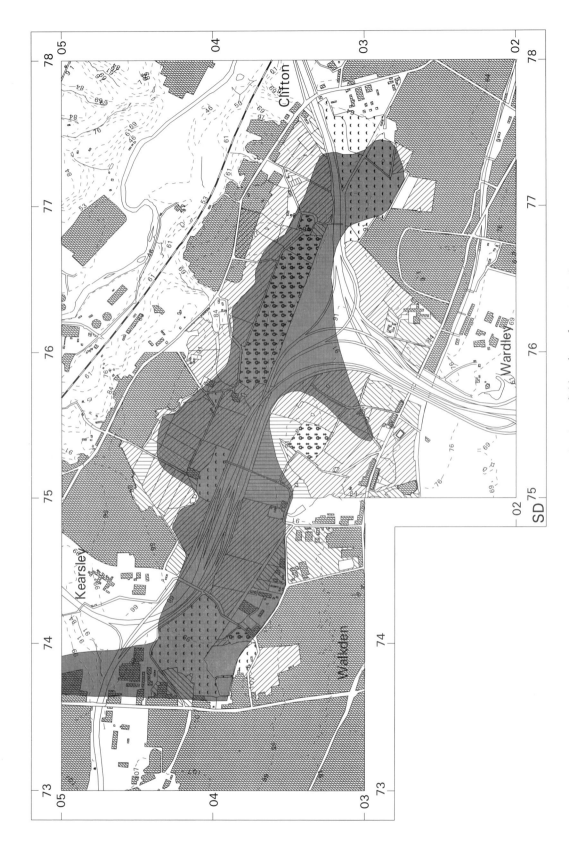

Fig 45 Kearsley Moss: land-use, fieldwork, and peat extent

Based upon the 1975 Ordnance Survey 1:10,000 map with permission of The Controller of Her Majesty's Stationery Office © Crown Copyright

The whole area has only two small fields of arable land and no finds were made. The extent of the peat is estimated using information from the Ordnance Survey First Series 6" map (Ordnance Survey 1845), taking into account the local topography.

Walkden Moss, the western part of the wetland, is almost completely built over in the area where a 'moor' was marked in 1845 (SJ 733 033). North-west of this, an apparent space of open ground, marked as moss on Yates' map (1786), contains colliery waste only.

Linnyshaw Moss, partly encroached upon by tips and industry on the western side, has peat surviving in a clump of trees at Moss End, east of Blackleach Reservoir at SD 741 038. The main peat area is a strip of pasture next to the south side of the M61, rising above the level of the surrounding land.

Morton Moss is all under pasture except for a slag heap planted with trees at the west side. A small area of peat survives west of New Hill Farm (SD 753 030) in the area that was still undrained moss in 1845 (Ordnance Survey 1845). Mosses named Wardley Moss and Red Carr would lie between Morton and Clifton. The only surviving open ground not covered by houses or motorways is high boulder clay, without peat, lying north-east of Wardley Grange School, separating Morton from Clifton. Any peat surviving has been destroyed by the motorway. Swinton presumably claimed rights between Wardley and Clifton. The only area that might be distinguished from those places not built over is a cemetery at SD 771 024. This appears to be skirtland, and is surrounded by housing of 1885 and c 1900, suggesting a fairly stable subsoil.

West of Clifton, the south-eastern part of the moss is an abandoned landscape, wild, with rubbish and a burnt and vandalised mill. The whole area is scrub or pasture, but the peat extent can be determined from molehills. Deep peat near the motorway falls to a valley of skirt on the east. The skirtland is boulder clay. Along the northern side of the M61 and M62 motorways, towards Kearsley, is a peaty dump of soil now planted with trees. At the north-east, the peat extent is confused by coal waste next to Clifton Moss Farm, and at the north a golf course has been 'landscaped' by strewing a coal tip about. The limit of peat was probably towards the A6 and Unity Brook, Kearsley. Near Ryders Farm the ground is skirt; a linear earthwork near this farm may be a colliery railway embankment (SD 7615 0387). A boundary stone, at SD 7661 0358, marks the division between Clifton and Kearsley.

Although there is no known archaeology from Kearsley Moss, further work should be done in view of imminent development threats. It is a little known moss with no previous palaeoecological study. Until recently there were 300 ha of peat, almost the same as Carrington Moss and more than Red Moss.

Palaeoecological assessment

Kearsley Moss was probably once a medium-sized (c 300 ha) raised or valley mire. It has subsequently been reduced in size to c 50 ha by agricultural reclamation and industrial development; the M62 and M61 bisect it. Today, deep peat deposits only remain in a small area to the immediate south of the M62 and north of Swinton, at SD 772 030, and in a slightly larger area south of the M61 and north-east of Walkden, at SD 753 037. The peats remaining in the Kearsley Moss system are mostly 1.2—1.8 m in depth, with a maximum of 2.5 m.

Stratigraphy

The base of the organic deposits is formed by a variety of silty clays and sands, probably representing a complex of fluvio-glacial deposits. A wood peat with much *Polytrichum* is found above these and is succeeded by *Eriophorum/Calluna* peats. In some places a short-lived wood/*Phragmites* peat preceded the switch to calcifuge assemblages. A final stratigraphic element is formed by *Sphagnum imbricatum* peats, but these have been removed from much of the site. Charcoal is often present in the peats, sometimes in relatively large amounts. As in other sites examined in Greater Manchester, there appears to be a trend for the greatest concentrations to occur in the lower stratigraphy, particularly in *Eriophorum*-dominated peats.

Interpretation

Brushwood with an acid *Polytrichum*-dominated ground flora developed over much of the fluvio-glacial deposits, with sedge-rich fen carr occurring on wetter margins. This phase was succeeded by an open vegetation characterised by *Eriophorum*. In some places the change was preceded by *Phragmites*, suggesting a brief flooding of the surface by nutrient-enriched waters. Burning of the vegetation occurred during this change and continued throughout the period of the *Eriophorum* domination. Finally this vegetation was replaced by *Sphagnum imbricatum*-dominated communities, indicating a change to wetter conditions. Peat removal of the upper stratigraphy has obscured the later development of the site.

Research potential

Kearsley Moss remains in a moderately good state of preservation (humification of the peats ranges from

H6—10, but mostly occurs around H7—8), and a substantial record of the stratigraphy remains in a few isolated places. However, it would seem possible that upwards of 2 m of peat is missing over much of the site where organic deposits remain and most, if not all, of the historic record is likely to be missing. A considerable amount of prehistoric peat probably survives and forms a valuable palaeoecological archive for study of local environmental change. The peats also provide another example from the region of burning disturbance at the changeover from minerotrophic to ombrotrophic communities.

The small area of peat in the eastern section of the site is under permanent grass and does not appear to be immediately threatened, but the western section is subject to arable cultivation and wastage is likely to continue at a steady rate.

Red Moss
(SD 638 100)

Red Moss, at 110 m OD, is located near the southern foot of the Rivington uplands, which rise to 456 m OD, and is near the A6 and south of Horwich (Fig 46 Plate 15). It now extends to about 65 ha, and was formed on glacial drift sands and gravels overlying a broad band of coal shales. Red Moss was a glacial mere site at the north-western end of a ridge overlooking the Mersey Basin, running towards Kearsley. The drainage falls west towards the River Douglas.

The best known archaeological find from the moss is a human skull, possibly Bronze Age, with some hair attached, found in 1942. It was then identified by forensic examination as belonging to a female aged c 30 who died 'some 3000 years previously' (M D Smith 1988, 4). The skull has now been lost. Nearby, and close to the M61, was found an antler pick with a broken shaft of 0.2 m length and a small branch near the base c 140 mm long. The pick is now in Bolton Museum (Plate 16), and is likely to belong to the Mesolithic period. A third find from Red Moss, now in the garden of Gibb Farm, is a quernstone of likely Roman date, found in a drainage ditch on the north side of the moss. It is c 400 mm wide and 150 mm deep. Other finds from the region have recently been listed and mapped, with the state of the moss as recorded on various Ordnance Survey maps since 1849 (GMAU 1993).

There appears to be no associated medieval settlement within the vicinity of the moss. The closest is Blackrod to the north-west, a place-name meaning 'black clearing' (Ekwall 1922, 45). Much of the area north of Red Moss came under Horwich Forest jurisdiction and there was little settlement in the area until the late fourteenth and early fifteenth centuries, when land was taken up as freehold.

Blackrod village lies on the line of a Roman road which runs north from Manchester to Chorley and joins up with the Wigan to Preston route. Watkin, citing a nineteenth century account by Whitaker, describes the road at Blackrod as a regular pavement of heavy stones found lying parallel with the present road. Watkin disagreed with a claim by Whitaker that a Roman station existed at Castle Croft in Blackrod, because there were no known remains of Roman date (Watkin 1883, 46—7).

The site of a twelfth-century castle, possibly erected by William Peverel, is located in the centre of the present village (SMR 1423; SD 6192 1065). The village was established by 1338 when Mabel de Bradshagh granted an endowment to the chapel of St Catherine, 'then newly built'. Blackrod mill on the River Douglas was recorded in 1606 (Farrer and Brownbill 1914, 5, 301, 303).

Blackrod township, in which Red Moss lies, and Lostock township were detached portions of Bolton parish. It is probable that the parish boundary along the Middle Brook tributary was the greatest extent of Red Moss. Park Hall (SMR 4318; SD 6275 0966), built in the late sixteenth or early seventeenth century on clay land, shows the southern limit of the moss at that time, although the peat may occasionally have spilled over to the Park estate. Recent work suggests that Barker of the Moss, recorded in Blackrod in the sixteenth century, is the source of the name of Barker's Farm on the west side of Red Moss (GMAU 1993).

The Blackrod Tithe Award of 1840 shows there was access across the moss from the south side, from Park Hall Farm, and from the eastern side, connected to a network of farms surrounding the moss: Moss House, New Close, Sefton Fold (a medieval moated site), Lower House, Old Harts, and Sharrocks Farm. There was access from Blackrod village via Moss Lane, on the west of the moss. Barker's Farm and Moss Farm are both situated on this lane (Bolton Archives Map of the Township of Blackrod in the Parish of Bolton in the County of Lancaster, surveyed 1840, 22/48/3). Moss Farm was renamed Gibb Farm before 1907, and is the site of the last hand peat-digging to be carried out on Red Moss (Tillotson 1907; Ordnance Survey Revised Edition 6" 1909, Lancashire Sheet 86). The Ordnance Survey First Series shows the western edge of the moss to have a small satellite coalpit owned by Victoria Main Colliery just south of the railway line (Ordnance Survey First Series 6" 1849, Lancashire Sheet 86).

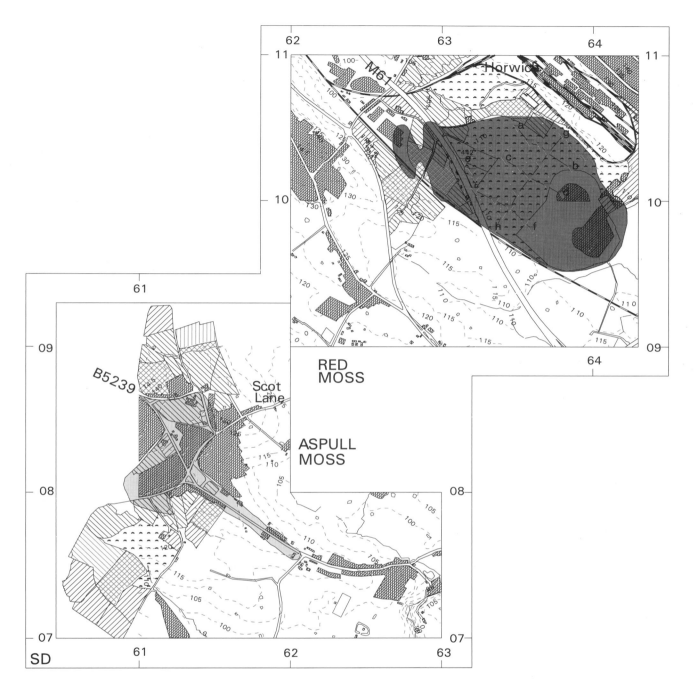

Fig 46 Red and Aspull Mosses: land-use, fieldwork, and peat extent with boreholes

Peat-firing experiments were carried out at Horwich by John Crowley and Co, Ironfounders of Red Moss (Drake 1861) to produce peat bricks for fuel as a coal substitute. The process is described:

> The peat, as it comes from the bog, is thrown into a mill arranged for the purpose, by which it is reduced to a homogeneous, pulpy consistency. The pulp is then conveyed, by means of an endless band, to the moulding machine, in which, while it travels, it is formed into a slab,

and cut into blocks of any size required. The blocks are delivered by a self-acting process on a band, which conveys them into the drying chamber, through which they travel forwards and backwards on a series of endless bands at a fixed rate of speed, exposed all the time to the action of a current of heated air... (Leavitt 1867, 45, 303).

The coal seam to the north of Park Hall Farm was abandoned by the end of the nineteenth century, and

Plate 15 Hand-cut peat blocks at Red Moss. This small-scale single-person operation was proscribed when the site was designated an SSSI

in 1901—2 Blackrod Cottage Hospital was set up on the site of the satellite coal pit (Tillotson 1899—1901, 1902—4), and named on the Ordnance Survey Revised Edition 6" map (Ordnance Survey 1909) as a smallpox hospital. By 1911 the Cottage Hospital was replaced by Richardsons Peat Moss Litter Co (Tillotson 1916, 1922), becoming the Red Moss Works by 1930 (Geological Survey of England and Wales, Lancashire Sheet 86, 6":1 mile (1:10,560), 1930). Afterwards the site was called Red Moss Farm (Tillotson 1932). The site of another Red Moss Farm was located to the east of Park Hall Farm.

Field mapping in 1992 and 1993
Red Moss (Fig 46) has a large rubbish tip at the south-east which is causing pollution, and sewage has been deposited on the west. The remainder of the moss is wild, drained by trenches up to 2.4 m deep. At least two 1 m cuts of peat were taken off *c* 1940, and a small quantity was still hand-cut at the north until designation of the site as a SSSI. The trenches and rooms should be regularly watched for finds.

One field is cultivated on the northern edge; no finds were made on the pebble and clayey soil where the Roman quern was found. West of the motorway two arable fields showed the western extent of peat. No finds were made on the high spots of pre-Flandrian soil in these fields, where the skirtland edge lies about 2 m above the present moss level, north of and next to the M61 motorway.

Red Moss, a SSSI, is threatened by continued rubbish tipping; it should be sampled with sondages to test for buried sites, and to collect further palaeoecological samples before the peat archive is affected. The sands known to exist underneath the moss may support early lithic sites that were later buried. Palynological work shows there was clearance of woodland around the moss from the Neolithic period onwards, but no evidence of agriculture was found (*see below*).

Palaeoecological evidence
Red Moss (SJ 638 100) is a relatively small mire famous as the North West's palynological regional type-site for the Flandrian (Hibbert *et al* 1971). At present the land-fill development, which is located in the south-east of the site, occupies perhaps a third of the original mire expanse. The current area occupied by peat covers some 0.75 x 0.85 km, in a roughly rectangular block stretching from the M62 in the south-west to the former locomotive works in the north-east, and from the landfill tip in the south-east to Gibb Farm in the north-west (Fig 46).

Previous research
Jones *et al* (1938) recorded borehole data from a site they termed 'Red Moss', but as their work was concerned with the underlying hard geology no details of peat stratigraphy were noted. However, they provided the first detailed interpretation of the late-glacial development of pre-peat surfaces in the region.

A major palynological study was undertaken on the site between 1968 and 1971, during the course of construction of the M62 along the moss's south-west boundary (Hibbert *et al* 1971). The investigation, aided by an extensive programme of radiocarbon dating, was carried out as part of a national series of palynological studies designed to define regional differences in the pollen record, and to facilitate correlation between vegetational changes in different areas. As a result of this work Red Moss came to be regarded as the palynological type-site for the Flandrian period in North West England. Because of the importance accorded to the regional record in Hibbert's study, little emphasis was placed on local or anthropogenic effects on vegetation in later prehistory. Accordingly, the main features of the pollen record arising from this research are described here and interpreted from a more archaeologically skewed perspective than that originally attempted.

Until the current study, subsequent stratigraphical work at Red Moss has been entirely of a non-palaeo-ecological nature, consisting largely of numerous borings undertaken by several utilities and public bodies to ascertain peat depths and underlying deposits for economic assessment purposes.

Sampling strategy

The current survey employed 24 boreholes placed across the site in north-west to south-east and northeast to south-west orientated transects (Figs 46—54). The dating of stratigraphic features in the following section relies on extrapolation of data from an extensively dated core taken in 1968 from the line of the M62 to the west of the site (Hibbert *et al* 1971). The dating of stratigraphic changes across the bulk of the site remains largely tentative, therefore, and may need revision should further research be undertaken.

Condition and survival of the peat archive

No part of the mire now retains any topmoss. The survey indicates that probably about a third to a half of the original mire expanse has been completely removed, the railway works, M61, and tip having taken its place. The motorway appears to have been constructed on some of the deepest peats, almost certainly where the 1968 core was taken (Hibbert *et al* 1971).

Elsewhere peat has been truncated to a varying degree as a result of a long history of domestic and commercial cutting. The moss surface is now marked by a regular pattern of wide (*c* 10—15 m) rectilinear cuttings separated by much narrower (*c* 2—3 m) baulks orientated north-east to south-west (Fig 46). These features dictate the distribution of most of the remaining deep profiles which are largely confined to the baulks, reaching depths of *c* 3—4 m in many places; probably 1.5—2 m of peat is truncated in the former cuttings. However, the very deepest surviving peats are located outside the area of former commercial exploitation in the south-west portion of the moss, adjacent to the M61. Here, *c* 6.5 m of well-preserved organic deposits occur in a small area under permanent grass.

Topography and mire classification

The highest points on the current mire surface occur on its eastern boundary between *c* 110 m and 112 m OD while the lowest parts of the moss are found around the 106—7 m OD contour.

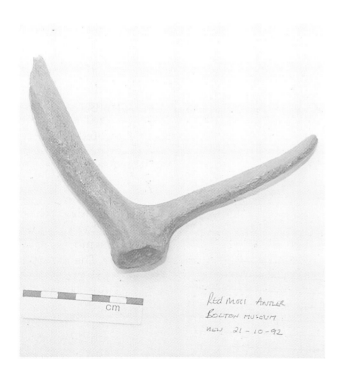
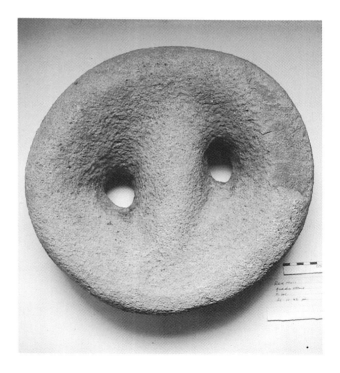

Plate 16 Antler pick and Roman quernstone from Red Moss (Bolton Museum)

The moss is underlain by an irregular mineral surface of clays and sands.

Generally speaking, this mineral substrate may be viewed topographically as a mosaic of shallow hollows, channels, and broad ridges enclosed within a larger, higher, ridged basin. A deep channel occurs to the west of the site, where it contains in excess of 6.5 m of organic deposit (SD 633 101). Elsewhere, shallower hollows occur in a more general plain of clays and sands occurring between the 101 m and 103 m OD contours. Several islands of higher mineral ground project from this topography, notably to the west, immediately adjacent to and either side of the deepest channel, and in the approximate centre of the current peat site. Superimposed on this general topography is a gradual incline of the surface from north to south and from north-west to south-east.

Although the earliest organic deposition appears to have occurred in the deep channel situated to the west of the site, the bulk of the mire probably began forming c 7000—6500 cal BC, with fen-carr peats over the wide flats of clays and sands in the central portion of the site. This probably built up over time to form a raised (ie shallow convex) profile characteristic of a raised mire, although the modification induced by years of disturbance means that this is not apparent at the present day.

The immediate pre-peat stratigraphy
The present survey did not investigate the underlying inorganic surface, but the studies of Jones *et al* (1938) indicate that the underlying glacially-deposited boulder clays of the region exhibit a tripartite division. A lower boulder clay, probably representing a period of intense glaciation, was recorded, followed by 'middle sands' (deposits interpreted as representing a temporary retreat of the ice) and finally upper boulder clays representing a terminal period of glaciation. The cessation of cold conditions was characterised by the formation of a complex of meltwater features including lakes, outwash fans, and channels. A complex kame attributed to this phase is situated to the north-west of the moss (Jones *et al* 1938).

Peat stratigraphy (Figs 47—54)
Isolated deep channels and hollows within the Red Moss basin complex allowed the early initiation (c 9000—8000 cal BC) of peat accumulation marked by bryophyte and monocotyledon communities, but over much of the mire the peat stratigraphy can be divided into five broad stages. Bottom deposits are often fen-carr assemblages comprising brushwood peats, wood peats with monocotyledon remains, and

pure monocotyledon peats. These deposits are usually succeeded by more solid wood peats, usually of *Betula*. In many areas the wood peat stage is followed by peats indicative of poor-fen conditions and characterised by *Eriophorum*, *Polytrichum* and sometimes *Aulocomnium palustre* and *S sect Acutifolia*. Following on from this, pure *Eriophorum/Calluna* or *Eriophorum/Calluna/Sphagnum imbricatum* peats are found. In many places these peats are interrupted by bands of purer *Sphagnum imbricatum* peats. A final major stratigraphical change occurs when *Sphagnum imbricatum* assumes domination. In some places the transition is marked by an initial short-lived phase of *S sect Cuspidata*-dominated pool peats.

Carbonised material
As at Chat Moss, burnt plant fragments, often of Ericaceous species, are a common feature of the Red Moss peats. The frequency of carbonised material varies across the stratigraphy and it is difficult to discern any obvious patterns. There is, however, a certain amount of evidence to suggest that initiation of burning often occurred immediately after the transition from wood peats to *Eriophorum*-dominated stratigraphy, and charred material can be at its most frequent occurrence in this stratigraphic context (*eg* SJ 6370 1033; SJ 6315 1027; SJ 6358 1022). Carbonised material seems to be most prevalent in the middle reaches of the stratigraphy, tailing off in frequency towards the top, although small amounts of carbonised material are associated with some of the lowest deposits too (*eg* SJ 6315 1027, SJ 6365 1025) indicating fire disturbance of varying intensity through most of the prehistoric period.

Palynology and regional vegetational change
The study carried out by Hibbert *et al* (1971) recorded a sequence of vegetation change dating from the immediate post-glacial possibly through to the early Iron Age (based on the presence of an obvious *Tilia* decline in the upper part of their diagram). In all, six major pollen assemblage zones were recognised. The evidence suggests that the regional vegetation cover during Flandrian Ia (beginning c 10,051—8418 cal BC (9798 ± 200 BP; Q-924)) was characterised by vegetation of an open nature with *Betula*, *Salix*, *Pinus*, *Juniperus*, *Populus*, Gramineae, and Cyperaceae. *Filipendula* was also particularly common in this early stage. By Flandrian Ib (9253—8098 cal BC; Q-924) *Betula* had increased its proportion of cover, forming a more closed community with a subsequent reduction in grasses, although fluctuations occurred throughout this phase, which also saw the appearance of *Corylus* pollen in small quantities. The beginning of Flandrian Ic (8091—7491 cal BC; Q-921) was characterised by the decline of *Betula* and the rise of

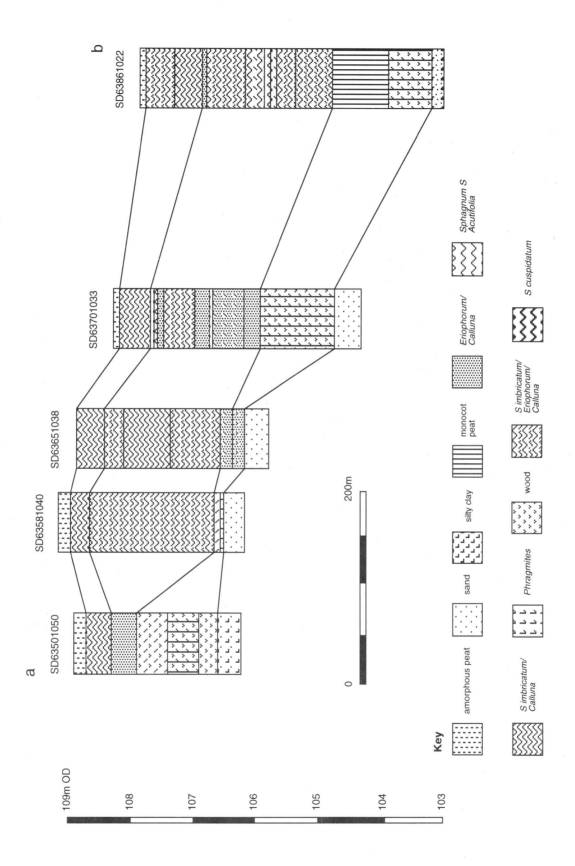

Key

amorphous peat	sand	silty clay	monocot peat	Eriophorum/ Calluna	Sphagnum S Acutifolia
S imbricatum/ Calluna	Phragmites	wood	S imbricatum/ Eriophorum/ Calluna	S cuspidatum

Fig 47 Red Moss: transect A

91

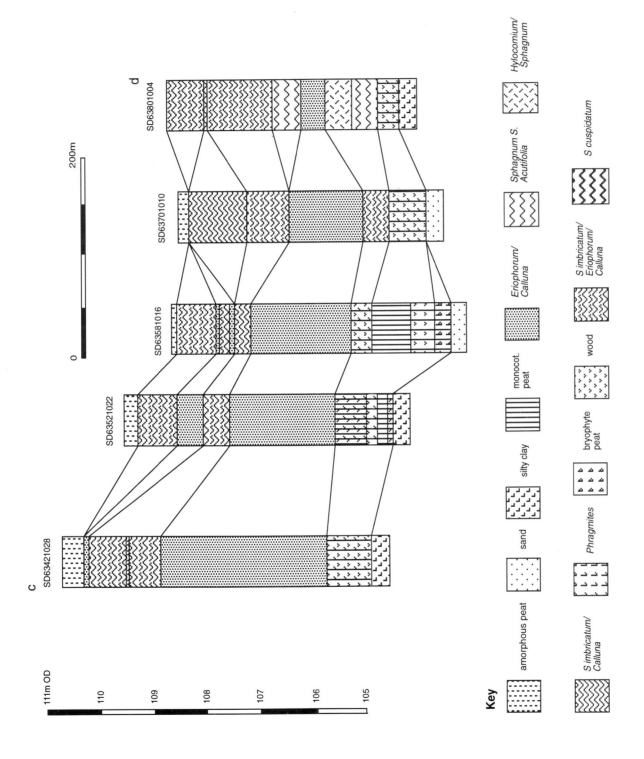

Fig 48 Red Moss: transect B

Key

amorphous peat

S imbricatum/ Calluna

sand

Phragmites

silty clay

monocot. peat

bryophyte peat

wood

Hylocomium/ Sphagnum

Sphagnum S. Acutifolia

Eriophorum/ Calluna

S imbricatum/ Eriophorum/ Calluna

S cuspidatum

92

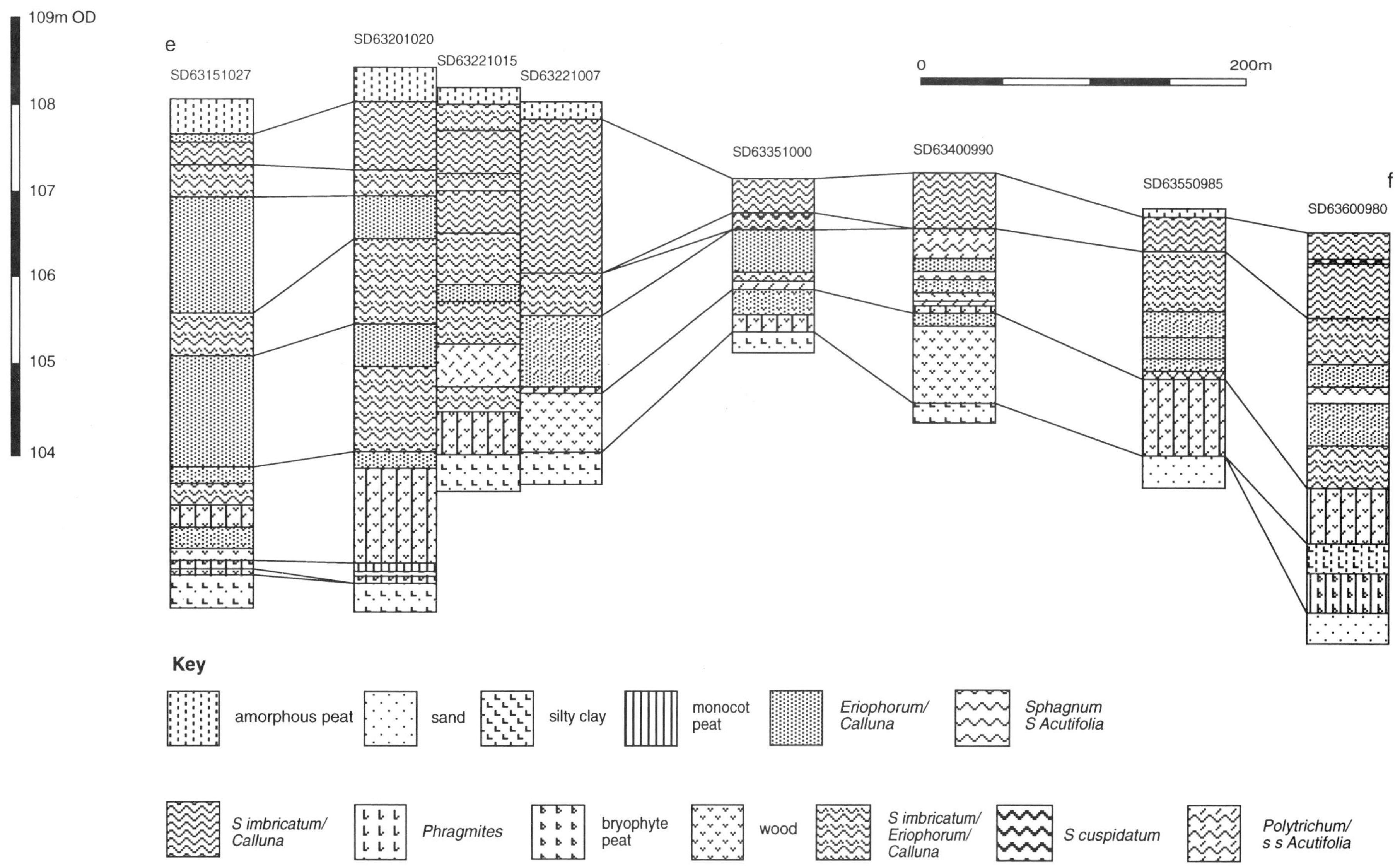
109m OD
108
107
106
105
104
e
SD63151027
SD63201020
SD63221015
SD63221007
SD63351000
SD63400990
SD63550985
f
SD63600980
0
200m
Key
amorphous peat
sand
silty clay
monocot peat
Eriophorum/ Calluna
Sphagnum S Acutifolia
S imbricatum/ Calluna
Phragmites
bryophyte peat
wood
S imbricatum/ Eriophorum/ Calluna
S cuspidatum
Polytrichum/ s s Acutifolia

Fig 49 Red Moss: transect C

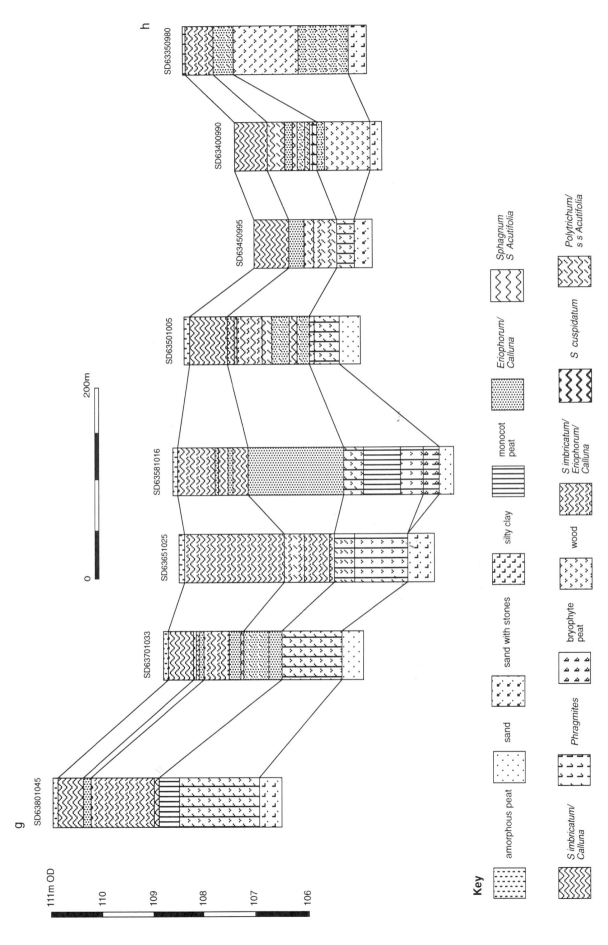

Fig 50 Red Moss: transect D

Key

amorphous peat	sand	*Phragmites*
S imbricatum/ Calluna	sand with stones	bryophyte peat
silty clay		
monocot peat	wood	
Eriophorum/ Calluna	*S imbricatum/ Eriophorum/ Calluna*	
Sphagnum S Acutifolia	*S cuspidatum*	
		Polytrichum/ s s Acutifolia

94

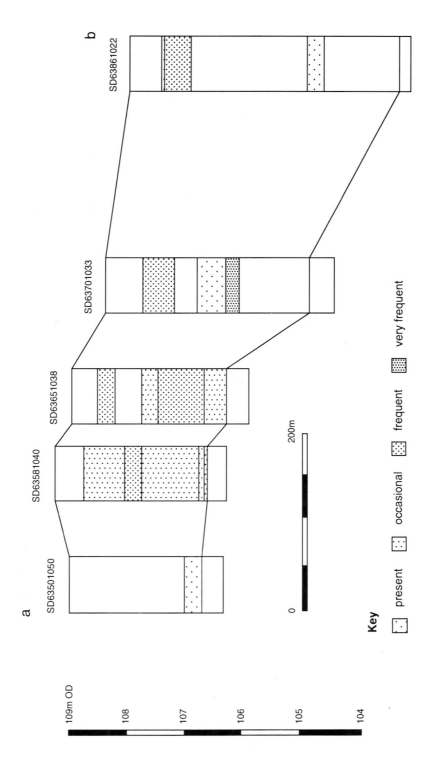

Fig 51 Red Moss: transect A charcoal distribution

Key

present   occasional   frequent   very frequent

a   SD63501050   SD63581040   SD63651038   SD63701033   b   SD63861022

200m

0

109m OD

108

107

106

105

104

95

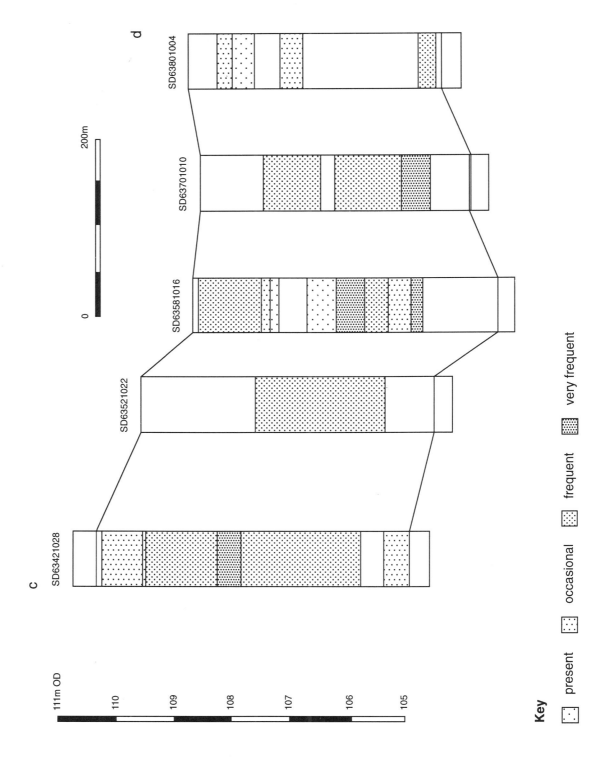

Fig 52 Red Moss: transect B charcoal distribution

Key

present occasional frequent very frequent

96

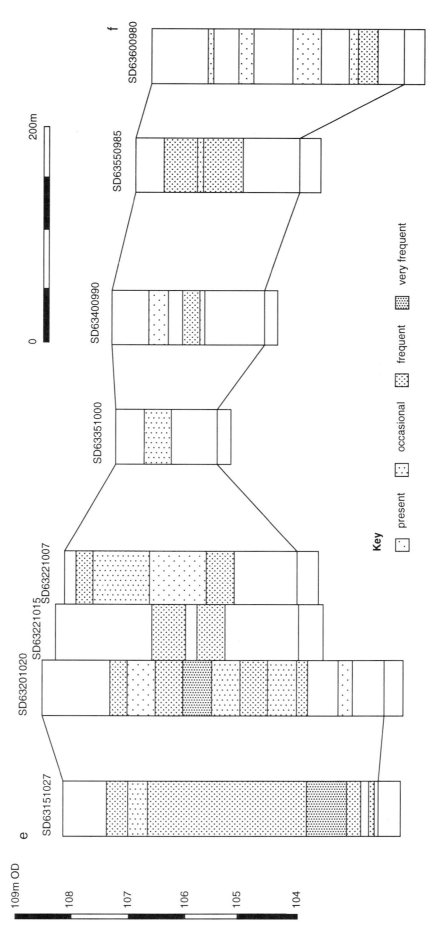

Fig 53 Red Moss: transect C charcoal distribution

97

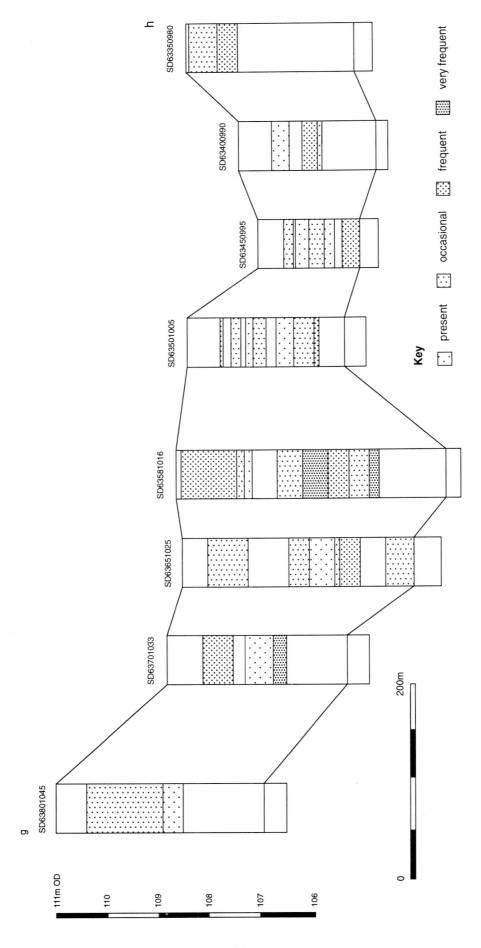

Fig 54 Red Moss: transect D charcoal distribution

Key

present | occasional | frequent | very frequent

98

Corylus. After a brief peak in representation, both *Pinus* and *Salix* pollen declined during this period. The change to Flandrian Id (7531—6646 cal BC; Q-919) was marked by a further decline in *Betula*, but a rise in *Pinus*. *Ulmus* maintained a steady but relatively subdued presence while *Quercus* began a steady rise in representation. *Corylus* maintained its position as a common and widespread species.

The opening of Flandrian II sometime between 6180 and 5720 cal BC (7101±120 BP; Q-916) saw the rapid rise to prominence of *Alnus* (the representation of the species at the Red Moss site is likely to be exaggerated by the wet nature of the habitat), and the further diminution of *Betula* as a major component of the regional forests, while *Pinus* tumbled from being a prominent species to an insignificant member of the woodland communities. *Quercus* and *Ulmus* rose to become prominent species as did *Tilia*, forming a closed mixed forest with a *Corylus*-dominated understorey. Flandrian III opened with the decline in the pollen of all tree species, but particularly *Ulmus* in characteristic elm decline pattern between 3990 and 3640 cal BC (5010±80 BP; Q-912). Later, other tree species apart from *Ulmus* recovered to some extent, but there followed an increasing trend to openness in the woodland cover, with light demanding plants becoming more and more common in the pollen record. Obvious anthropogenic clearance episodes also became frequent during the later part of Flandrian III, and were characterised by the occurrence of cultural indicator plants such as *Plantago lanceolata* and *Rumex* spp, along with reductions in tree pollen. Ericaceae also increased dramatically in representation. The expansion of trees such as *Fraxinus* during Flandrian III also point to disturbance of the forested landscape as the species is known to be an effective competitor in secondary woodland (Smith 1970). However, the clearance episodes remained relatively small and short-lived (probably <100 years), and the landscape remained predominantly forested. No further radio-carbon dates are available for the more recent levels of the Hibbert profile and so it is impossible to be precise about the timing of clearance episodes detectable in the pollen diagrams. However, it is probable that several such episodes span the Neolithic, Bronze, and Iron Ages and imply a continual human occupation of the catchment area throughout this period, similar to that recorded from other areas in the North West (Wimble 1986; Twigger 1988).

General overview and discussion of Flandrian environmental change

At the end of the last glaciation, the landscape of the Red Moss district is likely to have been a complex of fluvio-glacial features including moraine ridges, sinuous sandy banks, sterile impounded lakes, and dumps of gravels and silts. As the climate warmed, biological activity began softening the hard edges of the raw landscape. Part of this process included the colonisation by aquatic plants of the wet hollows in the Red Moss basin complex, marking the inception of the proto-moss. On the basis of the 1971 study, organic accumulation here is known to have commenced at least as far back as 9000 cal BC. The current survey, however, proved the existence of organic deposits over 6.5 m deep (at least 0.75 m deeper than Hibbert's core), and because of this it is possible that material of earlier date may exist in the deep channel discovered at the western end of the moss.

During the immediate post-glacial period the drier areas surrounding the site were colonised by a fairly open vegetation of birch, willow, pine, and grassland, with wet ground dominated by sedges. Birch continued its rise in importance between *c* 8900 and 7700 cal BC, and formed a major component of the extensive fen-carr habitat which was developing in the Red Moss basin complex. Some isolated burning occurred in the communities within the basin complex, but it appears to have remained infrequent until the later stages of the carr development. Hazel began to form a more and more important constituent of the continuing woodland succession as other forest trees, such as elm and oak, began to make their presence felt. Sedges and grasses went into relative decline as the surrounding forest began to take on a more closed aspect, except in the Red Moss basin area where a birch-dominated fen-carr continued to predominate in the wetter conditions. Between 7000 and 6300 cal BC pine reached its greatest abundance and thereafter went into rapid decline, while birch also declined in importance. The new dominants of the rapidly changing forests were oak, elm, alder, and lime, while hazel continued its importance, possibly as an understorey to the closed canopies or maybe as part of it (Rackham 1980). These changes, occurring around 6000—5700 cal BC, also coincided with a vegetation change in the Red Moss basin complex from minerotrophic fen-carr communities to ombrotrophic or poor fen communities dominated by *Eriophorum vaginatum*. The incidence of burning also seems to have increased at this transition stage, with relatively high frequencies of burnt fragments of Ericaceous species, particularly heather, recorded across the mire. It is uncertain whether this burning was of natural or anthropogenic origin. Hibbert *et al* (1971) did not record charcoal particles in their study, but the pollen data from this period outlined previously are ambiguous regarding evidence for any analogous human-induced disturbance in the catchment area. Alder increased at this time; this is a largely ubiquitous feature of British pollen diagrams from this period (the traditional 'Boreal—Atlantic transition'), and has often been regarded as indicative of a change to a wetter climate (Godwin 1975). This might lend weight to the argument that the burning was of human origin. Oak, however, also increases in representation, at the same time providing evidence inconsistent with major clearance activity. Similarly, hazel pollen also increases at this time. Again this may simply be due to an increase in average warmth of the climate, although it has been suggested that the plant is relatively resistant to fire and hence may be differentially encouraged by a moderate degree of burning (Smith 1970). The decline of pine and birch could be interpreted as implying deforestation in the catchment, although this feature is widespread in English and Welsh pollen diagrams referable to this period, and the cause is usually attributed to natural successional changes resulting from ameliorating climate as the Flandrian stage progressed (cf the behaviour of *Pinus* in other interglacials; eg West 1968). Heather, grasses, sedges, and *Sphagnum* all increase markedly, suggesting the creation of more open areas, but the data for these species are more likely to be reflecting local changes within the basin complex. Because of this it would seem more likely that the burning recorded at this time was probably localised and restricted to the immediate vicinity of the mire. A fuller discussion of the possible causes and effects of burning on the site, and the wider implications of burning on the Manchester mires as whole, will be found in the period discussion (*see Chapter 6*).

Flandrian II ends with the elm decline, dated at Red Moss to c 3990—3640 cal BC (5010±80 BP; Q-912). From this period onwards there is evidence for a number of forest clearance episodes marked by declines in tree pollen and increases in pollen from open ground indicators. Hibbert's study neglected the upper 1.5 m of stratigraphy and it is likely that most of the late and post-Iron Age record is lacking from the record in the published paper. Because of this it seems likely that most of the clearance episodes recorded span the Neolithic, Bronze, and Iron Ages. No cereal pollen was associated with any of this disturbance.

Changes in the stratigraphy also occurred during this period, with a switch from *Eriophorum*-dominated peats to peat with a high *Sphagnum imbricatum* content. There is evidence to indicate that at least in some locations the change was preceded by a short-lived *S sect Cuspidata* pool peat, an indication perhaps of a rapid switch to wetter climate.

There is some evidence for other phase changes in the upper stratigraphy, possibly indicating wet—dry switches in the climate, but they are ill-defined and may represent localised changes only. Much more detailed study would be necessary to elucidate detailed phase changes in the upper *Sphagnum* peats from Red Moss; in the absence of more detail, speculation about possible climatic effects on the upper stratigraphy can only be couched in the most general terms.

Recommendations for further work
Further work at Red Moss could profit from targeting the post-Bronze Age pollen record, thereby completing the record of vegetation change from the site as far as possible.

Detailed palynological study of the stratigraphy immediately associated with charcoal concentrations is also desirable, in order to establish whether any links exist between these features and anthropogenic indicators in the pollen record.

Plant macrofossil analysis is recommended as a means of gaining possible palaeoclimatic information and/or evidence for artificial disturbance to the catchment area.

Summary
Red Moss is a partly destroyed and reclaimed possible raised mire which now carries a semi-natural and degraded mire vegetation. Accumulation of organic sediments began in deep, isolated hollows within the mineral substrate by the beginning of Flandrian I and subsequently peat formation extended to cover the entire site. The stratigraphy follows a fen-carr wood peat—*Eriophorum* peat—*Sphagnum imbricatum* sequence, and there is possible evidence of wetter interludes within the ombrotrophic sections of the stratigraphy characterised by pool peats dominated by *S sect Cuspidata*. Most of the upper stratigraphy is truncated. There is evidence indicating fire disturbance of varying intensity through most of the prehistoric period.

Despite the damage to the Moss, the surviving peats remain in a good state of preservation and over much of the site probably contain a pollen record from the Mesolithic to the Iron Age or Romano-British periods. Isolated hollows containing deep deposits dating back to the late-glacial period also survive. Taken together with the status of the moss as the regional palynological type-site, these factors mean that Red Moss has considerable importance as a palaeoecological archive.

5

THE SMALL MOSSES

Only seven of the smaller mosses of Greater Manchester, described below, have any surviving peat and, except for White Moss, peat survival is minimal in all cases. The mosses are Chew Moor, Highfield Moss, Hopwood Hall Moss, Marland Moss, Royton Moss, Warburton Moss, and White Moss. These small mosses were accorded the same type of archaeological fieldwork as the larger ones; palaeoecological work was, however, limited to establishing the nature of the moss to see if further analysis was required, or if it would help in establishing the details of landscape history in the area.

Chew Moor
(SD 668 081)

Chew Moor is located near Chew Moor village, on the north side of the M61 as it crosses St John's Road (Figs 4, 55). The area at c 110 m OD overlies glacial drift and Coal Measures; surviving peat is not immediately obvious. The lower part of the moss lies in the Croal glacial drainage channel, running east—west from Red Moss to Kearsley. This relatively undeveloped, badly drained channel has been used for the M61 motorway.

No archaeological finds are known from the region; the place name 'Chow more' is first recorded in the sixteenth century (Ekwall 1922, 45). It is probable that the moor was used to mark the boundary between Westhoughton and Lostock townships. The Cockersand cartulary records for 1272 a boundary which was probably at Chew Moor (Farrer 1900, **2** 675). In the sixteenth century there was a dispute over the title of the wastes, moss, pasture, and turbary in Lostock Moss (Farrer and Brownbill 1914, 5, 79). An Enclosure Award was made in 1808 (Lancashire Record Office AE 6/10).

Chew Moor is shown on Yates' map of Lancashire (1786), with the village occupying the middle east side, along St John's Road (called Mirey Lane on the Ordnance Survey First Series 6" map (Ordnance Sur-

vey 1894 Lancs Sheet 16NE; surveyed 1888/92). Yates' map shows a watercourse flowing north towards a confluence with Middle Brook, springing from Red Moss. The Bolton to Bury line of the Leeds and Liverpool railway cuts the moss from north to south on the west side of the village. There was much coal-working activity in the area (Farrer and Brownbill 1914, 5, 20).

Field mapping in 1992 and 1993
The village lies on a ridge, with peat in various surrounding places on the high ground and in two slades that meet and run into Middle Brook. All the land is pasture. The M61 drains down one of the slades towards the village and peat was discovered in construction trenches. Deep peat occurs in this same valley near Wicken Lees Farm (SD 6683 0780), and between the old railway sidings to the north of the farm, towards the Middle Brook.

Only remnant peat survives on the high ground of Moss Hall Farm (SD 661 078), but more is in a slade to the south. Near Rosemount Farm (SD 6653 0764) there was 0.45 m of peat in some fields above the slades. The exact extent of peat cannot be plotted in the pasture, but it seems Chew Moor was a valley moss that spread widely on to high ground. The historical extent was about 80 ha, mostly on high ground (Yates 1786; Ordnance Survey 1848). More peat lies to the east of the historic moss area in the valley draining to Rumworth Lodge Reservoir (SD 675 076). On Figure 55 the extent of the peat is taken from historical sources. The area was very wet; nearly all the nearby slades and valleys have some peat in them.

There were no furlong boundaries or medieval ridge and furrow, showing that the area was not arable in the Middle Ages. A small amount of poorly preserved nineteenth-century narrow rig occurs at SD 6635 0742.

Palaeoecological assessment
Chew Moor is currently under flooded former pasture and is dominated by rushes. Organic deposits reaching nearly 3 m in depth survive in two distinct

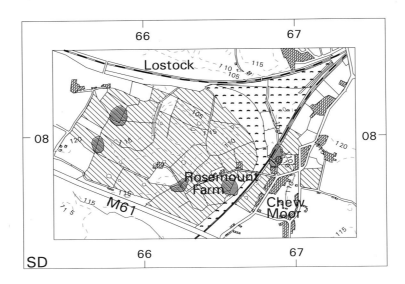

Fig 55 Chew Moor: land-use, fieldwork, and peat extent

basins separated by a higher ridge of mineral ground. The easternmost basin holds the deepest deposits.

Stratigraphy

Basal silts with a gyttja-like texture occur in the bottom half-metre of deposit and they are succeeded by either monocotyledon remains (*Schoenoplectus lacustris*, *Phragmites australis*, and *Carex rostrata*) or bryophyte assemblages with *S sect Cuspidata*. Sedge and wood peats form sequences above these, interspersed by clay and silt bands. Silts form a cap to the peats in the eastern deposits, but a wood peat forms the uppermost deposit in the western basin.

Interpretation

The sequences of deposits indicate that a minerotrophic system persisted on the site over a long period. They may well represent the remains of a former floodplain mire influenced by the Middle Brook. Several silt bands are recognisable in the stratigraphy which are likely to be the result of flooding episodes. No chronology can be ascribed to the deposits although it is likely that prehistoric peats are represented.

Research potential

The organic deposits at Chew Moor are in a poor state of preservation (humification ranges from H6 to H10), and may also suffer from contamination from recent flooding. It would probably be possible, however, to undertake some pollen analysis which would allow insight into local vegetation changes. Similarly, analysis of the silt bands may be worthwhile as they could be related to human disturbance of the catchment. The site forms a local palaeoecological resource and is not immediately threatened, except by contamination with polluted water.

Highfield Moss
(SD 613 959)

Highfield Moss, now an SSSI, is located south of Golborne and Lowton, near the edge of the present county boundary, and is cut by the Manchester Junction railway line (Figs 4, 56). Only a few hectares of peat remain, overlying, at *c* 20 m OD, glacial sands, gravels, and clays on top of Keuper sandstone. Lowton, as the name meaning 'hill' or 'mound', suggests (Ekwall 1922, 99), is situated on a raised sandstone ridge, with drainage in all directions (Worsley 1988, 1).

The site is interesting because of its proximity to a Bronze Age cemetery where four barrows were noted in the late nineteenth century, straddling a low-lying, undulating, sandstone ridge near the Winwick and Croft boundary (SJ 6193). The ridge is an important topographical feature in the locality, falling southwards to Warrington and providing one of the few fording points of the River Mersey between Runcorn and Manchester.

In 1826 a barrow near Kenyon Hall, Winwick Lane (SJ 6195 9480), yielded a bronze tanged awl, and a Beaker urn decorated with lozenge-shaped hachures and three rows of cord impression. The barrow was destroyed in 1903 during roadworks, revealing a circular bell-barrow, c 10 m in diameter and up to 1.2 m high, with a ditch c 2 m wide and 0.3 m deep. Grave goods of a food vessel urn, two Collared Urns, and an accessory cup, were dated approximately between the eighteenth and fourteenth centuries BC. The barrow had been reused for a windmill, although no remains were found (SJ 6194 9480; SMR 4719). A second barrow produced an unusual Beaker pottery form.

A flint knife of a type rare in the north of England (Plate 17), now held at Merseyside Museum, was found in the vicinity of this group of barrows. Also at Winwick, close to the Kenyon-Highfield Lane, several Bronze Age exotic metal and stone artefacts have been found. Among them are a bronze tanged dagger with rivet holes, a polished porphyry axe-hammer, a shouldered double-edged sword found in association with a cremation urn in 1859, a bronze socketed axe decorated with chevrons, a bronze palstave, and a ring (May 1903).

The place name Lowton is first mentioned in 1201, in the form 'Laitton' (Farrer and Brownbill 1914, 4). Common of pasture was claimed here in 1292 and also in 1356. There were disputes in the sixteenth century concerning rights of pasture (Farrer and Brownbill 1914, 4, 152 n16). Kenyon Hall to the east, dated to 1613, is now a farmhouse (SJ 6220 9494; SMR 924).

The 1762 Lowton Enclosure Award shows there were about 50 acres of open wetland, on which six landowners were allotted parcels of land and up to 30 people were given the right of turbary (Worsley 1988, 29). The Liverpool—Manchester railway, constructed in the 1820s, included an area of Highfield Moss. The watercourses at each end of the moss were channelled into siphons below the line. A shooting range was opened on Highfield Moss in the mid-nineteenth century (Worsley 1988, 35).

Field mapping in 1992 and 1993

The moss remnant, now drained and covered with semi-natural vegetation, lies in a low basin trapped on high ground. It is bisected by the railway, and excavated waste soil from the cutting has probably buried and preserved some of the moss in its 1830 condition. The moss has a surround of dark-stained and peaty skirtland, which goes up a slade on the north-east.

Plate 17 Large, bifacially-worked dagger from near Highfield Moss: length 175 mm (by permission of the Board and Trustees of the National Museum and Galleries on Merseyside, neg N93.59; acc no 65.98)

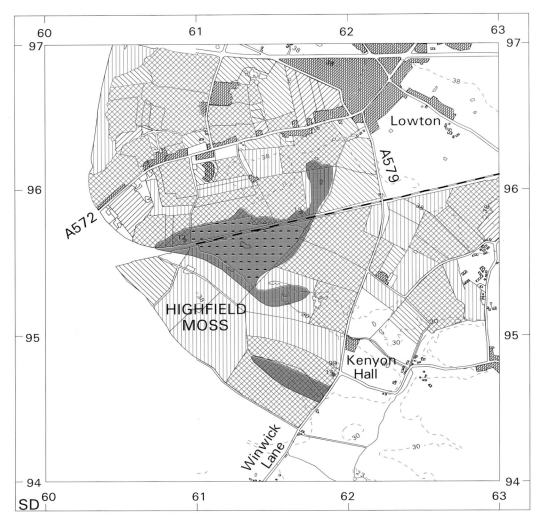

Fig 56 Highfield Moss: land-use, fieldwork, peat extent, and archaeological sites

The whole area around the moss (Fig 56) is arable, and most of it was in excellent weathered condition. Much of the ground is sandy, and two sandy areas adjacent to the peat yielded a few burnt stones, but there were no worked lithics (Fig 56; sites GM 12, GM 13). Nothing else was found of either Iron Age or Roman date, confirming other evidence that there are few sites in this region. If there were, they would surely be found at Lowton, in a good archaeological location with water and light soil, under optimum survey conditions. To the south, near the barrow excavated in 1903, a few burnt stones were discovered (GM 14) that probably relate to it. A separate valley peat lies south of the moss near Oven Back Cottage (SJ 6180 9455).

The present field boundaries north of the moss have earthen banks that are probably old enclosures, rather like the hedge banks of Devon. Early enclosure in Northamptonshire has marked hedgebanks about 1 m high, at Charwelton (1492) and Chacombe (1634) (Hall forthcoming 1995).

Ground away from the moss was also surveyed to obtain a wider context. On lower ground at Lowton House Farm all the soil is clay, some very heavy and waterlogged. Several small enclosures have hedges with aratral curves that conform with the topography as though they had furlongs, some of them long (400 m), similar to examples on the Yorkshire Wolds and at Carnforth, North Lancashire. No soilmarks or ploughed-over headlands were discovered, however. One field south of Lowton Heath (SJ 614 965) has several lands of wide ridge and furrow. To the east, near Morris's Farm (SJ 625 955) are linear banks, the boundaries of medieval furlongs, that underlie the present road and field hedge; Figure 56 shows the expected pattern of strip ploughing. It is evident that away from the low, wet Mersey Valley there was a considerable amount of arable land in the medieval period.

No palaeoecological assessment of Highfield Moss was made during the current phase of NWWS fieldwork as it is intended to cover the site as part of the West Lancashire survey.

Hopwood Hall Moss
(SD 878 086)

A peat bed at Hopwood is perched in a small basin to the north-east of Hopwood Hall, in an area called Lords Wood, at c 140 m OD (Figs 4, 57, Plate 18). A small tributary of the River Irk, called the Trub Brook, runs south-westwards cutting a narrow channel below the west side of the hill. The underlying drift geology is glacial sand on clay beds, covering solid rock of the Carboniferous series.

Hopwood was a township in Middleton parish; part of it was given to Heywood in 1863, and the rest divided, in 1894, between Rochdale and Middleton (Farrer and Brownbill 1914, 5, 170). The nearest large town is Middleton, approximately 2 km to the south, with Heywood slightly farther away on the north side. The Rochdale Canal passes north—south on the eastern side of the hillock. The Lancashire and Yorkshire (Manchester—Leigh Branch) railway line passed north—south of the east side, using the same valley as the Rochdale Canal. Modern developments include the construction of the M62, crossing west—east 1 km to the north of Hopwood, and the road improvement of the A664 c 0.25 km to the south of Hopwood; none of these directly affected the moss.

There have been no archaeological finds in the area. Hopwood manor was first recorded in 1277 (William de Hopwood) (Farrer and Brownbill 1914, 5, 170 n5), and seems to have remained in the possession of the de Hopwood family until 1773 when it passed to Edward Gregge of Werneth, who took the name Hopwood. A survey of Edward Hopwood's property in 1570 refers to 500 acres of moor, 'mossette' and turbary (LRO, D, DHP, 39/1). A lease between Edmund Hopwood and Isabella Schoffilde, dated 1587, refers for instance to the 'colefylde' at Siddal (Middleton Civic Society 1990).

By 1846 the property was in the care of R Greg Hopwood, who invested in agriculture. A long list of improvements given in an account made by the site agent Richard Dixon (previously agent at White Moss), refers to the manufacture of compost, using amongst other substances, '1900 cubic yards of bog soil', mixed on the principles proposed for making compost by the Royal Agricultural Society (Rothwell 1850, 26).

Latterly the Hall and grounds were bought by the monks of De La Salle (c 1955—85) and converted into a college. The college was recently taken over by the Local Education Authority, and the parkland is managed by the Healey Dell Nature Reserve, Rochdale Council.

Field mapping in 1992 and 1993

The surviving moss is small and lies in a high trapped basin only 100 m in diameter. There are a few drainage dikes showing that peat cutting has occurred, but the moss is now wild with cotton-grass. From the dark stain in the basin slopes above the moss it seems that about 2 m of peat have been removed. The surrounding area is pasture and no finds were made. To the north are many drainage ditches where the moss used to extend, according to the historical evidence. The moss is probably stable as long as there is no further peat cutting or drainage.

Palaeoecological assessment

Set in the grounds of Hopwood College (SD 879 087), this small basin mire less than 5 ha in area is situated at an altitude of c 140 m OD and covered by rough grazings dominated by *Molinia*. Some *Sphagnum* occurs in drainage ditches across the site. The peat is mostly 1—2 m deep, but a very limited area contains deposits reaching 4.6 m.

Stratigraphy

Over much of the mire the thin wood peats with an upper 0.5 m of *Eriophorum*-dominated peats occur. However, the deepest section consists of a basal wood peat (predominantly of *Betula* with bryophytes) overlying silty boulder clay. This is succeeded at 3.75 m by *Eriophorum*-dominated peats (initially preceded by *S sect Cuspidata* peat). At 3 m the assemblage changes back to a wood peat with *Polytrichum*, *Hylocomium splendens*, and monocotyledons present up to 2.1 m. Charcoal is also noticeable in this section of the stratigraphy. Following this a series of woody peats with varying amounts of *Calluna*, *Aulocomnium palustre*, *Hylocomium splendens*, *Sphagnum squarrosum*, *S sect Sphagnum*, continues to 0.5 m when an *Eriophorum/ Sphagnum* peat forms the final stratigraphic element.

Interpretation

The site is located in a series of small steep-sided hollows in hummocky boulder clays which contain some organic material. A date for the earliest truly organic deposits is difficult to estimate. However, by analogy with similar sites of known age, the fact that open water deposits are absent and that the sequence starts with wood peats suggests that they are likely to date from the Mesolithic period at the earliest, as opposed to late or immediate post-glacial times. The minerotrophic fen-carr deposits represented by the wood peats are succeeded by ombrotrophic deposits in the form of *Eriophorum/Sphagnum* peats, and represent a change to wetter conditions at the site with the vegetation becoming insulated against the effects of groundwater. A return to drier conditions is indicated by a switch back to woody peats above this. The

Fig 57 Heywood, Hopwood Hall, and Marland Mosses: land-use, fieldwork, and peat extent
Based upon the 1977 Ordnance Survey 1:10,000 map with permission of The Controller of Her Majesty's Stationery Office © Crown Copyright

Plate 18 Hopwood Hall Moss basin mire

latter stages of this second woody phase is marked by the presence of charcoal in the deposits, and ends abruptly with evidence for flooding of the surface in the form of abundant remains of *S sect Cuspidata*. The upper half of the mire is formed from plant remains, indicating an acid ombrotrophic system being maintained up to the surface, where a change to slightly drier conditions is marked by the assumption of dominance by *Eriophorum*. The current surface is dominated by *Molinia caerulea*, a characteristic species of drained peat surfaces.

The data are too limited to reach any firm conclusions, but it is worth observing that charcoal occurs in the deposits preceding a change to much wetter conditions. This allows the possibility that anthropogenic disturbance may have induced hydrological change at the site.

Research potential
Although the age of the deposits remains unknown, it is probable, on the basis of comparison with analogous systems in the region, that many of the deposits span much of the prehistoric period. The peats display an adequate state of preservation (humification values ranged from H5 to H8) for palynological pur-

poses. Because of the small catchment area of the site any such studies would be most likely to shed light on local vegetation changes.

Marland Moss
(SD 875 120)

Marland peat basin is located *c* 1.5 km west of the town of Rochdale, in the western Pennine foothills, on the southern side of a broad bend in the River Roch (Figs 4, 57). It lies on glacial sands and gravels (Geological Survey 1970). Marland Mere is now a lake and garden, with a golf course beyond. The wetland was probably once about 300 m in diameter, covering 7 ha. The place name may derive from *mere*, 'lake', but could also come from mere, 'boundary', denoting the boundary between Castleton and Bury (Ekwall 1922, 55). Fishwick prefers the place-name association with a waterhole (Fishwick 1889, 69).

No archaeological remains are known from the Marland area. The manor of Marland was granted to Stanlaw Abbey in *c* 1200, paying rent to Hugh de Eland, and 6d rent to the chief lord. The peat basin was used as a medieval fishery; in 1343, at Lancaster Assizes 100 shillings worth of fish had been taken from the mere of the Abbot at Merland (Fishwick 1889, 69).

At the dissolution of the monasteries there had been a court for the tenants of the abbey at 'Overland', believed to be 'Merland' (Farrer and Brownbill 1914, 5, 203 n10). In 1652 William Leach was fined for fishing in the mere at Marland manor court (Fishwick 1889, 70). A manorial survey of Rochdale taken in 1626 stated that the free tenant, Henry Ratcliffe, had bought all the land that was held by the abbots of Whalley; one of the tenants occupied a plot of land described as:

> Diverse closes; arable, meadow and pasture adjoining south east on Marland Mare and extending to the River Roch in Brierley's occupation, containing c 46 acres, worth £2 13s 4d; 2 closes of pasture and meadow adjoining to the fold on Marlan Mare [sic] containing c 7 acres, worth £3; and a large fishpond well stored with fish called Marland Mare containing c 7 acres, £3. The annual rent received from the estate was worth £229 17s 8d, and 120 acres of the total of 720 acres consisted of moss and rough ground (Fishwick 1913, 71, 17).

On Yates' map (1786) there is a small settlement at Marland on the north side of what is now the A58, lying at the crossroads with a minor road leading to the river, skirting a poorly illustrated feature that is probably the small basin of Marland Moss. On the Ordnance Survey First Series 6" map (1851) a small track leads to a sandstone quarry and across the river to a fulling mill. Marland in the mid-nineteenth century was a small village with about 13 buildings, including a smithy, clustered around what appears to be an open green.

Field mapping in 1992 and 1993

Marland Moss lies in a small deep basin, now partly excavated as a pond in a well-kept park, and surrounded by steeply rising ground. The small basin has sides steep enough to be easily appreciated. To the north the basin ridge falls deeply and sharply to the River Roch.

1—2 m of peat survive, and the park gardeners report that 'the ground shakes' when tractors are driven over. Black-stained sandy skirt soil lies on the hillsides. There was probably once 3 m of peat above the present pond level (Fig 57).

This is an interesting site, because the historical records show that there was open water present in the Middle Ages. Some of the peat is probably fairly recent, and there may be fishnet weights and other finds buried at the bottom. The moss is much disturbed by the pond and garden construction, but, if there are any developments here, the surviving peat should be sondaged and the outside skirtland ploughed for detailed field survey.

Palaeoecological assessment

The site at Marland is located adjacent to a boating lake and comprises c 2 m of amorphous peats and gyttjas overlying a sandy silt. The pond has been excavated from the organic deposits and so only the periphery of what may have been a small kettlehole deposit now survives. The degree of humification is high (c H9), but it is likely that limited pollen analysis could be undertaken at the site. It is possible that the basal deposits may be of considerable antiquity, perhaps late-glacial. The upper deposits seem to have been severely truncated and their worth as a palaeoecological resource is ambiguous. The site is under permanent grass and is not threatened.

Royton Moss
(SD 937 070)

Broadbent Moss and Royton Moss (Figs 4, 58) form one continuous wetland, located north of Oldham, and east of the small village of Heyside, on Higginshaw Lane (B6194). The wetland lies at c 190 m OD on glacial drift, boulder clay, sands, and gravels, which overlie flagstone in shale and Middle Coal Measures. Part of the moss is now conserved as a Site of Biological Interest; the total area was probably about 20 ha.

No archaeological finds are known from Royton. Heyside was associated with the Shaw family from at least the mid-seventeenth century, and may possibly be the moor referred to by Ralph Chetham, who died c 1538, leaving a 'take and farmhold and part of the moor hey in Royton' (Farrer and Brownbill 1914, 5, 114 n28). Broadbent House (SD 9415 0667; SMR 169), east of the moss, has mid-seventeenth-century features. Coal mining occurred in the area, and a nineteenth century cotton mill was built at Turf Lane End, just south of Heyside; a railway line bisects the site.

Field mapping in 1992 and 1993

Royton Moss is a valley mire hemmed in with hills on the east by Broadbent. On the Royton side (west) there is much encroachment by industry and tips.

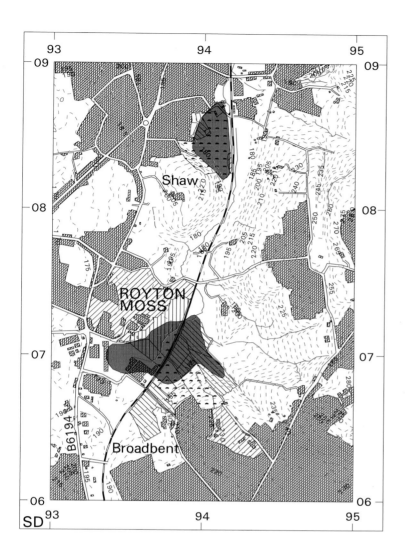

Fig 58 Royton and Shaw Mosses: land-use, fieldwork, and peat extent

Based upon the 1990 Ordnance Survey 1:10,000 map with permission of The Controller of Her Majesty's Stationery Office © Crown Copyright

Next to the railway a dike section revealed 1.5 m peat with logs at the base, overlying clay. All fields east of the railway are pasture, but molehills and dikes reveal the peat extent. On Figure 58 the extent of peat on the west is taken from the 1922 Geological Survey (1:10,560 Lancashire Sheet 97 NW).

The western side of the moss is threatened by further tipping and possible pollution from seepage. On the east the pasture fields appear to be stable.

Palaeoecological assessment
The vestigial mire at Royton Moss represents a relict portion of a valley mire which formerly occupied a channel in the boulder clays on the western fringe of the Pennine Moors. Situated at *c* 190 m OD the site, covering perhaps 5 ha, carries a peat-forming vegetation with *Calluna*, *Eriophorum*, and some *Sphagnum*; a maximum of 2.5 m of peats remain at the deepest point.

Stratigraphy
A light grey silty clay forms the inorganic substrate and is overlain by monocotyledon remains (possibly of *Scheuchzeria*) with many seeds of *Menyanthes trifoliata*. The assemblage is replaced at 2 m by wood peat which is in turn succeeded by *Phragmites* peat still rich in wood fragments. This assemblage predominates up to 0.5 m when it is succeeded by pure *Phragmites* peat; it is likely that the deposits are truncated.

Interpretation
The site seems to have originated as a shallow elongated basin with open water which was colonised by aquatic plants. This phase was relatively short-lived and was replaced by a more terrestrial fen-carr vegetation. Fen-carr communities with much reed continued to dominate the site through most of its history, until an increase in wetness caused decline in the carr element, leaving a reed bed. The later ontogeny of the site is likely to be obscured by peat removal.

109

Research potential

The peats at Royton Moss are moderately well pre-served (H6—9) and would prove suitable for palaeo-ecological analyses. The peats are probably mostly prehistoric in age and would allow insight into local vegetation changes and regional changes in the east-ern Manchester plain as well as the western foothills of the Pennines at this point. The change from fen-carr deposits to a pure *Phragmites* peat may well be worth pursuing to elucidate whether catchment disturbance may be affecting the hydrology of the site.

Warburton Moss
(SJ 720 895)

Warburton Moss (Figs 4, 59) is situated on boulder clays overlying Lower and Middle Coal Measures, at a height of 20 m OD. The site of the moss lies south of Sinderland Brook, near Carrington Moss, and is sepa-rated on the western side from the village of Warburton by a strip of arable land; it probably once extended to 200 ha (Ordnance Survey 1848). Warburton village is located on the south bank of the River Mersey, which has been subsequently diverted into the Manchester Ship Canal, leaving the old river channel intact.

Although there are no archaeological data, the place name and documentary evidence suggest that Warburton was of some importance in Saxon times. The River Mersey is thought to have been the early Saxon boundary between the kingdoms of Mercia and Northumbria to the north.

The medieval owners were called Warburton and had a park by 1469, which is marked on Saxton's map of Cheshire (1577). The boundary has possibly been fossilised by the Lymm to Dunham road, and the curved sweep of road north to Partington probably followed the eastern line of the park and moss bound-ary, although it was not laid out until the seventeenth century (Warburton court leet, 1646, Warburton Muniments, John Rylands Library Archives, Manchester).

Warburton Park Farm and Park Lodge Farm (SJ 7018 9024; SMR 383) are early buildings which still survive in the village. Near the moss, Granary Cottage and Overton Farm, buildings of c 1600—56, contain wattle and daub, and at Wigsey Farm a concealed cruck structure is preserved. The village was important because of the Mersey ferry crossing to Rixton, approximately half-way between the fording points at Warrington (downriver) and Barton (upriver). In 1836 a toll bridge was built, and the ferry

terminated when the bridge owners bought the ferry company (Faulkner 1989, 27). Most of the land and village remained the property of the Warburton fam-ily until its sale in c 1920 to the Cooperative Workers' Society (GMAU 1992c).

Field mapping in 1992 and 1993

The historic area of Warburton Moss is now mostly devoid of peat, and has an undulating pre-Flandrian surface in which it is difficult to determine the exact extent of skirtland. The surviving peat lies in the moss centre, at the highest part of a shallow slade that deepens in opposite directions (Fig 59). Skirtland seems to stretch to the north as far as the road. On the south is a central sandhill, and south of that another slade which also has dark peaty soils, hemmed in on the south by the Dunham Massey road. Hence the moss had two centres. South of the Dunham Massey road the ground rises higher and has light-coloured sandy soils that are not skirtland. Difficulties arise because so much soil in this area is dark and further darkened by nightsoil deposits. Probably most of the ground was moss and meadow land and not ploughed until relatively recently.

At Birch Farm the only real peaty area lies around Moss Wood, the northern of the two low centres, which appears to be a valley moss; its area has not changed since the nineteenth century. Peat at its western end contains many wood fragments. To the north-east and north-west the modern road appears to form the skirtland boundary.

Jack Hey Gate Farm, on the west, has some sandy soil, and one of the slades goes through it. All of the main central sandhill was searched in close detail when excellently weathered and clean, but noth-ing archaeological was recovered. No flints or other finds were made on any of the few other sandy areas.

Away from the moss, cropmarks have been re-ported. Two proposed sites (Caldwell Brook, *see Appendix 3 for details*) at Peterhouse Farm were not visited because the field was under set-aside. A third site (GM11 SJ 7344 8974) at Sinderland House Farm was under thin winter corn. The proposed cropmark lies on a small sandy knoll. There were no finds, but burnt stone occurred only on the knoll, and not in the remainder of the field, even though nightsoil debris was widespread. This may well be a prehistoric site; the original photograph has not been examined.

Palaeoecological assessment

Old maps suggest the former existence of a mire, perhaps extending to 300 ha, immediately to the

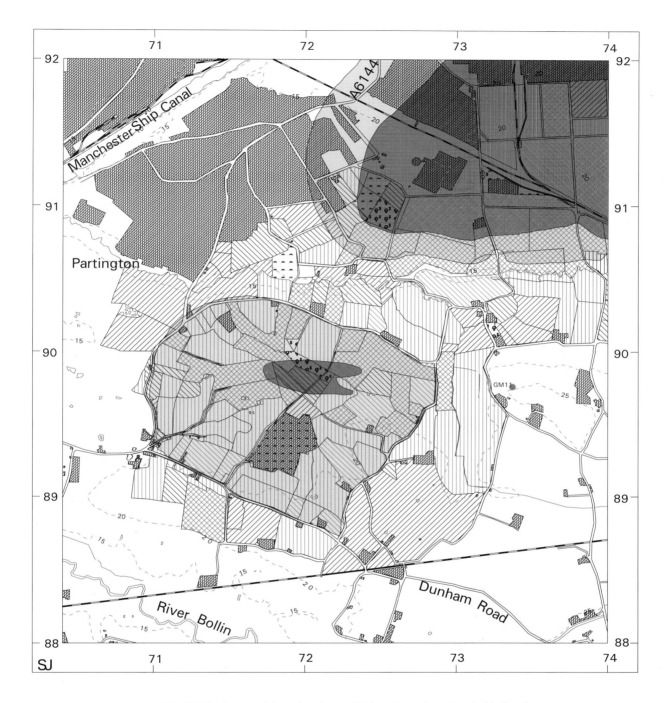

Fig 59 Warburton Moss: land-use, fieldwork, and peat and skirtland

south-west of Carrington Moss (SD 722 895). It would seem to have been bounded to the north and east by Red Brook and to the south and west by minor roads. Today there is only a vestigial central pocket of peat, reaching 0.75 m at the deepest point. The badly humified (H9—10) deposits record a basal *Phragmites* peat which is overlain by *Sphagnum imbricatum* peat. It is possible that originally the moss may have carried only shallow peats and

may have begun growth relatively late in the Flandrian period. In this respect it may be analogous to the Carrington Moss system immediately adjacent, although radiocarbon dating is needed to confirm this notion.

Limited palaeoecological analysis would be possible, although this could be difficult due to poor preservation of the deposits.

111

Fig 60 White Moss: land-use, fieldwork, and peat extent

Based upon the 1990 Ordnance Survey 1:10,000 map with permission of The Controller of Her Majesty's Stationery Office © Crown Copyright

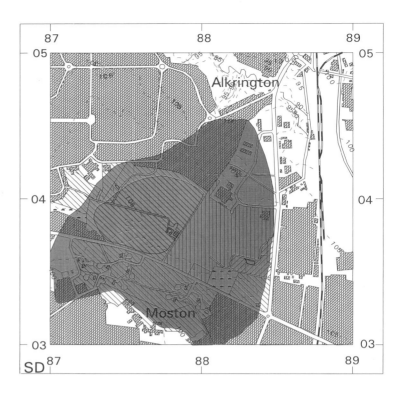

White Moss
(SD 879 036)

White Moss (Figs 4, 60) is the name latterly given to a large area of moss which incorporated the moss names of Theyle Moor and Nuthurst Moss, situated in the northern section of Moston township, the south-east corner of Alkrington township, and the southern border of Chadderton township, in Oldham parish. To the west of the moss is the township of Blackley in Manchester parish, and to the north the township of Chadderton in the parish of Prestwich with Oldham. The bounds of Chadderton and Moston divide White Moss into two sections. Blackley township also held a portion of 'moore' here, shown on the Booths Hall estate map of 1637 (Axon 1906, 30).

White Moss is sited on boulder clay overlying the lower and middle Coal Measures. The area to the south is fairly hilly, sitting on an outcrop of Permian rock, at a height of *c* 102 m. The township and the edge of White Moss are bounded on the northern side by the River Irwell, and on the southern side by the Moston main road and Boggart Hole Clough. The only known archaeological find is a barbed and tanged arrowhead, found to the south-west at Boggart Hole in 1959 (Bu'lock 1961, 21).

During *c* 1440—1503 litigation over the boundary between these three mosses occurred, and there was an eruption of White Moss in 1633 (*see Chapter 6: Historical descriptions and management of the mosses*). A manorial survey of Manchester in 1322 shows that most of Blackley was enclosed within the pales of a deer park (Harland 1861, 2, 36). Park pales were said to have been discovered buried deep in the mossland in an excavation of *c* 1834 'on the confines of Blackley and Moston', possibly north of Lumm Farm (Booker 1854, 8).

White Moss was improved in *c* 1840 (Fielding nd, 204, 205). The peat was 1.2—3.0 m deep over marl and sand, with tree stumps of oak, alder, beech, fir, and some yew. A farmhouse was built in the centre of the moss. To the south-west of Moss Farm, a marl and sand pit was utilised for dressing the mossland during reclamation, carried by tramlines (Ordnance Survey 1848). A tour of the area was described in 1858 by John Higson, who said of the moss:

> to the right a stunted tree or two and a couple of factory chimneys alone broke the horizon. The dykes and drains yielded a rusty scum or sediment of chalybeate tint... nearly all the moss was in tillage, and in a central situation was a farmhouse [Moss Farm] occupying a lonely site. On the left was a small piece of raw bog (Crofton 1907, 39).

Early industry in the area included the washing and bleaching of linen, hence the name 'bowker'

112

Plate 19 White Moss. A considerable depth of well-preserved raised mire peats are preserved under a golf course

which is probably the origin of the place name Boggart Hole (Crofton 1907, 36). Coalmining began in the sixteenth century, and wire manufacture was established at Moston (Farrer and Brownbill 1914, 4, 265). Moston is now largely urbanised with the exception of the cemetery. Alkrington has developed into a large housing estate, and the eastern side of the moss in Chadderton has grown into an even larger industrial sprawl of chemical works and a sewage farm.

Field mapping in 1992 and 1993

The north-west part of the historic moss area is now a golf course, and contains the only surviving peat, which rises to about 1.5 m above the general level. An open trench in the centre showed that peat was more than 1 m deep. Clay lay below the peat at the western edge. The golf course has some nineteenth century narrow ploughing ridges, showing that this type of cultivation is not medieval. Moss Lane probably defines the north-west boundary. The area south of the road is pasture and school playing fields, with sandy, terraced ground, probably disturbed, lying between tower blocks of flats. To the north-east is a football

ground and rubbish mound. On Figure 60 the outer line represents the area of moss in 1848 (Ordnance Survey 1848), and the inner area is the amount of peat surviving in 1992.

Any development should be preceded by sondage sampling, and fieldwalking the ploughed surround of the moss.

Palaeoecological assessment

Early maps suggest that White Moss once formed a raised or large basin mire some 400 ha in extent. Only about 50 ha remain (SD 875 040, Plate 19). Peats up to 4.35 m deep remain at the site.

Stratigraphy

Silty clays form the inorganic base of the moss and in the deepest deposits up to 0.75 m of organic muds or gyttjas overlie this. Following this stage, wood peats developed, to be succeeded by a more open brushwood peat with remains of *Polytrichum* and *Eriophorum*. This stage seems to have persisted for some time (up to 1 m of this peat type exists in places) and was followed by a sequence of *Eriophorum*- and *Sphagnum imbricatum*-dominated peats. The initial switch from *Eriophorum* peats to those with a significant *Sphagnum imbricatum* component is preceded in some places by *S sect Cuspidata* pool peats. The final stratigraphical element is formed by almost pure *Sphagnum imbricatum* peat. Time constraints precluded microscopic analysis of peats from the site and so no information is available regarding the presence of carbonised material in the deposits.

Interpretation

Deep basins within the boulder clays allowed open water deposits to accumulate forming gyttjas. By analogy with similar sites in the region it is possible that these may date from the late or immediate post-glacial period. In common with most of the other mires in the county a wood peat eventually developed as the hollows in the tills became terrestrialised by hydroseral succession. An open carr woodland with acidophilous vegetation persisted for some time before a change to more ombrotrophic conditions resulted in *Eriophorum/Calluna*-dominated communities establishing themselves. A sudden change to wetter conditions, suggested in some parts of the moss by the presence of *S sect Cuspidata* pool peats, led to a switch to *Sphagnum*-rich vegetation, particularly *Sphagnum imbricatum*. Following this there are at least another two phase changes between *Sphagnum imbricatum* and *Eriophorum* domination, possibly indicating climatically-induced switches.

Research potential

The remaining peat archive at White Moss forms an excellent reserve of material for palaeoecological analysis, and the peats are generally in a good state of preservation (humification values are H4—9, but are predominantly in the H5—7 range). As well as providing a valuable resource to facilitate study of local vegetation change, possibly through much of the Flandrian period, the stratigraphy displays well-defined changes and might provide useful data regarding climatic and/or humanly induced hydrological change at the site.

The remaining surface of the vestigial mire is under permanent pasture in the form of a golf course and deterioration of the remaining archive is likely to be slow.

6

THE ARCHAEOLOGY AND PALAEOENVIRONMENT OF THE MANCHESTER WETLANDS

This chapter describes the archaeology of the lowland mires of the Manchester region, together with their environs. It is split into conventional archaeological periods each with a regional introduction to give a general context. There then follows the evidence from the mosses, resulting from the current archaeological and palaeoecological survey. The chapter concludes with overviews of the archaeological and palaeoecological evidence for the development of the landscape. The details of the archaeological history, land-use, and the present condition of individual mosses are given in Chapters 3—5 and in the Gazetteer (*Appendix 1*).

Early prehistoric

Evidence for early Mesolithic activity in the Manchester area occurs as small lithic scatters which are mainly found in the Rossendale Uplands, and farther south along the western Pennine slopes, between Saddleworth and Marsden. Dense concentrations of finds have been made during the Tintwhistle Moor Survey (Garton 1991). Preliminary results indicate that the material collected is *in situ*. There are worked tools with utilised waste blades and flakes, suggesting that some sites were occupied. Upland sites, like March Hill, reveal multiple period occupation, but others appear to be transient temporary camps marked by a scatter of flints (Barnes 1982, 21, 23).

Lowland Mesolithic flints and cherts have been collected in Greater Manchester from Radcliffe, in the Irwell Valley (including an axe from Black Moss), Ashton Moss (Nevell 1991a, 23), Timperley (pers comm P Faulkner, South Trafford Archaeology Group), and Chat Moss. In the region of Hough Moss perforated pebble-hammers have been found. The current fieldwork has identified a lithic site at Chat Moss which has some small-blade late Mesolithic material. It is one of the largest sites known in the region and is undisturbed, being until recently buried and protected by peat. Further study will enable the full nature and context of the site to be better under-

stood. The current interpretation of the palaeoecological data is given below, and shows that the environs of the site were fairly open at the time.

The Pennine Mesolithic material falls into two major phases — an early, broad-blade phase, which in the middle of the seventh millennium BC changed to a later narrow-blade industry. The narrow-blade industry is subdivided into three groups: one with scalene triangle-dominated assemblages (*eg* from March Hill), a second dominated by straight-rod microliths, often blunted along one edge, and a third having trapezoidal microliths (Barnes 1982, 29).

The material from the upland area shows a chronological variation, the earliest generally having a dominance of a distinctive white and grey flint, probably originating from the Yorkshire Wolds or the Yorkshire coast. Better quality flint is introduced in the later phase of narrow-blade technology, probably also originating from Yorkshire. Chert was increasingly used, even though it is inferior in quality to flint. It occurs in several types: black chert of probable Derbyshire origin, brown chert perhaps from the Yorkshire Dales, and a flinty chert from upper Wharfedale (Faul and Moorhouse 1981, 79—81). The Chat Moss chert pieces are all made of brown material.

Accumulating evidence from outside Greater Manchester, in Merseyside and Cheshire, suggests that hunter-gatherers practised transhumance in upland and coastal lowland grazing areas; an early Mesolithic site at Tatton Park, Cheshire, is proposed as a temporary meeting site (Cowell 1991; Higham 1993, 16). For a recent account of work in the lower Mersey and the coastal region, see Cowell and Innes (1994).

The rather scant evidence from the distribution of known Mesolithic flints in the region suggests that some land was open and potentially habitable. The palaeoecological evidence shows that peat was beginning to develop in wet hollows with deciduous trees on the higher ground. There were possibly deliberate burning episodes during the period that may have been to keep the landscape open, so creating habitats attractive to game and grazing animals needed for food.

The environment during the Mesolithic

The end of periglacial conditions *c* 10,500—10,000 cal BC allowed the colonisation of the open landscape by trees, particularly birch and pine. In the wet hollows of the lowlands, which were to form the future mosslands, fen-carr vegetation and open water communities initiated the hydroseral process. By the middle of the Mesolithic, *c* 8000—7500 cal BC, deciduous trees such as oak and hazel were becoming prominent, while pine became common. Terrestrialisation of the wetter areas had advanced to the stage where there was a widespread distribution of carr and poor fen communities. Paludification of formerly dry ridges began, and repeated burning of the vegetation commenced over wide areas.

Towards the end of the Mesolithic period, true ombrotrophic communities became established in these areas and began the expansion of the mires, while increasing temperatures allowed the spread and predominance of deciduous trees such as alder, birch, hazel, oak, elm, and lime on the surrounding upland.

There is evidence to suggest that modification of the land and its soils by Mesolithic activity could have encouraged acidification and soil erosion which, along with a subsequent change in climate towards wetter conditions, may have contributed to the widespread growth of blanket peat (Tallis in Johnson 1985, 313—32). The present survey detected limited evidence that hints at the possibility of similar processes affecting lowland mires too, with widespread occurrences of carbonised plant material being present in peats referable to this period.

Virtually all of the mires surveyed produced evidence for *in situ* burning of vegetation during the prehistoric period. Although absolute dating evidence is limited, it seems likely, on stratigraphical grounds, that many of the burning episodes relate to the Mesolithic period. A similar situation has already been recorded by the NWWS in Merseyside (Cowell and Innes 1994) and other workers have established Mesolithic charcoal elsewhere in the region's peats (*eg* Cundill 1981; Tooley 1978).

The current evidence is insufficient to establish definitely whether some, or all, of the Greater Manchester Mesolithic burning evidence is anthropogenic in origin. A major problem of interpretation is posed by the probability that no completely unmodified ecosystem has existed in the lowlands of Britain for perhaps 5000 years, which means it is impossible to compare the Manchester data with a natural 'control'. Despite this, as explored earlier in the NWWS volume for Merseyside (Cowell and Innes 1994), a case for artificially-induced burning can be argued from the observation that British woodland is extremely resistant to natural combustion (Rackham 1986), and that wet, peat-producing carr communities might therefore be assumed to be similarly robust in the face of natural disturbance. Balanced against this view is the observation that swamp and fen vegetation is commonly ignited by lightning strikes in the Everglades of Florida (Komarek 1973). However, although this establishes the fact that it is possible for fen vegetation to combust naturally, the data refers to plant communities which, although analogous to British types, may not be entirely comparable, as they are not floristically identical and experience a different climatic regime.

If the fire record of the peats does reflect human activity, it could be marking either the passage of accidental isolated fires linked to domestic activity or perhaps deliberate attempts to manipulate the landscape on a broad scale. The widespread occurrence of charcoal in the Manchester peats would perhaps favour the latter suggestion. A motivation for such activity is plausibly suggested by the concept of the 'fragile ecosystem' proposed by Simmons and Tooley (1981). This argues that interference at relatively unstable ecotonal boundaries may have a major effect on the dominant plant species. Reedswamp and carr woodland can be accommodated within the notion of such 'fragile ecosystems', being seral in nature and forming a relatively open environment set amongst the more robust closed forest of the time. Controlled burning is frequently used today, both in England (particularly in the Norfolk Broads; B D Wheeler pers comm) and in North America, as a management tool for the maintenance of fen vegetation and to prohibit the invasion of woody species of plants (Komarek 1973). Although the antiquity of the technique is unknown, ethnographic parallels with North American Indians have led some to the conclusion that Mesolithic people may have attempted to manipulate reed and carr habitats by burning, in order to drive game (Mellars 1976) or to encourage vegetation attractive to grazing ungulates (Simmons and Tooley 1981).

Whatever the cause of such widespread burning of the wetland surface, there remains the possibility that ecological succession may have been deflected or altered in some way by it. Carbonised material is often found at the transition marking the onset of ombrotrophic or poor fen conditions. This observation is not restricted to Greater Manchester, having been recognised during NWWS fieldwork in all the other counties surveyed to date (*eg* Huckerby *et al* 1992; Huckerby and Wells 1993; Cowell and Innes 1994). The feature has been recorded in sites as far apart as Cheshire and the Scottish border, as will be reported in forthcoming NWWS volumes.

Many authorities now accept the probability that human activity affected the timing and extent of mire development in the uplands (*eg* Simmons and Tooley 1981; Moore 1986), and evidence supporting this view has been extensively documented from numerous blanket mires, usually associated with palynological indications of forest clearance (Moore 1975). Is it possible, therefore, that a similar process might have been operating in the lowlands and that the establishment of ombrotrophic communities may have been encouraged by the repeated burning of the initial carr and minerotrophic communities occupying these areas? Such an effect would be enhanced if combined with clearance of woodland in the surrounding catchment area (Moore and Willmot 1976). In Manchester, the evidence for such a coincidence is currently limited to Nook Farm, Chat Moss, where it seems that clearance was involved in the spread of peat over the underlying sands during the Neolithic and early Bronze Age. Near the sand island, fire, acting together with a rising water-table, may have prevented the regeneration of trees on the mire surface and encouraged the spread of peat. Elsewhere in the North West there are indications of this process from studies by Tooley (1978) and Innes and Tomlinson (1983), but more research is required to test these conjectures.

Neolithic and Bronze Age

In the Neolithic period, it is likely that the poor drainage of the Pennines and the spread of blanket peat at higher altitudes limited settlement. The few known sites produce small quantities of cultural material, mainly lithic, with some pottery and a few monuments (Barnes 1982, 39). Higham has suggested that the area lies at the periphery of regional communities (Higham 1993, 19).

Leaf and tranchet arrowheads have been collected on Saddleworth and Marsden Moors, which may be contemporary with Grooved Ware, *c* 2000 BC. There are isolated examples of the contemporary Seamer and Duggleby axes, one from Bacup and a similar find at Milnrow. The overall picture of axe distribution around the central Pennines shows a concentration on the flanks of major river valleys. Neolithic monuments close to the Mersey Basin include a chambered long cairn at Anglezarke, at 270 m OD. In Cheshire, a small building excavated in Tatton Park was radiocarbon dated to 3500—2945 cal BC (Higham 1993, 17).

In other parts of the lowlands a Langdale stone axe has come from Urmston and a flint dagger and other finds from Crumpsall. Fieldwork at Ashton Moss recovered eight flints assigned to the Neolithic period, and the Nook Farm site on Chat Moss (GM 7), discovered during the present survey, produced Neolithic worked flints and chert pieces. These flints are all that is known from the lowlands; they show a low level of activity consistent with previously recovered evidence from the region, which comprises a few artefacts of saddle querns and rubbers found near Rochdale (Barnes 1982, 40).

This scarcity of settlement evidence is reflected in the palaeoecological record, which suggests a low level of agricultural activity with small-scale clearings in the surrounding woodland, by this time consisting of a mixture of alder, birch, hazel, oak, and elm, with the latter declining, as elsewhere, in the early Neolithic period.

In the developing lowland mire ecosystems a mosaic of poor fen and ombrotrophic mire, separated by higher, drier ridges carrying light birch woodland, would have dominated the landscape. These remaining wooded areas were subject to small-scale clearance in the Neolithic, as seen from the evidence at Nook Farm (GM 7), and it is possible that the spread and development of mires in these areas was encouraged by this anthropogenic disturbance. Periods of climatic deterioration, marked in the stratigraphy by the presence of *Sphagnum cuspidatum* pool peats, may also have affected the region during the Neolithic and Bronze Age, perhaps increasing the difficulty of human exploitation of the landscape. Such peats have often been cited to provide proxy evidence for shifts to wetter conditions related to prehistoric climatic perturbations (Barber 1981; Haslam 1987). A degree of contemporaneity between these pool peats needs to be established, however, before such features can be regarded in this way. Stratigraphically, it would seem possible that at least some of the pool peats identified may be such climatic signatures, but much more extensive and detailed dating would be required to achieve any confidence about the identification of climatic phase-shifts (*sensu* Barber 1981) from the Manchester peats.

The limestone Peak District of the Pennines appears to have been one of the most favoured areas for occupation by early settlers, who, by the early Bronze Age were leaving material remains of a distinctive 'Peak Culture' (Barnes 1982, 43). Concentrations of burial sites on the western uplands, spreading into the lowlands, include poorly-recorded earthen round barrows, stone cairns, and ring monuments, most with urn cremations, food vessels, and Beaker-style pottery (Winwick Barrows and Castleshaw). Bronze Age flint material is very difficult to distinguish technically from Neolithic material (Cowell 1991, 48; Cowell and Innes 1994); no flints assignable to the Bronze Age were discovered during the present mossland survey.

Radiocarbon dates for features contemporary with early Bronze Age material begin in the mid-twenty-first century BC, and late Bronze Age material appears in the eighth century BC, slightly later than dates from lowland England. In the Pennine region, the main cultural remains are pieces of metalwork, approximately dated by typology and technology. In the Manchester area, a Beaker group was excavated at Pule Hill, east of Castleshaw, consisting of four food vessels and a pygmy vessel. A pygmy vessel was found at Rivington, and another at Clifton in the Irwell Valley. A pygmy vessel associated with a flat, riveted bronze dagger occurred with a burial at Haulgh, Bolton.

It is very difficult to locate funerary monuments in the lowland zone, and previous references have been vague or lack adequate record. The only excavated sites are at Haulgh, a cremation at Rivington, barrows at Winwick (*see above: Highfield Moss*) and the Wind Hill cairn at Heywood. The distribution of sites shows a link with the western flanks of the Pennines, at an average of 240 m OD. A flat bronze axe was found at Radcliffe, and other implements of the Bronze Age include a spearhead from Boggart Hole Clough, Middleton (Barnes 1982, 47, 54, 55).

There are many stray flint finds not associated with occupation sites, and small stone objects usually associated with grave assemblages, such as bone pins, jet, shale, and amber ornaments. Stone objects have been found in various locations in the Rossendale uplands, and in the Irwell and Mersey Basins (Barnes 1982, 65). From the wetlands come several exotic artefacts: a perforated macehead was found at Cheadle Heath, Cockey Moor produced a looped palstave, Bryn Moss a bronze spearhead, and Heaton Moor yielded a brown flint knife. Close to Chat Moss the Manchester Ship Canal works recovered an axe-hammer, a bronze socketed spearhead and a gold pendant (*see Irlam*). More enigmatic is the hollowed-out logboat from Trafford which may be Bronze or Iron Age, in spite of a much later radiocarbon date. These mossland finds show there was a little Bronze Age activity, but in the absence of any identifiable nearby settlement sites, it is difficult to interpret their significance. They are probably strays, for if they were

associated with burials the accompanying cremation ceramics or well-preserved bodies would most likely have been identified.

Towards the end of the Bronze Age a more general change to wetter, cooler conditions is indicated by the expansion of the mire systems and the rise to dominance of *Sphagnum imbricatum* from c 1700—1400 cal BC onwards. This climatic deterioration probably encouraged the abandonment of upland areas above 350 m, and these were covered fairly rapidly by blanket peat after c 1500 BC.

The Iron Age

Earthwork remains of hillforts and enclosures are found in the hinterland of the region. Overlooking the limestone district at 500 m, Mam Tor is the nearest late Bronze Age/early Iron Age hillfort to Manchester. The site has traces of late Bronze Age occupation, in the form of metalwork, pottery, and hut platforms, located on the southern slope. Enclosures are known in the region: at Castlesteads, Bury and elsewhere, and the postulated site of a Bronze Age hillfort at Buckton Castle (Barnes 1982, 72), although recent re-evaluation suggests that this site was a later monument, dating from the twelfth or thirteenth century AD (Nevell 1992b, 52).

The archaeological fieldwork results of 1992—3 support the notion of generally wet conditions affecting the Mersey Basin during this period: not a single Iron Age or Roman site was discovered. Ceramic remains of both these periods are normally easily identifiable on the surface of an arable field, yet none was discovered.

One Iron Age site near a wetland is known in the region, lying north of Great Woolden Hall, on a narrow belt of well-drained sandy soil between Chat Moss and the substantial stream of the Glaze Brook. The site was discovered by Higham as a cropmark in 1986, when a bivallate promontory enclosure was visible; it is of considerable interest, being the only authenticated pre-medieval cropmark near to a moss in the whole of Greater Manchester. It was subsequently excavated by GMAU during 1986—8 (Nevell 1989a, 35).

The enclosure, covering a hectare, lies on glacial sands and clays, with a high water-table; the top stratigraphy has been partially destroyed by ploughing and drainage trenches. A resistivity survey detected the ditches, but fieldwalking of the whole site yielded flints and prehistoric pottery. The two outer ditches on the north side of the enclosure were cut to a depth of 1.42 m into the water-table, revealing evidence of silting and re-cutting. Two circular structures, 13—15 m in diameter, lying within the enclosure on the west side, had U-shaped timber trenches; one of them appeared to have a southerly entrance and a hearth. Two other smaller circular ditched features, 8 m and 5 m in diameter, also appeared to be associated with postholes or stakeholes. The pottery recovered included a quartz clay fabric, one with surface burnishing, another with a frilly rim, as well as typical coarse local fabrics, containing heavy quartz temper with irregular firing. Roman pottery types recovered were Black Burnished ware, mortaria, and a local coarse fabric. A date of occupation from the third century BC into the Roman period is suggested (Nevell 1989a, 35—43; 1994).

Relatively wet climatic conditions continued into the Iron Age, with continuing paludification of the uplands as well as low-lying areas of the Mersey and Irwell valleys. Habitation was probably difficult above the 200 m contour. In the valleys, there was expansion of raised peat bogs and valley mires to form large *Sphagnum imbricatum*-dominated mire complexes. From the Iron Age onwards, these began to dominate the landscape around the middle sections of the Mersey Basin, forming, from a prehistoric agriculturalist's perspective, what amounted to wet deserts. The unattractive qualities which low productivity ombrotrophic systems offered in terms of economic exploitation may be reflected in the peats of the developing mires themselves, where reduced frequency of carbonised material in later sections of the profiles, if of human origin, attests to a continuing, but lower intensity of interference with the vegetation of these systems. Much of the carbonised material recorded from the ombrotrophic levels is Ericaceous, suggesting the clipping of leggy heather by light fires in a fashion similar to a modern muirburn (Muirburn Working Party 1977).

Once again, a problem lies in the impossibility of distinguishing between natural and artificial fires based on the available data. In a study of the incidence and causes of fires on the ecologically analogous moorlands of Derbyshire, however, Radley (1965) suggested that there were few major fires before 1900 and that it was not certain that lightning had ever caused a fire. When the fact that the study is primarily concerned with terrain which is drier than the prehistoric Manchester mires were ever likely to have been is taken into consideration, it can be seen that, although scanty, the evidence relating to the history of mire combustion leans towards the artificial cause as the most likely source of the burning.

The Roman period

Evidence for Roman activity in the region near the mosses is slight. Wigan may be the Roman *Coccium* of the Antonine Itinerary; a Roman settlement is suggested by chance finds and is now confirmed by excavation (Tindall 1985, 19—20; Jones and Price 1985). The region was consolidated early in the period, with forts established at Manchester, Castleshaw, Slack, and *Melandra*. A road from Chester through Northwich, Altrincham, and Stretford into Manchester, crossed the Pennines to York via Castleshaw and Slack. Another early route, almost entirely obliterated by later development, ran from Wigan east to Manchester, and on to *Melandra*. At the fort in Manchester there was an area of 5—10 acres used for ironworking, but by the third century industry appears to have declined, with increased building of domestic structures. It is possible that the mosses were a source of bog iron ore.

Most of the Roman roads have not been confirmed by archaeological means, although many are no doubt authentic, as judged by the usual geographic methods of identifying long, fairly straight lines linking known towns or military sites. The line of the road in Stretford was exposed and revealed a surface made of boulders 1 m below the ground surface, and another surface *c* 1 m beneath the first, being a 'wattle road of brushwood laid upon sand and covered by ling and gorse, with a ditch on either side' (Crofton 1899). The ditches very probably show that the road is Roman, and the brushwood was presumably laid in peat to stabilise the structure as it crossed the wet Mersey Valley. The site must still have been fairly wet in 1899 for the brushwood to have survived.

There are single finds of Roman coins around Manchester, and 17 coin hoards, such as that from Denton, and two groups from Boothstown, north of Chat Moss. Roman pottery outside the fort areas has been found at Rainsough and Stretford. Two quern fragments, with a distinctive double hopper manufacture, have been found at Red Moss and Kitts Moss, Bramhall, and are believed to be Roman in origin (drawn to our attention by E Wright).

Human skulls discovered in the mosses are of considerable interest, being known from Ashton Moss, Red Moss, and Chat (Worsley) Moss. Only the last is dated and falls into the Roman period; in none of these cases are the circumstances of the original deposition known.

The finds may be analogous to the better recorded recent finds at Lindow Moss, Cheshire, to the south, which have included an isolated skull, probably male (Stead *et al* 1986). All the finds at Lindow Moss have been dated early in the Roman period (Gowlett *et al* 1989; Hedges *et al* 1993, 154—5), and there may be evidence for a widespread 'bog-body cult' in the Mersey Valley region.

Truncation of the peat archives means that little can be related about the regional vegetation during the Roman period in Greater Manchester. There are likely to be isolated areas of peatland which contain profiles relevant to this period, for example the Botany Bay Wood area of Chat Moss and the remaining deep peats at Red Moss, but more intensive research is necessary to locate and analyse such stratigraphies. Evidence from the nearby mid-Lancashire site of Fenton Cottage (Wells and Huckerby 1991) indicates an extended period of drier mire conditions on the Over Wyre mosses during this time, suggesting the probability that the climate improved for much of the Romano-British period in mid-Lancashire. There is a strong possibility that this may also have been true for the Greater Manchester region.

Saxon

There is very little material evidence for activity during the immediate post-Roman period in the Manchester area. Historical records state that the Mercian Penda battled with the Northumbrians in AD 642, led by Oswald, at a site called 'Maserfelth'. Makerfield has been suggested as the location from the similarity of the name and in view of a church dedicated to St Oswald at nearby Winwick (Kenyon 1991, 77). The boundary between Mercia and Northumbria was very probably the River Mersey, since the name means 'boundary river'. A bank and ditch called the Nico Ditch, circling southern Manchester from Ashton Moss to Stretford, links the southern and eastern mosses and may relate to this period. It may be a defensive burgh work, although no date has been determined (*see Appendix 1: Hough Moss*).

From AD 700 onwards it is possible, from the place name evidence, that there was Anglo-Saxon settlement in the area at places such as Abram, Recedham (in Rochdale), Golborne, Makerfield, and Salford (Kenyon 1991, 82). Scandinavian settlement along the Mersey Valley is suggested by names like Flixton, Davyhulme, and Urmston, although the place-names may not originate from the earliest etymological source: Flixton may have been named after an Englishman with the Scandinavian name of *Flikkr* (Kenyon

1991, 127). In this part of the region, during the eleventh and twelfth centuries, there is frequent mention of Norse names such as Leysing and Sweyn of Barton (Farrer 1902, 63).

There is evidence for an early church at Eccles, as well as the name-form; the churchyard has a circular plan, and a fragment of Saxon carved cross-shaft was found close to the ford over the Irwell, during excavation for the Manchester Ship Canal, (Morris 1983, 9).

Other Saxon remains occur: a pre-Norman cross fragment at Bolton, and a possible Saxon cross in St Leonard's church, Middleton, although Edwards (in Morris 1983, 8—9) suggests this is of medieval date. The churchyard of St Chad's, Saddleworth, is circular. A few coin hoards of pre-Norman date have been recorded: two hoards of sceattas were found in Manchester, Stretford produced a ninth—tenth century coin hoard of 400 Saxon silver pieces, and a single Saxon coin came from the cemetery of Bowdon church, Trafford (Morris 1983, 13). A comb found in the Irwell may be Saxon (Roeder 1889—90, 204).

The limited evidence outlined above suggests that the Saxon period saw a widespread, if rather thin, settlement of the region. Although the material evidence is slight, the foundation of this part of medieval Lancashire was made. Early settlers picked out the best and driest locations in what must have been a wet and mossy Mersey Valley, and impact on the mosses may have been minimal.

Medieval

The Domesday Survey of 1086 refers to only 16 vills, and for most places no documents survive until the Great Inquest of 1215 (Farrer 1903, 1—114). Domesday Book records two churches in the Manchester area: St Mary's, Manchester, and St Michael's, thought to be the church at Ashton-under-Lyne. A minster church may have existed at Eccles rather than Manchester, but there is no direct evidence for this (Kenyon 1991, 143).

Parishes and townships are large compared to lowland England, and in many cases boundaries were not finalised until the late Middle Ages, or even later. Parishes in Greater Manchester usually comprise many townships, except for Ashton-under-Lyne and Radcliffe, the only single parish townships (Fig 61). Some townships have detached portions which almost certainly relate to exploitation of particular resources, such as Whittle, a detached portion of Bury parish enclosing Siddal

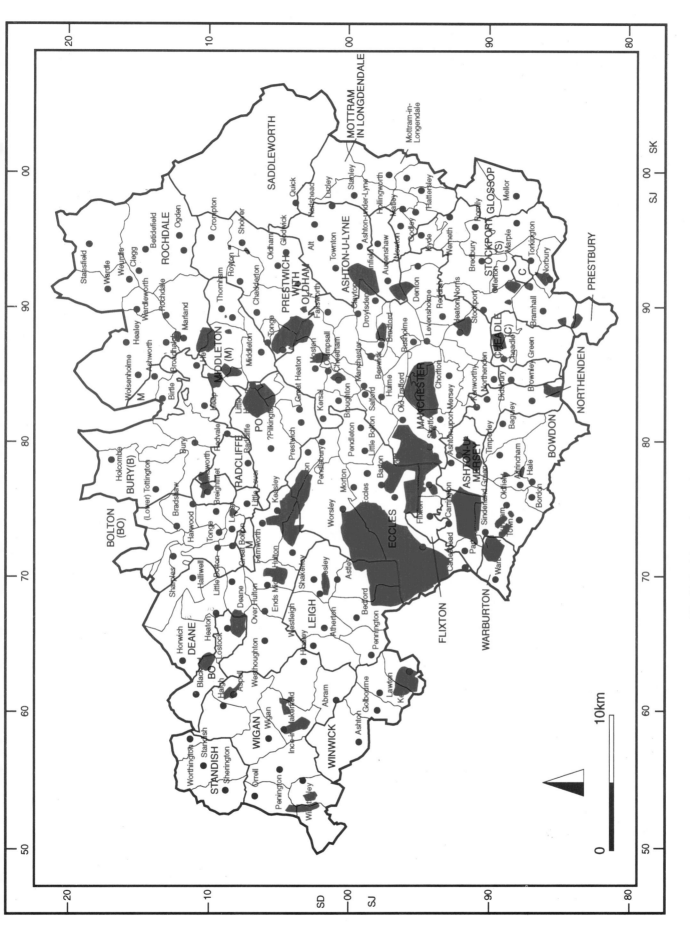

Fig 61 Parishes and townships in Greater Manchester

121

Moor. Other detached parts relate to estate division in the early Middle Ages (Tupling in Carter 1962, 116).

The settlement pattern of the region is characteristic of late development of marginal land. The place names north of Chat Moss are dominated by *leah*, 'clearing', names: Leigh (formerly West Leigh), Astley, Shakerley, Tyldesley, Worsley, and Kearsley. Few of the places are recorded in the Domesday Survey, not necessarily because they did not exist, but because they were small, of low value, and grouped together for taxation purposes. The size of a settlement has a direct effect on its later development. With the deteriorating climate after the thirteenth century, arable agriculture became more difficult, especially in areas of high rainfall. Small settlements were diminished with the general population decline, and were likely to be enclosed, being in the hands of one or a few owners. A change to animal husbandry in pasture closes was an economy well suited for the area from the fifteenth century onwards.

Many settlements consisted of very small vills that were little more than manor houses with appurtenances. This is clear along the banks of the Mersey and Irwell between Chat and Carrington Mosses. Here sufficient open landscape survives to show that only a narrow belt of dry land existed along the river banks, hemmed in and occasionally engulfed by the large untamed mosses. There was not room for any extensive arable that would support settlements of considerable size. The potentially arable areas of Cadishead, Irlam, Barton, Partington, and Carrington were small: about 450, 450, 860, 450, and 320 acres (180, 180, 350, 180, 130 ha) respectively (Fig 61), by comparison with Midland townships, usually 70—90% arable, which frequently cover 1500—3000 acres.

Manorial extents (valuation surveys made after the death of lords of the manor) support the physical evidence. The manor of Barton had 11 acres of demesne and other arable land not quantified in 1322. This compares with arable demesnes of upwards of 400 acres each for manors belonging to the Earl of Lancaster at Higham Ferrers, Raunds, and Rushden in Northamptonshire, in 1298 (*Cal. Inquisitions post mortem* 3, 296, from PRO E 133 81).

There is physical evidence from the present survey to show that medieval settlement and arable land was limited. The small amount of alluvium on the surface of the Mersey Valley between Carrington and Flixton shows that there was very little arable land causing run-off.

Agriculture

The published evidence for medieval arable in Lancashire suggests that it was not as extensive as in other parts of the country (Elliott in Baker and Butler and Baker 1973, 41—92). Much of the county was enclosed early and there are few open-field surveys from the seventeenth century, as occur for Midland counties, detailing precise land-use. There is little evidence for three-field cultivation systems, which is not surprising with so much pasture available. Organised cultivation systems with communal grazing on fallow only become necessary when there is insufficient pasture to support hay and grazing for animals. Most Lancashire townships have ample pasture near the Pennines, the mosses, or on other low areas of meadow and marsh unsuitable for cultivation. The percentage of arable land in townships near the mosses in the Irwell and Mersey Valleys would be low, leaving adequate space for grazing.

There are no broad strips of medieval ridge and furrow surviving in pasture fields, but in several places soilmarks have been observed, showing that they once existed (Glazebury and Bedford). Only farther out, on higher ground away from the Mersey, was there sufficient arable to develop earthwork headlands of the type familiar in the Midlands (Hall 1982, 25—8). These were seen at Kenyon, with one field of wide ridge and furrow at Lowton, and may have occurred under the urban areas near Leigh, but are not in evidence near the largest mosses, such as Morley in Astley or Warburton. The narrow grass strips common in various places, at Denton, Reddish, and elsewhere, are of nineteenth century date, as proved by their occurrence on White Moss, which was not drained and used for agriculture until 1840.

Plate 20 and Figure 6 show the likely boundary of arable and pasture in Astley and Bedford. North of Morley's Lane the Ordnance Survey First Series 6" map (1845) shows a hedge pattern that had almost certainly been set around open-field furlongs. Hedges with aratral (*ie* ploughing) curves near Morley's Hall, cut by the Bridgewater Canal, are particularly clear. It is in this area that soilmarks of strips were seen and where the modern ploughsoil is of a light colour, relatively low in organic content, characteristic of land long cultivated.

South of Morley's Lane the hedges are of large irregular enclosures, probably fifteenth or sixteenth century in date, subdivided with later straight hedges likely to be of mid-eighteenth century date. Such fields are typical of enclosed pasture land. Fieldwork supports this interpretation, in that no trace of medieval ploughing, in the form of either soilmarks of strips or headlands of furlongs, was found in the dark soils.

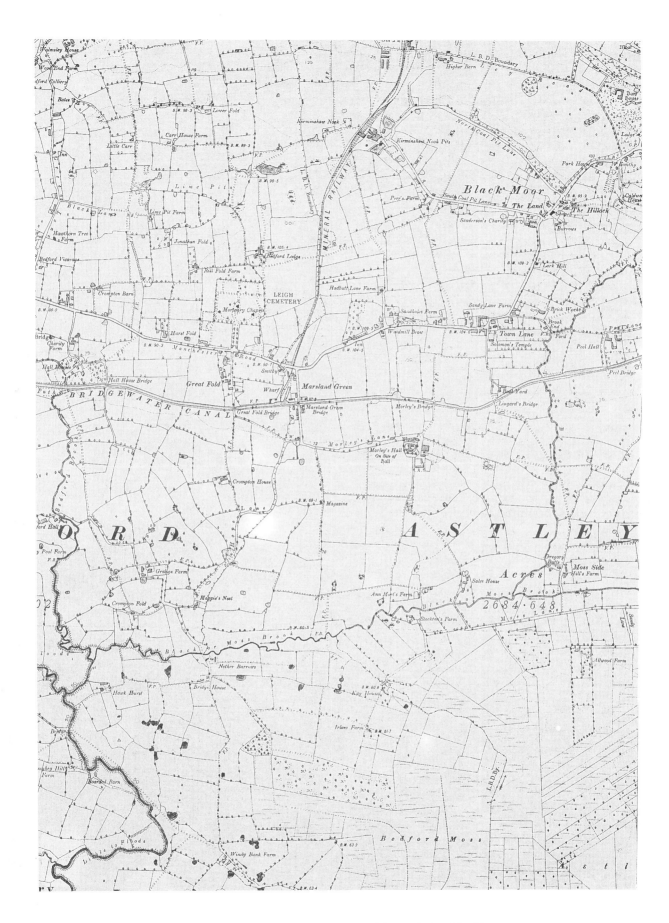

Plate 20 The Astley area in 1840 (First edition 6 inch)

123

The post-Conquest period saw rapid changes to the landscape caused, in particular, by assarting of forest, although few documents for the Manchester area describe what was happening in this early period. Only a few medieval fields survive, which is likely to be a true representation of the limited extent of arable in much of Manchester because of marshes and poor quality clay soils. The evidence suggests a mainly pastoral economy; deeds and other records reveal field names consistent with cattle raising, reclamation of marshland, and clearance ('riddings'). A lease of 1421 refers to an enclosure of wetslack at Rid Hall, near Timperley Moss (Farrer and Brownbill 1914, 4, 367 n43). At Barton, medieval documents show there was a small village lying by a ferry crossing of the River Irwell, with arable fields close to the centre. The tenants held pasture and turbary rights in nearby woods, heaths, and mosses. The manor was located outside the arable, and the lord of the manor leased an extra 30 acres of pasture and 12 acres at Boysnope. There were woods at Boysnope and Lostock, with pasture at Hagmoss, Whitmoss, Bromihurst, and Pullegrene. The resources of Chat Moss were shared by Barton, Worsley, and Astley (Harland 1861, 359).

The fourteenth century, as elsewhere, was a period of population reduction, with poor crop returns. This encouraged animal husbandry, which was a less labour-intensive economy. The shift to pastoralism continued in Lancashire, with arable agriculture practised in a small way on the borders with Cheshire.

The relative wealth of Lancashire and north-east Cheshire, compared to the rest of Britain, was below average in the mid-sixteenth century. The reasons for this include poor quality pasturage, and the competition that the Pennine foothills suffered from the Yorkshire woollen industry. The Mersey Basin pastoralists had competition from dairy producers in Cheshire who were closer to populated Midland markets. A comparison of the 1334 and 1524—5 lay subsidies for the Lancashire area shows two changes: the poverty of the Manchester embayment in the earlier period, compared to south-west Lancashire (Cheshire border), and the reversal of fortunes in the sixteenth century (Morris 1983, 22, 24).

One of the first markets documented was at Manchester, and a fair was held in Salford, granted by Earl Randle of Chester before 1232. The Manchester manor survey of 1320, listing the toll prices, shows the type of produce traded at the fair: livestock and hides, poultry and eggs, grain, provisions, beer, wax and honey, fish (including herrings which implies trade with the coast, probably along the Mersey Valley), pottery, cloth and linen, coals and bakestones. Cheshire salt was distributed at the fair, with purchasers coming from Yorkshire. Specialist commodities such as spices

and iron were listed and, by the sixteenth century, goldsmiths worked in the town. There was export of cloth to Ireland, and Irish weavers settled in the region. The developing industries in the Mersey embayment probably encouraged new markets at Ashton-under-Lyne (established in 1413), Bury (1440), and at Atherton and Leigh, and these areas show an increase in wealth continuing until 1524 (Morris 1983, 25, figs 13, 14). These markets show what was happening in the hinterland; none of the produce would have come from the mosses at this date.

Moss and peat

The wetness of the Mersey and Irwell Valleys is apparent on Figure 4, from the extent of the largest surviving mosses lying in close proximity. Chat Moss is not far from Rixton Moss on the west, and the fieldwork has shown that Warburton and Carrington Mosses almost joined up. Farther east, Sale Moor and the Stretford Moss complexes were once dominant, and possibly even larger. There are no longer any obvious topographical reasons why there should be moors or mosses in much of this region. There are no well-defined hollows as at Marland or Hopwood, or steep-sided valleys as at Chew Moor or Royton. The Mersey Valley was extremely wet, and was almost entirely covered by moor, pasture, and laterally spreading moss, apart from small high and dry spots on which late Saxon and medieval villages became established. In this terrain vills could only be small with limited amounts of arable land.

Historical descriptions and management of the mosses

The earliest account of mosses is a Manchester survey of 1320, and a second was commissioned by the Crown two years later, both probably in response to a statute of 1275 requiring manorial surveys (Harland 1861, 293, 359, map showing the extent and detached parts; Farrer 1907, 48—75).

The boundary interpreted for Heton by Harland is much the same as the township boundary. The 1320 survey describes *Heton Mos*, with 30 acres of moor turbary, of which the free tenants had *housebote* (the right to gather fuel) leaving some for commoners, and fuel worth 6s 8d was to be cut for the Lord. 'Moor turbary' probably refers to heath turf rather than moss peat. There is reference to moors and turbaries at Openshaw, Crumpsall, and Denton, all of which were exhausted or reclaimed in the Middle Ages. At Openshaw there were 100 acres of moor turbary exhausted by the tenants and lords of Gorton,

Openshaw, and Ardwick. The lord of Ancoats, Sir John de Byron, had misappropriated 40 acres. At Crumpsall there were 40 acres of moor pasture, and at Denton 398 statute acres were allowed for use by the lord of Manchester and others (Harland 1861, 328).

The second survey lists Openshaw still with 100 acres of moor turbary, Crumpsall with 40 acres, and 70 acres of moor on Heton Moss. Also listed in the survey was Har-Moss (possibly Ashton Moss, Crofton 1907, 36) with 20 acres of moor, Whit-Moss 10 acres, Bromihurst 120 acres, and in Hal-Moss 12 acres of turbary of the lord's soil with rights to the tenants of Barton. The lord had no right to make a charge on the turbary, but he could receive higher rents for arable in recompense. Hal-Moss is claimed by Crofton to be Hal-Moss in Blackley (Crofton 1907, 36—7), although it is difficult to believe that the tenants of Barton, on the other side of the parish, were allowed rights of turbary in Blackley. More likely Hal-Moss refers to a section of moss in the vicinity of the village of Barton on the Irwell, probably a section on the southern edge of Chat Moss. Chat Moss is described as the soil of the lords of Barton, Worsley, Astley, Workedley, and Bedford, and a wood with turbary rights was granted to the tenants of the lords.

There are no early documents to describe the ownership and use of the mosses elsewhere in the county. Less precise accounts are given in deeds and wills, and in two cases, depositions relating to disputes over White Moss, Ashton Moss (Crofton 1905a, 313; 1907, 44—5), Warburton Moss (Warburton muniments, John Rylands Library Archives) and Cockey Moor (Raines 1856).

White Moss (Theyle Moor) is mentioned in a 1427 *Inquisition Post Mortem*, and in other related records of 1465 and 1500—50, which describe the boundary separating Theyle Moor and Nuthurst (Crofton 1907, 37, 39—40). Litigation concerning the exact bounds of the moss appears to have been started by Thomas Chetham of Nuthurst sometime between 1440 and 1503. The disagreement was about the rights and claims of commoners who opposed the claims of the Chetham and Chaderton families: how much pasture the commoners should have, rights of turbary and enclosure, and the demands for tithes from new-born animals. The litigation was recorded on a 'platt or carde' (plan), dated 1556, which conveniently identifies the boundaries at that date. An amendment made in 1594 seems to have solved the disagreement.

Wills and inquisitions sometimes confirm ownership or use of a moss, such as an item in the inventory of the goods of Thomas Leyland of

Morley's Hall (at the north-west of Chat Moss) in 1564: 'iij hundrethe loode off turves xls' which is presumably turf extracted from Chat Moss (Piccope 1857).

General comments on the state of the mosses in the sixteenth century were made by travellers such as Leland (1533), who wrote the well-known description of the eruption of Chat Moss in 1526 (Crofton 1902, 140—1; *see Chapter 3: Previous archaeological finds*). Leland also noted that Sir John Holcroft's house, Hurst Hall, 'stood in jeoparde with fleting of the mosse'.

An eruption of White Moss in 1633 was also described by Earl Cowper of Derbyshire (Crofton 1902, 140):

> We have forewarning signs of God's judgements many. A great earthquake, immoderate rains, and great inundation of waters, strange lightnings on Sabbath day last. The report is that Ribble water stood still for two hours together. There is, some three miles from us about Blakeley and Mosson, the White Moss. Saturday night this ground brake forth, and, by the violence of the wind and the force of the water which was within, it removed itself; it came in height four or five yards, and in breadth near twenty yards and some-times more, and it went violently till it came to a place of descent which we call a clough, and then went down along such place for the space of a mile and a half, until it came into that river [Irk], and did raise it as high again as it was before, and so putrefied the water that our water [at Manchester] was as black as a moss pit, and at Hunt's Bank it left I think near a hundred load of moss earth behind it; how much then may we suppose it left in other places? A strange work of God it is…

This report of an apparent 'bog-burst', along with the one reported by Leland from the west end of Chat Moss in 1526, is perhaps understandable when the evidence for a period of climatic deterioration in North West Europe centred around the sixteenth to eighteenth centuries (the so-called 'Little Ice Age'; Lamb 1972) is taken into consideration. The mires would probably have been reaching their greatest extent at this period, just prior to the commencement of large-scale drainage and reclamation operations. Increases in precipitation, which evidently characterised this era, are more likely than at any other time in the historic period to have tipped the hydrological balance of certain systems into instability, causing these spectacular catastrophic failures of the mire structure.

Management of the mosses is illustrated in documents of the sixteenth and seventeenth centuries. At White Moss, the early plan of Theyle Moor shows the plots, on the western side of the moss, west of Moss Farm in the area then known as Hall Moss, resulting from an early enclosure agreement. The plan also shows, in stylised form, moss extraction 'rooms' (square pits) lying on the northern boundary of the present moss, near Hayfield and alongside the canal. The moss rooms are pictured as irregular intakes along the moss boundary (Crofton 1907, 51)

A court leet for Warburton, in 1647, records orders for trenching, or guttering, by all those who had moss rooms upon the moss. An estate map of c 1757 shows irregularly-sized moss rooms, and haphazard encroachment onto the moss as peat digging exhausted the supplies. Warburton Moss, then called Whitehead Moss, was enclosed in 1778 (Warburton Muniments, Catalogue Box 21 (original map held at Arley Hall, Cheshire)).

There are a few archaeological remains of the extraction industry: moss rooms occur at Red Moss, where small quantities of peat were still removed by hand until recently; large-scale commercial ditching occurred at Woolden Mosses on Chat Moss, with iron railway tracks used to transport turf sods, and turf-cutting machinery still survives at Woolden and Red Mosses.

In the early industrial period, there was interest in the extraction or 'winning' of peat, on a fairly large scale. A description of 1842 gives great detail of kiln construction at Cadishead, for the efficient carbonising of peat turf to make a pure charcoal. The article shows that peat was still used and gives the price of turf: for top or 'white' moss, considered the 'worst' of peats, the cost was 4s or 5s 'the Load'; brown peat was 5s 6d, and black peat, called 'iron turf', was sold at 6s (Albert 1842).

This economic differentiation undoubtedly related to the differing qualities of the upper and lower stratigraphies of the mires being exploited. The upper, 'white moss' peats referred to are the fresh *Sphagnum imbricatum* peats, while the 'brown peat' is the more humified *Eriophorum/Calluna* and wood peats, and the black peats are probably the woody basal peats. The latter types, being denser, provided a fuel with a higher calorific content than the more saturated *Sphagnum* peats (caused by the water-retaining hyaline cells of *Sphagnum* species). The colours described the appearance of the cut material when dried. Paradoxically, when the major use of peat changed from fuel to litter for livestock in the late eighteenth and early nineteenth centuries, the relative values of the two peat types reversed, with the more absorbent upper *Sphagnum*-dominated material becoming the more useful of the two.

Drainage of the mosses

The Duke of Bridgwater seems to have been the first in Manchester to drain a moss for land reclamation, on his Worsley estate. By 1762 a branch of the Worsley canal extended into the moss and was used for tipping waste from coal mines (Hadfield and Biddle 1970, 22).

Several accounts of the late eighteenth century record intensive drainage of the mosses in Manchester. Holt refers to the 1793 improvement of Trafford Moss, leased to Messrs Wakefield and Roscoe of Liverpool. Drains were cut at six-yard intervals leading into drains connected to the River Irwell a hundred yards away, causing a natural consolidation of peat (Holt 1795, 94). A detailed description is given of the form of ditch, and the method and style of cutting. Sod drains were used, in which the spit section was covered with a dried turf sod to form a protected natural watercourse or drain. During 1830 to 1840 tileries produced the first type of ceramic drain, of horseshoe-section, at a low cost suitable for large-scale operations (Lobban 1973, 38).

Before successful cultivation could be achieved it was necessary to prepare a fertile peat. It was common at that time to pare and burn peat bog surfaces, but such fertility was a temporary measure, and after the first crop peat soils deteriorated. Marl and sand were the chief manure additives, and a marl pit at the southern end of Trafford Moss was used. Marl was transported on movable iron tracks to carry trolleys of about five hundredweight each (Aikin 1795, 382), pulled by horses. Nightsoil from Manchester was added as a fertiliser. Similar tracks are still in existence at Woolden Moss and part of Chat Moss.

The progress of the Trafford programme can be followed in subsequent accounts, and in particular one by Roscoe (Steel 1826, 312—19). He thought the method of draining 300 acres of Trafford Moss was tedious and expensive, but by 1826 he was able to report that Trafford Moss had produced crops of potatoes, clover, oats, and wheat.

Previously, in 1805, Roscoe had taken a 2500-acre tract of land on Chat Moss, and laid down drains, based on a central road (called the 'twelve-yard road', and still in existence), at a cost of 1d per cubic yard. He made an attempt to affect directly the natural drainage of the moss, and succeeded in reducing or curtailing the flow from the sites called 'ring pits' (spring sites). After paring, burning, and ploughing, the land was manured with marl carried along rails, as at Trafford Moss, in wagons weighing up to 15 hundredweight. The crops laid were partly influenced by the productivity of the land and the prices obtainable at markets. Generally, Roscoe grew crops of mixed roots and cereals; in 1825—6 only 160 acres of the total available on Chat Moss had been brought under cultivation. In the mid-nineteenth century Chat Moss was still being described as a black swamp:

> during six or seven months of the year the hue of the general surface is nearly black; in wet seasons, and after long-continued rain, much of it converted into swamp…in hot seasons, on the other hand, and ordinarily at midsummer, walking is easy and comparatively dry… (Grindon 1867, 29)

In the late nineteenth century open areas of mossland attracted the attention of the Board of Health Committee for Manchester, which was interested in finding sites for dumping waste from the city. With a choice of three areas of raw bog at Trafford Moss, Carrington Moss, and part of Chat Moss, Carrington Moss was selected for its ease of access. The operation was successful, as witnessed by the Manchester Natural History and Archaeology Society in 1892:

> Up till about four years ago this great piece of land was untouched moss and bog. Today it is a smiling and prosperous farm…here before our eyes we have waste products producing great results of wealth and beauty…last year the value of the produce grown by themselves was £3,043…Great loads of hay were being led to the huge barns to be transferred in due time for food for the 450 horses which the Committee work with (Anon 1892, 27).

Elsewhere in the county, drainage proceeded rapidly once transport reduced the cost of manure supply and of taking produce to markets. At Hopwood, the estate of Robert Gregge Hopwood, managed by Richard Dixon, had tiled horseshoe drains in 1846, and the soil was fertilised. Rothwell's account (1850) noted that Dixon mixed a compost of nightsoil from the cotton mills with 'bog soil'.

The agricultural usage of parts of Carrington and Chat Mosses continues to the present time. The exact extent of cultivation since the nineteenth century is similar to that shown on Figures 6 and 34, and can be seen on various editions of Ordnance Survey maps. Other mosses were left wild, although Kearsley was used for agriculture until the southern part was left derelict following motorway construction in the 1970s.

Conclusions of the Greater Manchester survey

The archaeology

Archaeological fieldwork has been restricted in the Greater Manchester region because of urban coverage and the prevalence of pasture on open ground. Sandy soils occur south of the Mersey, and west of the region towards Golborne, but are much rarer in the north, where clay predominates. The modern land-use follows the soil type; the light sandy soils at Warburton and Carrington are mainly arable, but pasture is almost universal in the north on the claylands. Detailed survey of the available arable, however, produced only a limited number of sites.

Even when sites are discovered the number of finds is very low throughout all archaeological periods. Cowell and Innes (1994) have recently discussed the problem for the Merseyside region, and it is common to the other counties in the North West (Middleton *et al* 1995; pers comm R Middleton and M Leah), presenting a serious difficulty when evaluating the significance of any results compared to the remainder of England. Nevertheless, the density and extent of settlement seems to be much less than in lowland England. The medieval evidence best illustrates this since the historical data are not affected by the extensive subsequent urbanisation; for other periods it may be argued that evidence has been destroyed without record, and that there could have been as much activity as anywhere else in the country. Recent work in Cheshire and Merseyside has documented that Iron Age and Roman sites are more common than previously assumed in this region (J Collens and R Philpott pers comm).

Aerial photographic techniques, as with ground survey fieldwork, are hampered by lack of arable and the widespread occurrence of heavy clay soils. Only a single pre-medieval cropmark, the Iron Age site at Great Woolden Hall near Chat Moss (*see above*), has been discovered so far in Greater Manchester near a moss. Possible cropmarks in the Glazebrook region (many on the Cheshire border) have not been confirmed by further study, and ground survey produced no results.

The archaeological evidence for the region shows that there was early prehistoric activity on the Pennines spreading down into the Mersey Basin. The details previously known are limited, and the present survey has added only little in terms of artefacts or conventional sites represented by lithic scatters.

The most significant new site is that found in the peat workings at Nook Farm, Chat Moss (*see Chapter 3: Nook Farm — a small Neolithic clearance?*). Not only has the site produced the greatest number of lithics, but it is of interest in that it was buried by Chat Moss peats and so, with the environmental evidence, gives a view of a much earlier landscape and its exploitation. Further work is planned to sample and perhaps excavate the site (J Walker pers comm 1995). The present results show that during the late Mesolithic to early Neolithic periods major portions of the Chat Moss area were relatively dry. There had been clearance in the immediate vicinity of the Nook Farm site, with some evidence for arable agriculture being practised. Significantly, this site was only identifiable due to intensive peat cutting which has partially exposed the underlying sands. Other prehistoric sites are sparse and identified with difficulty because of the small number of flints, and in many cases only burnt stones were found, making dating impossible.

The Iron Age and Roman periods are poorly represented in the region. The only Iron Age site identified recently, the enclosure at Great Woolden Hall which continued into the early Roman era, was discovered by aerial photography.

By the early Middle Ages most of the villages of the region had been formed. Taxation returns and manorial surveys show that settlements were relatively small and dispersed, with limited arable, while place names north of Chat Moss show there had been inroads into marginal land.

The survey was able to show that the extent of potential arable land relating to many townships around the large mosses was small, and that there was much meadow land as well as the 'waste' of the mosses. The unploughed lands and mosses were valuable for grazing, with turbary rights claimed by many manors in the region, peat forming an important resource until coal was readily available.

The wetlands of Greater Manchester have proved to be a rather artificial area to study as an economically balanced region before the Industrial Revolution. They are divorced from the wider region of southern Lancashire and northern Cheshire, where there is more archaeological activity and better survival of remains. Cowell and Innes (1994) note that the number of sites in Merseyside falls off in the east of that region. Manchester, lying at the foot of the Pennines, has a local climate unsuited for corn production and so was unattractive to early settlers with primitive agricultural techniques. East of the Pennines in Yorkshire and Derbyshire (Hart 1984), where a better agricultural climate is found, there is an abundance of archaeological activity. Modern agriculture reflects these differences; there is little arable land in Manchester or eastern Lancashire, by comparison with lowland Yorkshire and Derbyshire.

The palaeoecology

In contrast to the relative paucity of archaeological material in the region, the palaeoecological resource proved to be unexpectedly bountiful, given the long period of industrialisation and development in the county. Despite two centuries of intensive reclamation activities, the Greater Manchester mires still hold considerable reserves of peat of value to archaeologists and palaeoecologists. The four main mire complexes retain large areas of stratigraphy spanning the prehistoric period, and many minor basin mires, which would prove invaluable for local studies, survive in adequate states of preservation. All sites are, however, affected by drainage, and severe damage by development is imminent at Ashton and Red Mosses.

Two new findings relevant to the palaeoecology and classification of the major mire complexes in Manchester have emerged from the survey. The first is that the large mosses should be regarded as intermediate or ridge-raised mire types rather than simply raised mires. Secondly, it seems that fire disturbance was a constant feature of the systems throughout their history. How much this influenced the ecological succession of these systems and its cause remains an area of speculation.

Most of the mires in Greater Manchester are now known to have origins of considerable antiquity. Two sites, Chat Moss and Red Moss, are definitely known to have begun accumulating organic material in the late-glacial period, and there is circumstantial stratigraphic evidence to suggest that Ashton Moss may also contain deposits ascribable to this early period. White Moss is another candidate for possible late-glacial origin. Carrington Moss may prove to be the exception to this early origin, the shallow peats pointing to the Mesolithic or Neolithic period being a more probable time for its origin. In addition, many of the small basin mires might produce pollen from the early Flandrian period. Further investigative work is required to test these conjectures.

All of the major mires display a broad hydroseral succession which, when simplified to its basic form, follows a progressive change from fen-carr wood peats to *Eriophorum/Calluna* peats to *S imbricatum* peats. It is unclear whether all of these biostratigraphic elements are approximately contemporaneous from mire to mire, or whether they vary widely in their dates of inception and duration. It seems, however, that an expansion of the mire systems occurred across the region during the late Mesolithic to early Neolithic periods, with the widespread development of ombrotrophic or poor fen communities, dominated largely by *Eriophorum*, replacing carr—fen communities. Embedded within the *Eriophorum*-dominated ombrotrophic peats of this phase are widespread *S sect Cuspidata* pool peats variously dated to periods between 6000 and 1440 cal BC.

The final major biostratigraphic element found in the major systems is the switch from *Eriophorum/Calluna*-dominated peats to *S imbricatum* domination in the late Bronze Age/early Iron Age, marking a major expansion of mire communities in the area leading to their dominance of the lower-lying parts of the Mersey and Irwell Valleys.

Peat wastage

As noted earlier, estimates of peat wastage over the last 30 years from Chat Moss suggest a loss of 30—50 mm per year. These rates are significantly higher than the 10 mm per year loss estimated by the Soil Survey for Lancashire (Burton and Hodgson 1987), and could imply that Chat Moss is losing its peats faster than other areas in the region; or alternatively previous figures seriously underestimate the rate of regional peat loss under agriculture. Either assertion has considerable implications for modelling future peat distributions in the North West.

It is currently difficult to predict peat wastage accurately, due to the relative paucity of archive peat stratigraphical information necessary to act as a benchmark against which to compare the modern data. The establishment of a large body of stratigraphical information accrued as part of the NWWS should go a long way to fulfilling this need in the decades ahead.

The survey has been successful in identifying the present condition of the mosses and drawing attention to those where there are sufficient remains to warrant further study. Maps have been prepared to show the current extent and land-use of the former wetlands. Of the five mosses with significant amounts of surviving peat, three — Ashton, Kearsley, and Red Mosses — are subject to imminent serious threats that will substantially destroy them. It is essential that a full palaeoecological record be made, and that the buried old ground surface be sampled, to find any early prehistoric sites that flourished when the climate was drier. The two remaining large mosses, Chat and Carrington, are also unstable because of drainage and agriculture. These need further work and regular checking of all peat disturbance. Other minor and remnant mosses should be monitored as part of planning control, and action taken when any development occurs. The plans prepared during the present survey make it clear precisely which areas have peat and are under threat, so making monitoring easier than was previously possible. In all cases more radiocarbon dates and palaeoecological studies are needed to confirm the dating of the wetlands and to provide the evidence for changes of land-use and climate that is so well preserved. A complete view of the development of the regional landscape will then be obtained.

The NWWS palaeoecological survey has demonstrated the great potential for adding to the sum of knowledge of environmental history within the Greater Manchester region. It reveals the extent of the surviving deposits and provides a basis of knowledge for further research. Perhaps most importantly it also identifies the most threatened areas. The data accrued will greatly facilitate modelling the future of this wetland resource, thereby enabling appropriate management measures to be identified.

Glossary

Basin mire
Mire developed in enclosed waterlogged depressions which have become colonised by peat forming vegetation. Most often found in areas of glacial deposition where local irregularities of relief such as kames and kettle holes provide the necessary geomorphological environment of hollows with enclosed drainage. Usually characterised by relatively small surface area (*ie* < 50 ha) compared to depth (often > 5 m).

Biostratigraphy
Stratigraphy based on changes in biological assemblages in the sediment (*eg* pollen zones).

Blanket mire
Peat developed directly over mineral ground up to a considerable angle of slope (10—25°) normally in an upland environment.

Bog
General term for ombrotrophic mires.

Bog burst
Catastrophic failure of structural and hydrological integrity of a peat bog. The cause remains unknown but is always associated with a rapid increase in amount of water entering the system (usually caused by prolonged heavy rain). Bog bursts are rare events but may have been commoner before modern agriculture improved drainage of mire systems.

BP
Before present. In terms of radiocarbon dates present is defined as AD 1950.

Bryophyte
Member of the major group of the plant kingdom comprising the mosses and liverworts.

Cereal type
Large grass pollen grain of between *c* 40 and 60 microns in size.

Chronozone
A chronostratigraphic unit, defined as the time-span of a selected stratigraphic unit of another kind (*eg* a pollen zone) the boundaries of which have been radiocarbon dated (*eg* Flandrian II).

Cropmark
Variations in the growth of crops caused by disturbances in the soil due to features completely buried beneath. Cropmarks are most readily identified from the air under favourable lighting conditions.

Dendrochronology
Tree ring analysis. In prehistoric examples the dating of some species of tree remains can be achieved by matching variations in ring patterns with examples of known age. Variations in tree ring widths can also be used to shed light on climatic fluctuations in the past.

Devensian
The most recent glacial stage, culminating in a brief interstadial (Windermere interstadial LDeII) followed by a final cold period (The so-called 'Younger Dryas', LDeIII). The end of the cold conditions at around 10,000 radiocarbon years BP marked the opening of the Flandrian stage.

Ecosystem
An ecological unit comprising distinct plant and animal communities together with their abiotic environment (*eg* saltmarsh, ombrotrophic mire).

Ecotone
Transitional boundary between different types of ecological community.

Elm decline
A fall in elm pollen frequencies as percentages of tree pollen, dated throughout North West Europe to around 5000 radiocarbon years BP. It defines the boundary between the Flandrian II and III chronozones and Godwin's pollen zones VIIa and VIIb. Its exact cause

131

is controversial but it may be related to a complex interaction between disease, human activity and possibly climate.

Fen	General term for minerotrophic mires.
Flandrian	The present interglacial stage in Britain, dating from *c* 10,000 radiocarbon years BP to the present.
Flood-plain mire	Minerotrophic mire frequently developed in lowland alluvial plains traversed by sluggish rivers. Characterised by fen communities.
Fluvio-glacial	Descriptive term for deposits or processes connected with the hydrology of glaciers.
GIS	Geographic Information System. A map-based computerised archive and modelling system for geographical and related information.
Grenzhorizont	Literally, 'boundary horizon'. A prominent recurrence surface found in many European ombrotrophic mires. First identified by the German stratigrapher CA Weber and initially thought to be synchronous, reflecting climatic change. Recent research has indicated a wide range of dates for major humification changes but many cluster between *c* 1000— 500 cal BC.
Gyttja	Fine detrital mud, usually deposited in open freshwater conditions. Synonym: *Nekron mud*.
Humification	Degree of decomposition (of peat).
Hydrosere	A plant succession commencing in waterlogged sites.
Intermediate mire	Mire characterised by a stratigraphic profile indicating the overgrowth of a peat reservoir to cover low amplitude water partings causing coalescence with other peat reservoirs in adjacent basins. Intermediate because they form a type of mire morphology transitional between true raised mires and blanket mires. Synonym: *Ridge-raised mire*.
Interstadial	Relatively short-lived warmer period occuring within a full glacial stage (*eg* Windermere interstadial which lasted perhaps 500—1000 years at the end of the Devensian glacial stage).
Macrofossil	Usually taken to mean plant macrofossil. A term which covers everything that cannot be considered a microfossil. It effectively includes all plant remains which can be recognised by the naked eye or with the aid of a low-power microscope. The semi-decomposed plant remains are not true fossils, in the conventional geological sense of the term (*ie* having petrified tissue), but are sub-fossils. Cell contents are generally replaced by water in the peat sub-fossil material.
Minerotrophic	Used to describe a mire whose surface receives water from outside that mire's own limits.
Mire	Peat-producing ecosystem which develops in sites of abundant water supply.
Mire macrotope	Mire complex which has been formed by the fusion of isolated mire mesotopes.
Mire mesotope	Mire system developed from one original centre of peat formation.
Monocot	Abbreviation of *monocotyledon*, group of flowering plants possessing only one cotyledon. In the context of macrofossil assemblages, often used as a convenient shorthand to encompass undifferentiated remains of grasses, sedges and rushes.

Monolith tin	Open sided stainless steel rectangular container used for bulk sampling of sediments from exposed faces.
Moss	Local (northern England/Scottish lowlands) term for mire, usually, but not exclusively, of the acid, ombrotrophic variety.
Nightsoil	Victorian euphemism for raw sewage.
Ombrotrophic	Literally, 'fed by rain'. Used to describe a mire which receives water only directly from the atmosphere in the form of precipitation.
Ontogeny	Biological or ecological development through time.
Palaeoecology	The study of past interactions between plant and animal communities and their environment.
Palaeoenvironment	An ancient environment.
Palynology	The science and practice of pollen analysis.
Paludification	Initiation of mire formation over mineral ground by the lateral expansion of peat.
Peat	Partly-decomposed organic material formed in areas of permanent waterlogging.
Phase-shift	Change in macrofossil assemblage indicating a change in surface hydrological conditions (*ie* from a wetter to a drier situation or vice versa).
Pollen zone	A body of sediment with a consistent and homogeneous fossil pollen and spore content which differs in type and frequency from that of adjacent bodies of sediment.
Post-glacial	Used to denote the temperate period since the end of the Devensian at *c* 10,000 BP.
Radiocarbon dating	A method of dating ancient organic material. The technique involves the measurement of amounts of radiocarbon (which is subject to decay after death) remaining in organic matter.
Raised mire	Ombrotrophic mire characterised by a very low amplitude convex profile and usually occupying topographical situations such as level floodplains of river systems and alluvual deposits of estuaries. The ontogeny of raised mires usually indicates that ombrotrophic plant communities have developed over minerotrophic and/or aquatic ones as upward growth of peat has insulated the surface from groundwater influence.
Recurrence surface	A stratigraphic feature found in many ombrotrophic mires consisting of a boundary between two peat layers of markedly differing humification, the upper layer being less humified. Thought to be related to climatic changes (particularly increased rainfall) affecting the wetness of the mire surface leading to an increased rate of mire growth and peat accumulation. Recently this view has been challenged by Chambers (1991) who has suggested that human disturbance might also play a role. Recurrence surfaces have been classified into a sequence in Europe by the Scandinavian palaeoecologist Granlund and are often given the prefix 'RY' (from Recurenstyor), *eg* RYIII being the 'Grenzhorizont'.
Ridge-raised mire	Synonym for *intermediate mire*.
Sand island	An area of mineral ground exposed by the wasting of overlying peat. Sand islands sometimes reveal evidence for prehistoric human activity pre-dating the paludification of the surface by mire expansion.
Sere	Plant succession.

Skirtland	Dark-stained soil derived from decomposed peat. Skirtland represents the edges of former mossland which have retrenched due to disturbance (usually by agriculture).
Slade	Small or shallow valley.
SMR	Sites and Monuments Record.
Sphagnum	'Bog moss'. Genus of mosses often dominant in ombrotrophic mires which are characterised by their water retentive hyaline cells.
Tilia **decline**	A permanent decline in pollen of *Tilia cordata* (small-leaved lime) is a feature of a number of pollen diagrams from lowland sites in the British Isles, often associated with clearance phases. Various suggestions have been made to account for the phenomenon including climatic deterioration and selective felling by humans. Extensive radiocarbon dating of the decline indicates that it is metachronous, with a range from the Neolithic to the Iron Age. However, many dates cluster around the Bronze Age (Turner 1962).
Topmoss	Local term used to describe untruncated peat deposits.
Type site	In a palynological context, a locality chosen to be a standard for comparison of biostratigraphic units.
Valley mire	Mire occurring in small, shallow valleys or channels which are not enclosed so that movement of water along the long axis is possible even with only a slight gradient. Typical of wet, elongated depressions in acidic heathland areas of lowlands.
Wetland	A term which has still to achieve a commonly accepted meaning in common usage. Its ecological meaning remains vague but is usually defined as an area characterised by a water table which, for a significant part of the year, lies close to the substrate in which vegetation is rooted. In an archaeological context it is usually taken to refer to any large area of terrestrial ground containing extensive waterlogged or organically dominated terrain. For the purposes of the NWWS, however, it is restricted to lowland mire systems (raised mires, valley mires, flood-plain mires, basin mires) and their mineral ground hinterlands.

Glossary of botanical names used in the text and pollen diagrams

Alnus glutinosa (L.) Gaertner	Alder	*Ilex aquifolium* L.	Holly
Arctostaphylos uva-ursi (L.) Sprengel	Bearberry	*Juncus* L.	Rushes
Artemisia L.	Mugwort	*Juniperus* L.	Juniper
Betula L.	Birches	*Lycopodium*	Clubmoss
Betula nana L.	Dwarf Birch	*Melampyrum* L.	Cow-wheats
Betula pubescens Ehrh.	Downy Birch	*Menyanthes trifoliata* L.	Bogbean
Calluna vulgaris L.	Ling Heather	*Molinia caerula* (L.) Moench	Purple Moor-grass
Carex L.	Sedges		
Carex cf acutiformis Ehrh.	Lesser Pond-sedge	*Phragmites australis* (Cav.) Trin ex Steudel	
			Common Reed
Carex diandra Schrank	Lesser Tussock-sedge	*Pinus sylvestris* L.	Scot's Pine
		Plantago lanceolata L.	Ribwort
Carex echinata Murray	Star Sedge		Plantain
Carex elata All.	Tufted-sedge		
Carex paniculata L.	Great Tussock-sedge	*Polypodium* L.	Polypodies
		Populus L	Poplars
Carex rostrata Stokes	Bottle Sedge	*Potentilla* L.	Cinquefoils
Cerealia	Cereal type pollen	*Potentilla erecta* (L.) Raeusch	Tormentil
		Pteridium aquilinum (L.) Kuhn	Bracken
Chenopodiaceae	Goosefoot family	Pteropsida = Filicales	Ferns
		Quercus L.	Oak
		Quercus petraea (Mattuschka) Liebl.	Sessile Oak
Cladium mariscus (L.) Pohl	Great Fen-sedge	*Ranunculus* L.	Buttercups
Corylus avellana type (*Corylus* type)	Includes Hazel and Bog Myrtle	*Rhynchospora alba* (L.) Vahl.	White Beak-sedge
Cyperaceae	Sedge family	Rubiaceae	Bedstraw family
Dryas octapetala L.	Mountain Avens	*Rumex* L.	Docks and Sorrels
Dryopteris Adans	Buckler-ferns	*Rumex acetosa* Raf type	Includes Common Sorrel
Empetrum nigrum L.	Crowberry		
Erica tetralix L.	Cross-leaved Heath	*Salix* L.	Willows
		Scheuchzeria palustris L.	Rannoch-rush
Ericales	Pollen of Ericaceae plants	*Schoenoplectus lacustris* (L.) Palla	Common Club-rush
Eriophorum angustifolium Honck	Common Cottongrass	*Succissa pratensis* Moench	Devil's Bit Scabious
Eriophorum vaginatum L.	Hare's-tail Cottongrass	*Taxus baccata* L.	Yew
		Tilia L.	Limes
Fagus sylvatica L.	Beech	*Ulmus glabra* Hudson	Wych Elm
Filicales= Pteropsida	Ferns	Umbelliferae = Apiaceae	Carrot family
Filipendula ulmaria (L.) Maxim	Meadowsweet	*Urtica* L.	Nettles
Fraxinus excelsior L.	Ash	*Vaccinium vitis-idea* L.	Cowberry
Gramineae = Poaceae	Grass family	*Vaccinum oxycoccus* L.	Cranberry
Hedera helix L.	Ivy		

Appendix 1: Gazetteer of mosses

This Gazetteer contains descriptions of all the mosses in a single alphabetic list. The lists below arrange mosses in different orders for convenience; List A helps to locate mosses on Figure 4, and List B places mosses in their modern districts.

List A

1	Winstanley; Far Moor, Little Moor	SD 5302
2	Bryn and Ince Moss	SD 5903
3	Amberswood Common	SD 6004
4	Aspull Moss	SD 6108
5	Highfield Moss	SJ 6196
6	Black Moor, Tyldesley	SD 9601
7	Red Moss, Horwich	SD 6310
8	Chew Moor (Runworth Lodge)	SD 6607
9	Hollins Moss (Hulton Moor)	SD 7005
10	Kearsley Moss; Linnyshaw, Clifton	SD 7504
11	Black Moss	SD 7607
12	Cockey Moor (Ainsworth)	SD 7610
13	Unsworth Moss	SD 8307
14	Siddall Moor	SD 8508
15	Pilsworth Moor	SD 8409
16	Heywood Moss	SD 8709
17	Marland Moss	SD 8712
18	Hopwood Hall	SD 8708
19	Slattocks Peat	SD 8908
20	Shaw Moss	SD 9408
21	Broadbent/Royton Moss	SD 9407
22	Moss Grove	SD 9103
23	Hale Moss	SD 8903
24	White Moss	SD 8703
25	Ashton Moss	SJ 9298
26	Denton Moor	SJ 9095
27	Brinnington Moor	SJ 9192
28	Cheadle Moor	SJ 8789
29	Little Moor	SJ 9189
30	Woods Moor	SJ 9080
31	Stockport Moor	SJ 9188
32	Bramhall Moor	SJ 9192
33	Norbury Moor	SJ 9186
34	Kitts Moss	SJ 8883
35	Hall Moss	SJ 8883
36	Shadow Moss	SJ 8385
37	Hough Moss	SJ 8394
38	Turn Moss	SJ 8093
39	Stretford/Trafford Mosses	SJ 7895
40	Urmston E'es	SJ 7693
41	Sale Moor	SJ 7992
42	Timperley Moss	SJ 7889
43	Hale Moss	SJ 7787
44	Warburton Moss	SJ 7289
45	Carrington Moss	SJ 7491
46	Chat Moss	SJ 7197

Moss List A in alphabetic order

Moss name	Moss number
Ainsworth	12
Amberswood Common	3
Ashton Moss	25
Aspull Moss	4
Black Moor	6
Black Moss	11
Bramhall Moor	32
Brinnington Moor	27
Broadbent	21
Bryn and Ince Moss	2
Carrington Moss	45
Chat Moss	46
Cheadle Moor	28
Chew Moor	8
Clifton	10
Cockey Moor	12
Denton Moor	26
Far Moor	1
Hale Moss	43
Hall Moss	35
Heywood Moss	16
Highfield Moss	5
Hollins Moss	9
Hopwood Hall	18
Hough Moss	37
Hulton Moor	9
Kearsley Moss	10
Kitts Moss	34
Linnyshaw	10
Little Moor	21
Marland Moss	17
Moss Grove	22
Norbury Moor	33
Pilsworth Moor	15
Red Moss	7
Royton Moss	21
Runworth Lodge	8
Sale Moor	41
Shadow Moss	36
Shaw Moss	20
Siddall Moor	14
Slattocks Peat	19
Stockport Moor	31
Stretford	39
Timperley Moss	42
Trafford Mosses	39
Turn Moss	38
Tyldesley	6
Unsworth Moss	13
Urmston E'es	40
Warburton Moss	44
White Moss	24
Winstanley	1
Woods Moor	30

List B

The full list of mosses in the Greater Manchester area, listed according to the district in which they occur, the numbers refer to List A.

BOLTON

11	Black Moss	SD 768 075	SMR 4335
8	Chew Moor (Runworth Lodge)	SD 661 078	SMR 4336
	Darcy Lever Moss	SD 746 083	SMR 4340
	Halshaw Moor	SD 738 057	SMR 4332.1.1
9	Hollins Moss (Hulton Moor)	SD 700 055	
10	Kearsley Moss	SD 750 040	SMR 4332.1.0
7	Red Moss	SD 638 100	SMR 4341

BURY

12	Cockey Moor, Ainsworth	SD 768 100	SMR 3873
11	Black Moss	SD 768 074	
	(notes on this moss are included with Bolton, Little Lever District)		
13	Unsworth Moss (Back o' th' Moors)	SD 8307	SMR 3878

MANCHESTER

	Chorlton E'es	SJ 810 930	SMR 2025
	Crumpsall Moss	SD 840 005	SMR 2140
	(boundary unidentifed)		
37	Hough Moss	SJ 830 948	SMR 2026
	(Oose Moss, also includes Grindlaw Marsh, both boundaries unidentified to date)		
	Openshaw Moor	SJ 885 975	SMR 4479
36	Shadow Moss (Moss Nook)	SD 8385	SMR 2027

OLDHAM

21	Broadbent Moss	SD 942 071	SMR 6212.1.0
23	Hale Moss (Theyle Moor)	SD 896 032	SMR 7495.1.1
22	Moss Grove (Hollins)	SD 035 913	SMR 6214
21	Royton Moss (with Broadbent)	SD 937 070	SMR 6212.1.1
19	Shaw Moss	SD 940 085	SMR 6215
24	White Moss	SD 879 036	SMR 7495.1.0
	(parts of this moss are in Rochdale and Manchester Districts)		
	(Wey Moss and Saddleworth Mosses are blanket peats)		

ROCHDALE

16	HeywoodMoss	SD 870 098	SMR 5041
18	Hopwood Hall	SD 878 086	SMR 5045
	Little Moss	SD 870 050	
17	Marland Moss	SD 875 120	SMR 2311
15	Pilsworth Moor	SD 840 096	SMR 5055
14	Siddal Moor	SD 851 088	SMR 5076
19	Slattocks Peat	SD 894 088	SMR 5085

Some Rochdale peats may have developed from recent waterlogging (mapped from 1930s).

SALFORD

46	Barton Moss	SD 7397	SMR 3033.1.1
46	Cadishead Moss	SJ 703 933	SMR 3033.1.2
10	Clifton Moss	SD 768 033	SMR 4332.1.2
46	Great Woolden Moss	SJ 700 935	SMR 3033.1.4
46	Irlam Moss	SJ 7196	SMR 3033.1.5
46	Chat Moss		SMR 3033.1.0
	(general name for moss north of the Manchester Ship Canal)		
46	Little Woolden Moss	SJ 691 950	SMR 3033.1.6
10	Linnyshaw Moss	SD 745 038	SMR 4332.1.3
10	Morton Moss	SD 7503	SMR 4332.1.4
10	Red Carr Moss	SD 766 027	SMR 4332.1.5
10	Swinton Moss	SD 771 030	SMR 4332.1.6
10	Walkden Moss	SD 746 032	SMR 4332.1.7
10	Wardley Moss	SD 758 036	SMR 4332.1.8
46	Worsley Moss	SJ 715 975	SMR 3033.1.3

STOCKPORT

32	Bramhall Moor	SJ 912 871	SMR 2586
27	Brinnington Moor	SJ 913 923	SMR 2589
28	Cheadle Moor/Heath	SJ 872 892	SMR 2590
34	Kits Moss	SJ 888 847	SMR 2591

35	Hall Moss	SJ 885 837	SMR 2594
	Heaton Moss/Moor	SJ 882 915	SMR 2595
33	Norbury Moor	SJ 918 860	SMR 2597
29	Stockport Little Moor	SJ 911 896	SMR 2598
31	Stockport Moor	SJ 912 882	SMR 2600
30	Woods Moor	SJ 905 880	SMR

TAMESIDE

25	Ashton Moss	SJ 920 985	SMR 7472
	(Little Moss and Droylsden Moss)		
26	Denton Moor	SJ 904 950	SMR 3471

TRAFFORD

45	Carrington Moss	SJ 740 915	SMR 7922
	Dumplington Moss	SJ 775 918	SMR 7923
	Edge Moss	SJ 800 957	SMR 7924.1.1
43	Hale Moss	SJ 773 876	SMR 7925
39	Hennetts Moss	SJ 7894	SMR 7924.1.2
39	Longford Moss	SJ 798 950	SMR 7924.1.3
39	Lostock Moss	SJ 783 957	SMR 7924.1.4
39	Low Moss	SJ 781 952	SMR 7924.1.5
39	Lower Moss	SJ 805 959	SMR 7924.1.6
41	Sale Moor/Moss	SJ 794 927	SMR 7926
39	Stretford Moss	SJ 7997	SMR 7924.1.0
39	Stretford E'es	SJ 785 935	SMR 7927
39	Timperley Moss	SJ 780 894	SMR 7928
39	Trafford Moss	SJ 780 975	SMR 7929
38	Turn Moss	SJ 804 939	SMR 7924.1.7
40	Urmston E'es	SJ 760 935	SMR 7930
	Urmston Moss	SJ 750 950	SMR 7931
44	Warburton Moss	SJ 720 895	SMR 7932

WIGAN

3	Amberswood Common	SD 606 043	SMR 4844
4	Aspull Moss	SD 612 082	SMR 4845
46	Astley Moss	SJ 7198	SMR 3033.1.7
46	Bedford Moss	SD 688 973	SMR 3033.1.8
2	Bryn Moss	SJ 586 031	SMR 4847.1.0
	Diggle Flash	SD 6302	(caused by mining subsidence)
	Fir Tree Flash	SD 640 010	(caused by mining subsidence)
5	Highfield (Golbourne)	SJ 613 959	SMR 4848
2	Ince Moss	SD 590 028	SMR 4847.1.1
6	Tyldesley Moss	SD 690 016	SMR 4852
1	Winstanley Moor	SD 5302	SMR 4908

Gazetteer

AMBERSWOOD COMMON

Project Number	3
SMR Number	4844
Parish	WIGAN
District	WIGAN
NGR	SD 606 043

The original extent of Amberswood Common wetland is unknown; it is shown on Yates' map of Lancashire, and the place-name Moss Hall on the north side of the open common area shown on the First Series Ordnance Survey map (Lancashire Sheet 94, 6":1 mile, surveyed 1845—6, published 1849) indicates the whereabouts of the moss. The site lies at *c* 35 m OD west of Hindley, in an angle formed by the Westhoughton—Platt Bridge Road (A58), and the Atherton—Wigan Road (A577). It overlies glacial drift on Coal Measures, shales and sandstone.

A Manchester to Wigan Roman road was alleged to cut across the Common (Sibson 1846, 532, 552; Watkin 1883, 62). There is a moated hall site on the south side of Amberswood Common, called Lowe Hall, the ancient residence of the Langtree family, now mainly destroyed with only part of the moat surviving (SMR 4930, SD 6060 0325).

The moss has been largely destroyed by coal mining and ancillary industry. The First Series Ordnance Survey map shows the abandoned remains of Moss Hall Colliery, the Lancashire and North Western Railway (Lancashire Union) line running north—south on the east edge of Amberswood, the Lancashire and Yorkshire Railway (Pemberton branch) running west—east on the north edge of the common, and three mineral rail lines criss-crossing it.

1992—3 field mapping

The area is now a wasteland of levelled slag heaps with no archaeological potential.

ASPULL MOOR (Fig 46)

Project Number	4
SMR Number	4845
Parish	WIGAN
District	WIGAN
NGR	SD 612 082

Aspull Moor lies at the crossing of the B5238 and B5239 roads, 1.5 km south of Blackrod, Bolton, on the southern side of a low ridge running north—south from Blackrod to Kearsley. It overlies glacial clay drift on Middle Coal Measures, at *c* 125 m OD.

The site is shown on Yates' 1786 map of Lancashire, with a large community located on the fringes of the moor, and a moated site at the western edge, called Manor House Farm (formerly Moor Field Hall, SMR 596, SD 6071 0798). The moat partially survives, with a post-medieval farmhouse built in stone and brick; Richard and Thomas Gerard of Highfield are mentioned in the early eighteenth century (Farrer and Brownbill 1914, 4,122).

By the date of the First Series Ordnance Survey map (Lancashire Sheet 94, 6":1 mile, surveyed 1845—6, published 1849), the site had been industrialised, with a mineral railway crossing east—west, and a 'cannel pit' (Bradshaw House coal and cannel works) at the east end of the moor.

1992—3 field mapping

Little of the moss area survives, having been built over and mined, and all that remains is pasture. The site lies on a hill-top and spur, so was probably a blanket peat. An undisturbed part may be at SD 611 082 where there are rushes, but no peat is marked on the geological survey. Much mining has occurred to the south. The moss is remembered by the *Moorgate* public house at the beginning of Scot Lane. There is narrow-rig ploughing at SD 608 078; and cross narrow-rig ploughing at SD 614 088.

No further action is needed.

ASTLEY and BEDFORD MOSSES

Project Number	46
SMR Number	3033.1.7
Parish	LEIGH
District	WIGAN
NGR	SJ 7198

Astley and Bedford Mosses form the north-western corner of the large wetland complex called Chat Moss. The west side is bounded by Glaze Brook, the north by Black Brook; the township and parish boundaries square off the south-east corner from the rest of the moss. The site overlies alluvial sands and gravels and glacially deposited boulder clay, with Triassic Bunter Sandstone underneath.

The manorial history of the moss is given under Chat Moss. There are a number of ancient halls in the vicinity, for example Light Oaks Hall, now a brick built farmhouse, probably on the site of a building mentioned in 1356 (Walker and Tindall 1985, 192, SMR 4138, SJ 6758 9672). The site of Hope Carr Hall, from 1291 the residence of the Sale family, has now little trace of two possible moats (SMR 4948.2.0, SJ 6649 9871). Medieval Bedford Hall is now a farmhouse with a reputed tunnel leading to the moated site of Morleys Hall. The hall was possibly the early manorial site, although the manor had been divided by 1291 when it is first documented; the present hall dates from the early seventeenth century (SMR 4745, SJ 6742 9892). Morleys Hall, a sixteenth/seventeenth-century moated site on the north side of the moss, is associated with a Benedictine monk Ambrose Barlow. William de la Doune, tenant in 1301, appeared at a manorial court accused of felling up to 300 trees in Bedford wood, allegedly to repair the hall at Morleys (Farrer and Brownbill 1914, 3, 431 n17, SMR 4064, SJ 6896 9927). A mill close to Bedford Hall, demolished in the fourteenth century, was reconstructed on the other side of Glaze Brook (now Cheshire) with the agreement of the tenant of Light Oaks Hall; the remains of the medieval mill were thought to be visible in the 1970s (SMR 4745, SJ 672 989).

Astley Moss was one of the earliest to be enclosed; an Act dated 1763—4 saw the construction of Great Moss Road (WRO). The Bridgewater Canal skirts the northern edge of this moss. Astley Hall Estate, including part of the moss, was put up for sale in 1889 (WRO PC 5a/3). In 1908 Astley Green colliery sank the first shaft through unstable organic soils by hydraulics (Preece and Ellis 1981, 46)

The **1992—3 fieldwork** is described under Chat Moss in Chapter 3.

BARTON MOSS

Project Number	46
SMR Number	3033.1.1
Parish	ECCLES
District	SALFORD
NGR	SJ 7397

Barton Moss lies at the south-eastern end of Chat Moss, and is now largely drained by early reclamation and by motorway construction. It is situated on alluvial sands and gravels and glacial clays overlying sandstone. The manorial history is discussed with Chat Moss, Chapter 3.

The seat of the Bartons, Booths and Leighs is the site of the moated Barton Old Hall south of the moss, on the east side of Salteye Brook and north of Liverpool Road. A medieval cross, now at Eccles church, was rescued from one of the external hall elevations when demolished in 1897. A sixteenth—seventeenth century coin hoard was also discovered at this period (SMR 1904, SJ 7540 9794). The Manchester—Liverpool rail line, constructed under the direction of Stephenson, opened in 1830 (Ashmore 1982, 22), transects Chat Moss east—west, on the Barton side of the township boundary. By 1894 a cemetery had been laid out on the southern fringes of the moss (Second Series Ordnance Survey map, 6″:1 mile, Lancashire Sheet 103, surveyed 1894, published 1896).

The **1992—3 fieldwork** is described under Chat Moss, Chapter 3.

BLACK MOSS (Fig 62)

Project Number	11
SMR Number	4335
Parish	RADCLIFFE
District	BURY
NGR	SD 768 074

Black Moss site is approximately located by the place-name marked on the First Series Ordnance Survey map (Lancashire Sheet 95, 6″:1 mile, surveyed 1844—6, published 1850), midway between the river and the now-dismantled Lancashire and Yorkshire Railway, at *c* 100 m OD. The original extent of the moss is not known, although the place-names 'Black Moss' and 'Moss Shaw Fold' on the Ordnance Survey map close to the line of the railway suggests that the moss extended northwards. The site overlies alluvial gravels and sands, glacial drift, and shales and gritstones.

The waterfront archaeological site of Radcliffe E'es, located *c* 2.5 km east of the moss site was investigated in the 1950s and *c* 1960. Excavations are said to have revealed a double row of post-holes covered by brushwood, allegedly associated with Mesolithic flint implements and flakes. Pieces of worked wood were identified by Frank Sunderland (librarian at Radcliffe) as post-conquest, probably medieval (letter dated 18.05.61, GMAU archive Radcliffe E'es). An ard, dating probably to the first century, was similar to a Tommerby-type plough (letter dated 10.05.61; identification by Alan Aberg, Southampton Museum). The Ancient Monuments Laboratory determined two radiocarbon dates of no more than 200 years old (letter dated 29.9.72). Samples of hide and red raddle have been subsequently lost (SD 7980 0720, SMR 77). Other finds from the site include a cast bronze flanged-axe.

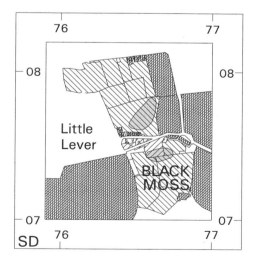

Fig 62 Black Moss: land-use, fieldwork, and peat extent

Based upon the 1980 Ordnance Survey 1:10,000 map with permission of The Controller of Her Majesty's Stationery Office © Crown Copyright

The Roman road from Manchester to Ribchester is thought to have crossed the River Irwell at Radcliffe (Watkin 1883, 53; Just, 1842, 409). Also in the vicinity of Radcliffe E'es was the 11—12th century Radcliffe manor-house with a crenellated fifteenth-century tower (demolished before 1844; SMR 354, SD 7958 0751).

The north bank of the Irwell has been frequently used for gravel extraction. Coal mining and brick pits operated, and on the south is the large site of the Mount Sion Printworks (shown on the First Series Ordnance Survey map). The Bolton and Bury Canal runs east—west, parallel with the River Irwell and south of the moss.

1992—3 field mapping
The area available for field examination lay east of Little Lever and was all pasture on boulder-clay soil. Two small rushy wet areas with slightly peaty soil are all that remain of this moss. The ground falls sharply to the Irwell to the south lying at a much lower level; the moss was therefore trapped in a small high basin that had at least two centres.

No further action except collection of dating samples is needed.

BRAMHALL MOOR

Project Number	32
SMR Number	2586
Parish	STOCKPORT
District	STOCKPORT
NGR	SJ 912 871

Bramhall Moor is shown on a map of the environs of Stockport *c* 1800 (A new and accurate map of the environs of Stockport, surveyed by William Stopford, held at Stockport Local Heritage Library), and the First Series Ordnance Survey 1″ map, although the extent of the moss is not clear from either. The site overlies glacial morainic drift at *c* 80 m OD, on top of Coal Measures, shales and mudstones.

The moor was reclaimed *c* 1825 (Dean 1991, 20). The village to the north-west of the moss, called Bullock Smithy, documented in the late sixteenth century, was at a boundary point of the Forest of Macclesfield. It is shown on the *c* 1800 map, and was re-named Hazel Grove *c* 1860, (Dodgson 1970, 256). By the late nineteenth century, the moss had been cut on the northern side by a branch of the London and North Western Railway (First Series Ordnance Survey map, Cheshire Sheet 19, 6″:1 mile, surveyed 1872, published 1882).

1992—3 field mapping
What survives of the area is now left as an open rough space that was a golf course until *c*.1986. It lies on sandy soil and was probably a heath, with drainage to the west; some peaty soil remains at the low part at SJ 907 875.

There are parts of several furlongs of medieval ridge and furrow of low profile (*ie* wide ridge, not the more common narrow ridge). Peat may have been dug from the slades (valleys) between the furlongs.

A raised earthwork of elongated oval form occurs at SJ 9114 8718, 40 m short diameter and *c* 1.5 m high. It overlies the ridges and is probably recent (of Home Guard date or related to the golf course).

BRINNINGTON MOOR

Project Number	27
SMR Number	2589
Parish	STOCKPORT
District	STOCKPORT
NGR	SJ 913 923

Brinnington Moor lies south of the River Tame above the confluence with the River Mersey, and on the east side of the M63 at the junction with Brinnington Road. It is crossed by the Manchester—Sheffield rail line, and lies at *c* 90 m OD, on glacial drift and Coal Measure sandstones. The extent of the moss is not known, it is shown as an irregular area on a survey of *c* 1800 (A new and accurate map of the environs of Stockport William Stopford, Stockport Local Heritage library), and is noted as a place-name on the First Series Ordnance Survey 1" map.

The name 'Brinnington' was probably the name of a Saxon settlement, and has been suggested as the original Saxon town of Stockport, destroyed in the mid-11th century (SMR 7528). No remains have been found. The area has had many coal-quarrying pits in the vicinity.

1992—3 field mapping
The area is built over. There is no obvious topographical reason for peat to have developed, with no sign of peat in gardens. Two-storey houses nearby seem stable.

BROADBENT and ROYTON MOSSES
Project Number	21
SMR Number	6212 .1.1
Parish	PRESTWICH WITH OLDHAM
District	OLDHAM
NGR	SD 942 071

The history and fieldwork of these mosses is described in Chapter 5.

BRYN MOSS and INCE MOSS
Project Number	2
SMR Number	4847.1.0, 4847.1.1
Parish	WIGAN
District	WIGAN
NGR	SD 586 031

Bryn Moss has now subsided due to mining, creating a water-filled hollow, commonly known as a 'flash'. It is located to the south of Wigan town centre, with the Leigh branch of the Leeds and Liverpool Canal on the west, and enclosed on the east side by the London and North Western railway. Before the canal was built, Ince Brook, a tributary of the River Douglas, flowed on the west side of the moss. It overlies boulder clay on Middle Coal Measures, shales, mudstones and sandstones.

A bronze spear-head, with a central rib shaft, 6.75 inches long, was found at Ince in 1892, between Ince and the Weaver. It lay below 6 m of 'peat, sand and marsh clay', and is now preserved at Warrington Museum (Warrington Museum Catalogue).

Close to the moss are the mid/late sixteenth-century New Bryn Hall (SMR 4792, SD 5881 0165), and near to Landgate Farm the moated site of Old Bryn Hall, built on a mound with a drawbridge in swampy ground (SMR 4655, SD 5779 0150). One of three halls in Ince, of possible late sixteenth-century origin, is located on the north-east side of the moss, at Ince Green Lane (SMR 4228, SD 5922 0430), and Hawkley Hall, on the site of a demolished sixteenth century mansion built on the site of a late fourteenth century hall, is located to the west of the moss (SMR 919, SD 5752 0310). Saxton's 1577 map of Lancashire depicts an enclosed deer park at Bryn. In the late eighteenth century, bulls were bought and pastured in the marshes (Slater 1910, 69)

Subsidence in the area was caused by the Pearson and Knowles coal mine, owned by the Ince Moss Colliery, sunk in 1863 (Rymer 1922). The landscape of the area in 1843 before any disturbance, is shown in Plate 1.

1992—3 field mapping
The only area not destroyed by mining and the wasteland of a large active rubbish tip, is Wigan Borough Cemetery which lies to the east of the estimated moss extent. Part of the cemetery lies low and floods. No peat survived in flower beds and the soil is sandy.

CADISHEAD MOSS
Project Number	46
SMR Number	3033.1.2
Parish	ECCLES
District	SALFORD
NGR	SJ 703 933

Cadishead, one of the many parts of Chat Moss complex, is found at the south-west corner of Chat, between Glaze Brook and Irlam Moss. It overlies alluvial sands and gravels on glacial clays with a solid geology of Permo-Triassic sandstone.

Cadishead Moss was separated from Woolden Mosses and Lower Irlam Moss by a straight drain, shown on the First Series Ordnance Survey map (Lancashire Sheet 110, 6":1 mile, surveyed 1845, published 1849). A method of carbonising turf was described in 1842 by Dominique Albert, who built an oven at Cadishead to take advantage of the proximity of the moss (Albert 1842, 399)
The general history of the moss is found in the account for Chat Moss.

1992—3 field mapping; see Chat Moss, Chapter 3.

CARRINGTON MOSS
Project Number	45
SMR Number	7922
Parish	ECCLES
District	TRAFFORD
NGR	SJ 740 915

The history and fieldwork studies of this moss are fully described in Chapter 4.

CHEADLE MOOR or HEATH
Project Number	28
SMR Number	2590
Parish	CHEADLE
District	STOCKPORT
NGR	SJ 872 892

The extent of Cheadle Moor is not known, although shown on a map of c 1800 (A new and accurate survey of Stockport, William Stopford), and the First Series Ordnance Survey, 1" map. It overlies glacial drift and Coal Measures at c 50 m OD, and is located near the junction of two main roads, Stockport—Cheadle road (A560), and Edgeley road (B5465). The nature of the peat type is uncertain.

A perforated stone macehead was found close to Cheadle Heath in 12 feet of gravel. The stone is thin, 5 inches diameter, with a hole 1.5 inches wide, and a rough edge; it is lost (SMR 799, SJ 8695 8975).

1992—3 field mapping

The whole area is built up.

CHEW MOOR (Fig 55)
Project Number	8
SMR Number	4336
Parish	BOLTON
District	BOLTON
NGR	SD 661 078

The history and fieldwork studies of this moss are fully described in Chapter 5.

CHORLTON E'ES
Project Number	
SMR Number	2025
Parish	MANCHESTER
District	TRAFFORD
NGR	SJ 810 930

Chorlton E'es is located at c 10 m OD on the north bank of the River Mersey, between the river and a tributary called Chorlton Brook (now mostly drained or culverted), close to Turn Moss. It overlies glacial alluvial terrace and Bunter sandstone. The area was probably prone to flooding, so creating a small wetland. The name e'es is thought to derive from Old Norse ey or Old English eg, meaning 'island'; rather than Old English ea meaning 'river' (Ekwall 1922, 10, 31).

Barlow Hall lies to the east, and appears to have been the site of a house from the fourteenth century, although it is now largely rebuilt and converted to a farmhouse (SMR 39, SJ 8225 9205).

1992—3 field mapping

The ground falls sharply from the built-up area to an alluviated plain next to the river and the late sewage farm. Where not scrub or disturbed, the site is all river alluvium with pasture fields. No finds.

CLIFTON MOSS

Project Number	10
SMR Number	4332.1.2
Parish	ECCLES
District	SALFORD
NGR	SD 768 033

The site overlies glacial-drift covering Middle Coal Measures. It is a part of the Kearsley complex of mosses lying between the Wigan—Manchester road on the south side, and the Manchester—Bolton road on the north. Farther north, the river Irwell has carved a deep valley with steep sides through Triassic sandstone.

The general history of the site is given with Kearsley Moss.

On the north side of the River Irwell, opposite Clifton Moss, a high promontory of land overlooking the river, called Giants Seat, is possibly a prehistoric fortification. Recent excavations have failed to confirm early activity (GMAU 1990, SMR 2910, SD 7740 0470).

On the south side of the river was an early mining engineering site at Wet Earth Colliery, one of the earliest to be mined on a moderately large-scale. Started in c 1750 by the Fletcher family using a steam-powered hydraulic drainage system installed by the engineer James Brindley, it was finally abandoned in 1928 (Ashmore 1969,106, pers comm Alan Davies, Buile Hill Mining Museum). A number of other coal mines and shafts occur where Coal Measures are close to the

surface and easily worked, *eg* Clifton Moss colliery begun *c* 1840 (Hayes 1986, 132).

The **1992—3 fieldwork** is decribed under Kearsley in Chapter 4.

COCKEY MOOR (Fig 63)

Project Number	12
SMR Number	3873
Parish	MIDDLETON
District	BURY
NGR	SD 768 100

The site of Cockey Moor is bisected by the Bury—Bolton road (B6196), at *c* 160 m OD. Ainsworth (also called Unsworth) village lies on the western edge. The moss overlies glacial drift, shale, mud-stones and Coal Measures; on the First Series Ordnance Survey map (Lancashire Sheet 87, 6":1 mile, surveyed 1844—7, published 1850), shale and sandstone quarries lie less than 1 km to the south.

A bronze looped palstave was found in a quarry at Cockey Moor in *c* 1839. The cutting edge measured *c* 6 cm, length *c* 15 cm, with a central rib (illustrated by French, 1894, 49, SMR 3633, SD 771 103). The alleged line of the Roman road 'Watling Street' is shown on the First Series Ordnance Survey clipping the eastern edge of the moss as it passes south—north from Manchester to Ribchester (SMR 14).

In the sixteenth century there was a dispute over rights of turbary on Cockey Moor (Raines 1856).

1992—3 field mapping
The whole site consists of pasture fields on boulder clay. There is no indication of any peat left, with only a few rushes in wet places. Some fields are as left by glaciers, rough and uneven, others have been ploughed leaving traces of furrows and some narrow rigg, with a much smoother surface. All around the ground falls sharply, there is no basin and Cocky Moor was probably a high 'blanket' moor. A deep stone quarry at NW has fine walls around it, and cottages with

Fig 63 Cockey Moor: land-use, fieldwork, and peat extent

eighteenth-century dates show there was no peat problem for building.

CRUMPSALL MOOR

Project Number	—
SMR Number	2140
Parish	MANCHESTER
District	MANCHESTER
NGR	SD 840 005

The boundary of Crumpsall Moor has not been identified because peat was exhausted and the soils reclaimed at an early period. The local geology is glacial drift overlying Coal Measures. A suggested site for the moss is close to the east bank of the River Irk in Crumpsall township, lying at *c* 50 m OD.

Several archaeological finds come from the vicinity. To the southwest, a ridge of land runs north-south between bends in the rivers Irk and Irwell, and flints have been found at Cheetham Hill and Crumpsall. Among them a Neolithic flint dagger came from a brick croft in 1908 (SMR 267, SD 841 001) and flint implements from Beech Hill (SMR 268, SD 8481 0077). Flint implements were found at Smedley Hill (SMR 269, SD 8481 0077), including a flint axe of Scandinavian type with a square distal end; black flints came from Brideoak Street, Cheetham Hill (SMR 270, SD 840 008), and a Neolithic scraper and flint knife were also found at Cheetham Hill (SMR 266, SD 8401). West of Cheetwood at Broughton, a cremation urn was found in 1873, in a small mound in the grounds of Broughton Old Hall (SMR 519, SD 832 021).

The ridge carries the probable line of the Roman road, Manchester to Bury/Ribchester. A Roman fort site has been proposed on Castle Hill (called 'Low-castor' on a deed of 1322); the area is now built up and the earthworks recorded by Whitaker remain undated (SMR 54, SD 8304 0249). Roman coins were found on Cheetham Hill (SMR 265, SD 8401).

A 'moss dyche' at 'Curme-sale' is mentioned in the 1320 survey of the bounds of Manchester, when it is also recorded that at Crumpsall there were 40 acres of 'moor-pasture' (Harland 1861, 305, 358). Close by is the site of Crumpsall Hall, demolished in 1825, and probably built when the Chetham family settled in Crumpsall in the mid-sixteenth century (SD 8461 0199). The area was extensively excavated for brick clay in the nineteenth century, and has since been built over (First and Second Series Ordnance Survey maps, Lancashire Sheet 104, 6":1 mile, surveyed 1845/1894, published 1848/1896).

1992—3 field mapping
Not visited because the area is completely urbanised.

DARCY LEVER MOSS

Project Number	—
SMR Number	4330
Parish	BOLTON
District	BOLTON
NGR	SD 746 083

Darcy Lever Moss lies east of Long Lane, between the Lancashire and Yorkshire railway (Bury—Bolton branch line (now dismantled), and on the north side the Bury and Bolton turnpike road (now the A58). The extent of the site is not known; on the First Series Ordnance Survey map (Lancashire Sheet 95, 6":1 mile, surveyed 1844—6, published 1850), the area had been drained and reclaimed. The place-names of Moss Lane, and Top o' th' Moss give an approximate location. The site lies on alluvial sands and gravels over glacial clay and Coal Measures.

A Bronze Age burial site lay near Crompton Fold, Breightmet, on the north-western corner of the moss. Twelve inurned cremations were discovered in 1790; originally attributed to the Roman period, they have subsequently been re-evaluated as Bronze Age. A bronze axehead is also known from Darcy Lever (Tindall 1986, 84). The area is pitted by many coal shafts (see Geological Survey of Lancashire Sheet 95 NE 6" 1910).

1992—3 field mapping
Not visited as all the area is mined.

DENTON MOOR

Project Number	26
SMR Number	3471
Parish	ASHTON UNDER LYNE
District	TAMESIDE
NGR	SJ 904 950

Denton Moor, now built over, is located south of Gorton and Audenshaw reservoirs, at *c* 65 m OD, and bounded to the north by the turnpike Manchester—Hyde road, and to the east by the Central and North-Western railway line. The site is underlaid by glacial boulder clays on top of sandstones, shales and marls.

In 1969, at the edge of the moor, a hoard of Roman 20 coins was discovered dating from the late 3rd century to early 5th century (Nevell 1992b, 75, SMR 8236, SJ 908 948). The waste of Denton is mentioned in the manorial survey of Manchester, extending to 200 acres (389 statute acres), with pasture and turbary rights (Harland, 1861, 351). Disputed common lands at Denton (292 acres) were enclosed about 1597 (Harrison, 1888, 112).

The area was divided into small fields on the Ordnance Survey 6" map (1850) with no mention of 'moor', except at the railway.

1992—3 field mapping
Most of the moor site is built over and no peat survives anywhere. At the south towards a tributary of the Tame, a small area of ground is open pasture, lying on clay soil, covered with narrow rigg (SJ 9094). Possible cropmark sites in fields E of the railway here were not examined, being under pasture.

More narrow rigg lies on open land at SJ 903 951, and on the large open area of Denton Golf Course, to the north of the old moor area (SJ 9096). The ridges are likely to be nineteenth century. At the N they do not lie parallel to the walls of King's Road. There is more similar ridge and furrow on another golf course at Reddish SJ 887 941.

DUMPLINGTON MOSS

Project Number	
SMR Number	7923
Parish	ECCLES
District	TRAFFORD
NGR	SJ 775 918

Dumplington Moss, a small area of peatland at *c* 20 m OD, is now completely redeveloped. It was probably part of Trafford and Stretford Mosses. It lies on the southern bank of the River Irwell, separated from Trafford Moss in 1761 by the construction of the Bridgewater Canal.

A plan of 1782 shows the division of the moss, and reclaimed fields called 'moss field' and 'little moss field' (reproduced in Crofton 1906, 21). The Holcroft notebook, a diary of events discussed by Crofton, records the early eighteenth-century use of wooden piles, and a fender 'cop' embankment used as flood protection. The upkeep of a 'weare' (Crofton 1906, 30), and procurement of 'cannel' (a form of coal) from Dixon Green, Worsley, are detailed. Dumplington and the associated village of Bromyhurst on the river bank, were both mentioned with the Fee of Barton in the thirteenth century (GMAU 1992a).

1992—3 field mapping
The whole area is a wasteland of scrub partly covered with dumped soil. West of Urmston Road is a modern rubbish mound, and east of the road, next to the M63 motorway, one area of peaty soil is exposed in the area of moss names of 1782 (SJ 772 962). Sondages would be required to identify buried archaeology. The map of 1782 by Bennet shows that 'moss' field names were mostly in the area of the industrial estate next to the Bridgewater Canal; the other side was Trafford Moss.

EDGE MOSS

Project Number	
SMR Number	7924.1.1
Parish	MANCHESTER
District	TRAFFORD
NGR	SJ 800 957

Edge Moss is situated, at *c* 30 m OD, on the north-west side of the A56, Manchester—Chester road, between Lower Moss and Longford Moss north of Stretford town centre. It is a parcel of the larger mosses that lie in Stretford, overlying glacial sands and gravels, boulder clays and sandstones. The moss name was perhaps suggested by being on the fringe of the main Stretford moss (to the west). Also see Stretford Moss.

1992—3 field mapping
The whole area is built up and industrialised; not visited.

GREAT or STOCKPORT MOOR

Project Number	31
SMR Number	2600
Parish	STOCKPORT
District	STOCKPORT
NGR	SJ 912 882

Stockport Moor was located near the road junction of the north—south A6 Manchester—Stockport Road and Dialstone Lane. The extent of the moor is not known; shown on a survey of *c* 1800 (*A new and accurate survey of the environs of Stockport*, William Stopford), it had been reclaimed and built over by 1859 (First Series Ordnance Survey map, Cheshire Sheet 19, 6":1 mile, surveyed 1872, published 1882). The site lies at 75 m OD, over glacial drift and Coal Measures, and is close to the alleged Manchester—Buxton Roman Road (SMR 28). Black Lake on Great Moor, drained before 1875, was the site of the last gibbet in Stockport (Anon 1991, 6). The first significant enclosure of commons was in 1712, and in 1805 an Act of Parliament was granted to enclose the remainder (Anon 1988, 5).

1992—3 field mapping
All built over except for a levelled playing field. A little peaty soil occurs nearby on the west. The housing dates from 1920 to 1935, and the school probably a little older. The site lies on the top of a hill and slopes down it somewhat. Slight dips in the streets at the west where the moor line is drawn on the map. Sandy soil; probably a blanket heath or moor, now destroyed. No further action is necessary.

HALE MOSS, Prestwich

Project Number	23
SMR Number	7495.1.1
Parish	PRESTWICH WITH OLDHAM
District	OLDHAM
NGR	SD 896 032

Hale Moss is part of the larger complex of White Moss (also known as Theyle Moor), located on the extreme eastern edge in an area now bounded on all sides by roads, the M66 on the north, Hollinwood Avenue (A6104) on the east, and Broadway (A663) on the south. It lies at *c* 95 m OD, on glacial boulder clay and middle Coal Measures. The Rochdale Canal crosses the moss north—south.

1992—3 field mapping
The site lies at White Gate End with only school playing fields left; not visited. Still a 'Turf Lane' to the E.

HALE MOSS, Bowdon

Project Number	43
SMR Number	7925
Parish	BOWDON
District	TRAFFORD
NGR	SJ 773 876

Hale Moss lies at *c* 25 m OD, on the eastern fringes of Altrincham town centre, south of Timperley Brook and clipped by the Cheshire Line railway, overlying boulder clay and sandstones. The site was still wetland on the First Series Ordnance Survey map (Cheshire Sheet 18, 6":1 mile, surveyed 1872—6, published 1882).

A flint arrowhead came from Altrincham (SMR 369, SJ 7698 8760), and other flints were found during recent excavations at Timperley Moated Hall site, less than 500 m to the north, (although the flints may be redeposited (pers comm to A Mayer, South Trafford Archaeology Group).

Altrincham is an early settlement close to the Manchester—Chester Roman Road. A first—second century Roman coin was found in 1984 *c* 0.5 km east of Hale Moss (SMR 1474, SJ 7815 8686). The moated site at Timperley had a thirteenth-century hall (SMR 3704, SJ 7770 8811, dated by radiocarbon to cal AD 780—1021 (1110±60 BP) and cal AD 1260—1437 (620±80 BP) (Beta Analytic Inc, Beta-63539, Beta-60502) (information kindly supplied by STAG).

1992—3 field mapping
Partly built over by late nineteenth-century housing, and partly converted to Stamford Park. The ground falls slightly towards the north, to the Timperley Brook, with soils of peaty sand; no arable land was available, no finds made. There is no obvious reason why there should be a moss here, other than it being part of an extensively waterlogged Mersey Valley.

HALL MOSS

Project Number	35
SMR Number	2594
Parish	PRESTBURY
District	STOCKPORT
NGR	SJ 885 836

The extent of Hall Moss is unknown; only a small section remained as a bog on the First Series Ordnance Survey 6" map (Cheshire Sheet 19, 6":1 mile, surveyed 1872, published 1882). The moss was called 'Gilbent', on a survey of *c* 1800 (*A new and accurate survey of the environs of Stockport*, William Stopford). It is located south of Bramhall, on the northern side of Hall Moss Lane and at the boundary of Cheadle parish and Woodford township in Prestbury parish. The site, at *c* 80 m OD, overlies glacial drift and Coal Measures. A watercourse running through the moss flows towards Handforth and joins the River Dean.

1992—3 field mapping
The moss site lies on high ground in a depression, where the valley of a brook widens. No peat survives, but there are peaty soils at the lowest part by the brook (SJ 8882 8367). Only two fields are 'arable', one a market garden (SJ 8840 8360) and the other currently used for a potato crop, neither available for examination. Clay soils predominate. The peat was probably 2—3 m deep, judged from the topography.

The area to the south (to Wilmslow Road, Woodford) was examined for general nature of the landscape. Nearly all the land is pasture, but has been ploughed as part of a long-period convertible system for intensive milk production. There are many quarries (for sand?) and long, deeply-scarred natural gullies. One field has ridge and furrow, probably nineteenth century, that appears to cross a slade.

HALSHAW MOOR

Project Number	10
SMR Number	4332.1.1
Parish	DEANE
District	BOLTON
NGR	SD 738 057

Halshaw Moor is an area of peat moss or moor, near the Kearsley Mosses. It is located at the north-western corner of Kearsley Moss, and shown on Yates' map and the First Series Ordnance Survey map (Lancashire Sheet 95, 6":1 mile, surveyed 1844—6, published 1850). It is now enclosed by the M61 and A666 junction on the south and east side. The area overlies glacial drift on top of Coal Measures, shales and mudstones.

The name of the moor is said to derive from the claim of a shoemaker, Alexander Shaw, who, before enclosure, laid claim to a 'squatters' right of ownership (Barton 1887, 5). Silver coins of the reign of Henry IV were discovered at Dixon Green, which is a small uncultivated tract of Halshaw Moor (*ibid*, 7).

In 1787 William Hulton of Hulton Park laid claim to the lordship of the waste and prevented commoners from taking coal from Halshaw Moor (Farrer and Brownbill 1914, 5, 35).

1992—3 field mapping
Now built over; not visited.

HEATON MOOR
Project Number	—
SMR Number	2595
Parish	MANCHESTER
District	STOCKPORT
NGR	SJ 882 915

Heaton Moor, referred to as both moor and moss, is located west of the Manchester—Stockport road (A6) and east of Errwood Road. The Stockport and Crewe branch of the London and North Western railway cuts the moor on the north-east edge. The site at *c* 50 m OD, overlies glacial drift and Coal Measures. The original extent is not known.

Near the south-east corner of the moss, a brown flint knife was found during the 1970s (SMR 794, SJ 880 909). The main road (A6), is alleged to be a Roman road from Manchester to Buxton. On the north side of the moor, a partially surviving moated site called Peel Hall, is said to have had a square fortified tower (SMR 64, SJ 8748 9248).

The manorial survey of Manchester, 1320, records 30 acres of moor turbary where the free tenants of Heton had 'housebote', and the lord could sell annually 6s 8d of turbary (Harland, 1861, 328).

1992—3 field mapping
Now built over; not visited

HENNETTS MOSS
Project Number	39
SMR Number	7924.1.2
Parish	FLIXTON
District	TRAFFORD
NGR	SJ 7894

Hennetts Moss is adjacent to Stretford E'es on its northern side, lying at *c* 10 m OD on an alluviated glacial terrace with stratified levels of peats and sands overlying sandstones.

The field pattern on the First Series Ordnance Survey map (Lancashire Sheet 110, 6":1 mile, surveyed 1845, published 1848) indicates a circular depression. A township boundary cuts through the centre of the moss, continuing a ditched line from Urmston Moss, and draining at the southern end into the River Mersey overflow channel. The overflow channel was also known as the 'ousel or Kickettye' Brook, and appears to have been embanked in the nineteenth century for use as a Mersey overflow (Crofton 1899, 3).

A general history of the area is discussed with Stretford Moss.

1992—3 field mapping
The greater part of the site is a grassed mound of a modern rubbish dump. At the far west is pasture meadow ground of the River Mersey, with a slight rise of higher land (an 'island') at Hillam Farm. The housing of Sands Lane is sited on the river valley bank. At SJ 781 942 the low ground consists of dark peaty soil, recently planted with trees. The moss appears to have been a valley type; no finds.

HEYWOOD MOSS (Fig 57)
Project Number	16
SMR Number	5041
Parish	MIDDLETON
District	ROCHDALE
NGR	SD 870 098 SMR 5041

The site lies at *c* 140 m OD. The local geology is sand and gravels with bedded clays to the south, overlying Coal Measures, shale and rock. Other moss names associated with this site are Hills Moss and Mados Moss. The Heywood Branch Canal passes to the south of the area. The Liverpool, Bolton, and Bury rail line (Lancashire and Yorkshire Railway) cuts across the southern part of the wetland and there was a late nineteenth-century engineering store based on the east side of the moss, with railway wagon works on the west side. There is map evidence of coal shaft sinkings across the area (Ordnance Survey map Lancashire Sheet 35, 6":1 mile, surveyed 1889, published 1891).

1992—3 field mapping
This is a valley moss lying either side of Roeacre Brook, east of Heywood. South of the railway are unspoilt pasture fields with sandy soil. Peat occurs in marshy ground, formed in a tributary to the main brook. North of the railway, some land is disturbed by sand quarries, tipping and industry. Peat in the valley is only about 0.3 m deep, visible in drainage trenches.

If threatened, the area needs ploughing for field walking; no action otherwise.

All the fields marked on the First Series Ordnance Survey map, named 'Hills Moss' are on high ground to the south-east.

HIGHFIELD MOSS (Fig 56)
Project Number	5
SMR Number	4848
Parish	WINWICK
District	WIGAN
NGR	SD 613 959

Highfield Moss is fully described in Chapter 4.

HOPWOOD MOSS (Fig 59)
Project Number	18
SMR Number	5045
Parish	MIDDLETON
District	ROCHDALE
NGR	8780 86

Hopwood Moss is described in Chapter 4.

HOUGH MOSS OR OOSE MOSS
Project Number	37
SMR Number	2026
Parish	WITHINGTON
District	MANCHESTER
NGR	SJ 830 948

Hough Moss lies south of Manchester city centre, between Rusholme Brook on the north side and Hough End Clough on the south, at *c* 20 m OD. The area lies over glacial-clay deposits and Bunter Sandstone.

An outburst of Hough Moss was wrongly located by Baines (Baines 1836, 268), and, as Crofton notes, occurred at White Moss, Middleton (Crofton 1902, 142). The site was mostly drained by the late eighteenth century, and the boundary, as shown on the historic mosses map (Fig 4), is taken from place-name evidence and the extent tributaries of the River Irwell and other natural boundaries.

Several prehistoric implements associated with the mossland were found when landscaping Alexandra Park; one perforated stone was possibly a macehead or net-sinker, and another may have been a round stone axe (SMR 80, SJ 8394). At Withington, in 1915, a fine-grained, polished stone axe was found, in sand 0.75 m below the ground surface (SMR 392, SJ 846 922). Another oval sandstone artefact, with a perforated centre, was discovered close to Alexandra Park in 1937 (SMR 66, SJ 8346 9435).

There is a north—south road to the east of this moss, from Manchester through Rusholme (a place-name suggesting a wet environment), south to Didsbury, which then approaches two fording points of the River Mersey at Northenden (mentioned in Domesday; 'Norwardine' then waste; Tait 1916, 215) and Cheadle. This road is a western branch of the Roman Manchester to Stockport road, forking at Ardwick.

Around what was probably the early medieval edge of the moss, a ditched, and once banked, boundary line runs south and east of Manchester. It begins in Ashton Moss curving west towards Rusholme, breaking at the edge of Hough End Moss, and re-forms at Stretford across to the edge of Urmston Moor. A full discussion on the ditch, known as the Nico Ditch, is given by Nevell (1992b, 78—83), and Arrowsmith and Fletcher (1993). It was possibly built to create a boundary between Saxon tribes, pre-dating the Norman Conquest. The boundary seems to have respected the then large wetlands of the southern Manchester Mosses, including Ashton Moss, Hough End Moss and Urmston Mosses.

Other sites in the vicinity include the moated Withington Old Hall, probably not a medieval site, now covered by a housing estate (SMR 398, SJ 8405 9368). Hough End Hall, the only Elizabethan mansion remaining in Manchester, is located on the south side of Hough End Clough (SMR 32, SJ 8248 9325). A sixteenth-century document records trespass of two mares, 10 cows, and six heifers on to 100 acres of Mosse Green and on to moss at Withington (Tait 1917, 49).

1992—3 field mapping

The region is built up. Open ground at Whalley Range and nearby (a school to the south and a football pitch, west) have sandy, slightly dark soils. There cannot have been much peat here by c 1900; the school, of three storeys, is dated 1910 and houses of c 1875 lie west of St Bede's College (SJ 832 950).

HULTON MOOR (HOLLINS) (Fig 64)

Project Number	9
SMR Number	4402
Parish	PRESTWICH WITH OLDHAM
District	BOLTON
NGR	SD 700 055

Hulton Moor overlies glacial gravels and clays on top of sandstones. It is indicated on the First Series Ordnance Survey map (Lancashire Sheet 97, 6":1 mile, surveyed 1844—5, published 1848) by the place-names Moss House, Moss Fold and Moss Field on the north side of Highgate Lane. Circular field boundaries on the map indicate the bowl of the moss.

A Roman road bisects the township running west—east, said to have been laid bare in c 1880 and found to be paved, c 3 m wide (Watkin 1883, 51). The place-name Hulton derives from the 'town' on the hill. There were three seats at Hulton, the main estate being that of the Hultons of Hulton, situated in the southern portion of Over Hulton (Farrer and Brownbill 1914, 5, 26). In 1556—7 Richard Brereton and Adam Hulton were recorded as holders of Hulton Moor, and summoned to answer the accusation of seizure of cattle owned by Robert Grundy of Runworth on Rumworth Moor (Farrer and Brownbill 1914, 5, 29 n58).

1992—3 field mapping

The whole area lies pasture except for two fields north of Moss Hall Farm. High ground to the south consists of pink boulder clay; the two arable fields have similar clays except in a narrow valley where a small area of peat survives with some skirtland. This was a valley moss. There is probably enough peat for dating.

IRLAM MOSS

Project Number	46
SMR Number	3033.1.5
Parish	BARTON ON IRWELL
District	SALFORD
NGR	SJ 7193

Irlam Moss is part of the southern complex of Chat Moss, lying between the Manchester—Liverpool Railway line and Irlam town. It is situated on alluvial sand and gravel terrace, and glacial clays, overlying the Upper Mottled Sandstone. The general history and utilisation of the moss is discussed under Chat Moss.

It was not until 1908 that Ordnance Survey maps showed the body of the moss divided into moss rooms and the site inhabited by farms and a network of roads and tram lines.

KEARSLEY MOSS

Project Number	10
SMR Number	4332
Parish	KEARSLEY
District	BOLTON
NGR	SD 750 040

Kearsley Moss lies in Bolton and Salford Districts. The name 'Kearsley' is used for the whole complex with the component parts listed below. For some details of each moss see this Gazetteer, for an overall account of the complex see Chapter 4.

Clifton Moss, Salford, SD 768 033 SMR 4332.1.2
Kearsley Moss, Bolton SD 750 040 SMR 4332.1.0
Linnyshaw Moss, Salford, SD 745 038 SMR 4332.1.3
Morton Moss, Salford, SD 7503 SMR 4332.1.3
Red Carr Moss, Salford, SD 766 027 SMR 4332.1.5
Swinton Moss, Salford, SD 771 030 SMR 4332.1.6
Walkden Moss, Salford, SD 746 032 SMR 4332.1.7
Wardley Moss, Salford, SD 758 036 SMR 4332.1.8

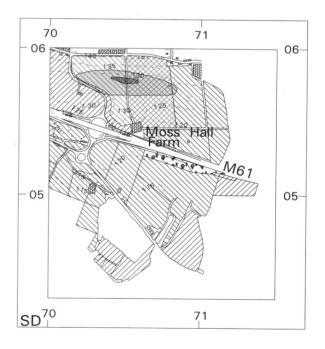

Fig 64 Hulton Moor: land-use, fieldwork, and peat extent

Based upon the 1990 Ordnance Survey 1:10,000 map with permission of The Controller of Her Majesty's Stationery Office © Crown Copyright

KINGS MOSS

Project Number	
SMR Number	
Parish	
District	WIGAN
NGR	SD 498 016

The present county boundary crosses the north side of this moss. It has therefore been considered in the Merseyside Wetlands volume (Cowell and Innes 1994, 123). The site lies at *c* 60 m OD on glacial till with Coal Measure geology. King's Moss was shown on Yates' 1786 map of Lancashire and also on the Ordnance Survey First Series map.

1992—3 field mapping
The village of King's Moss lies on sandy ground in a hollow along the Black Brook; lithic sites are likely to occur in this terrain. The area to the north-west, marked as a moss on the 1:10,000 Ordnance Survey map, is now a large rubbish dump in current use. It lies in the County of Merseyside and no fieldwork was undertaken near it.

KITTS MOSS

Project Number	34
SMR Number	2591
Parish	STOCKPORT
District	STOCKPORT
NGR	SJ 888 847

Kitts Moss is located to the west of Bramhall village centre, bordered on the south by Moss Lane, and on the north by Hack Lane (A5102). It overlies, at *c* 75 m OD, glacial drift over Sandstone. The small settlement of Bramhall is marked on the First Series Ordnance Survey map (Cheshire Sheet 19, 6":1 mile, surveyed 1872, published 1882).

At a house called Siddall, lying at the south-east corner of the moss, a fragment of a 'double-hopper' rotary quernstone was discovered in *c* 1890. The original artefact has been lost, but an illustration of 1892, shows that it was similar to a piece found in the 1940s at Red Moss (Heginbotham 1892, 135). The querns are probably Roman (Plate 16).

The First Series Ordnance Survey map shows the moss divided into thin strips — presumably moss 'doles'. A path, called Lumb Lane (now built over) to the south, lying parallel with the railway (Manchester—Macclesfield branch of the London and North Western Railway), was described on the Ordnance Survey map as a Roman road.

1992—3 field mapping
The moss lay in a basin located on high ground, 1—2 m deep, with the lowest part picked out by a 75 m contour on the Ordnance Survey 1:10,000 map. Moss Lane, forming the southern boundary, has a ridge rising just south of it. The eastern (Ack Lane East) and northern roads (Kitts Moss Lane) also follow a ridge around the basin edge; houses on the north side are dated 1876.

The basin was built up with bungalows in 1958—9. There is no peat left, and clay lies underneath the topsoil, according to a resident living at the lowest central part. Garden soils are still peaty, although some of the dark colour is caused by nightsoil. The drainage exit of the basin was probably at the west where the moss edge cannot be defined.

No further action is needed.

LINNYSHAW MOSS

Project Number	10
SMR Number	4332.1.3
Parish	DEANE
District	SALFORD
NGR	SD 745 038

Linnyshaw Moss lies over glacial boulder clay, lying on top of Middle Coal Measures, sandstones, shales and mudstones. It forms the south-western part of the Kearsley Mosses, partially reclaimed by the late eighteenth century (Bridgwater Estate map, Irlam Record Office). The general history of the site is described with Kearsley Moss.

The Wigan—Manchester road running east—west, south of this moss is alleged to be a Roman road (Watkin, 1883, 46). A preserved beehive (undated) made of willow basketwork four chambers high, similar to one found on Chat Moss, was found *c* 1710 at Linnyshaw Moss (Whittaker, 1773, 316)

The field mapping of 1992—3 is described in Chapter 4.

LONGFORD MOSS

Project Number	39
SMR Number	7924.1.3
Parish	MANCHESTER
District	TRAFFORD
NGR	SJ 798 950

Longford Moss is one of the moss parcels of what is here considered to be Stretford Moss. It is located on the east side of Stretford. The site overlies glacial sands and gravels, clay on sandstone.

A general history of the site is given with Stretford Moss.

By the time of the Second Series Ordnance Survey map (Lancashire Sheet 111, 6":1 mile, re-surveyed 1892, published 1895), Longford Brook had been canalised, and on the southern side there was significant urban encroachment.

Longford Hall is a mid-nineteenth century structure (SMR 17, SJ 8073 9448).

1992—3 field mapping
The site is built over; not visited.

LOSTOCK MOSS

Project Number	39
SMR Number	7924.1.4
Parish	MANCHESTER
District	TRAFFORD
NGR	SJ 783 957

Lostock Moss is part of the larger Stretford Moss, located on the north-east side of Stretford, north of Barton Road and south of the Bridgewater Canal. It overlies glacial sands, gravels and clays on sandstone.

A general history of the site is given with Stretford Moss.

On the First Series Ordnance Survey map (Lancashire Sheet 111, 6":1 mile, surveyed 1844—5, published 1848) Moss Lane meanders northwards across the site towards Stretford Moss proper, from the town centre.

1992—3 field mapping
The site is built over; not visited.

LOW MOSS

Project Number	39
SMR Number	7924.1.5
Parish	MANCHESTER
District	TRAFFORD
NGR	SJ 781 952

Low Moss is part of the complex of Stretford Moss. It lies at *c* 30 m OD, south of Barton Road and north of Urmston Lane, on the north-east side of Stretford town centre. What is now Moss Park Road, was called Low Lane on the First Series Ordnance Survey map (Lancashire Sheet 110, 6":1 mile, surveyed 1845, published 1848).

A general history of the site is given with Stretford Moss.

LOWER MOSS

Project Number	39
SMR Number	NONE
Parish	MANCHESTER
District	TRAFFORD
NGR	SJ 805 959

Lower Moss was a parcel of the larger extent generally referred to as Stretford Moss. It lies *c* 30 m OD, north of Stretford town centre on glacial gravels, sands and gravels, overlying Sandstone. The site is shown on a Trafford Estate map (Crofton 1905b), at the edge of the moss area, adjacent to Edge Moss, and the Manchester—Chester Roman road (A56).

Next to the moss lies the Great Stone, a glacial boulder which marked road mileage. On the First Series Ordnance Survey map (Lancashire Sheet 104, 6":1 mile, surveyed 1844—7, published 1848), there were gravel pits opposite the moss.

The history of the site is given with Stretford Moss.

1992—3 field mapping
The site is built over; not visited.

MARLAND MOSS (Fig 59)

Project Number	17
SMR Number	2311
Parish	ROCHDALE
District	ROCHDALE
NGR	SD 875 120

Marland Moss is fully described in Chapter 4.

MORTON MOSS

Project Number	10
SMR Number	4332.1.4
Parish	ECCLES
District	SALFORD
NGR	SD 7503

Morton is a small moss located within the Kearsley Moss complex, lying on the southern edge close to Wardley Moss and the Wigan—Manchester road. There is industrial activity nearby. The general history and fieldwork are given with Kearsley.

MOSS GROVE

Project Number	22
SMR Number	6214
Parish	PRESTWICH WITH OLDHAM
District	OLDHAM
NGR	SD 913 035

The Moss Grove site is located to the south-west of Oldham town centre, at the junction created by the present Manchester—Oldham road (A62), and Hollins Road (A6104). It lies at *c* 145 m OD, over glacial drift and Middle Coal Measures.

The alleged line of the Manchester—Castleshaw Roman road is shown on the First Series Ordnance Survey map (Lancashire Sheet 97, 6":1 mile, surveyed 1844—5, published 1851) as a slightly fragmented line on the southern edge of the moss, and even now called 'Roman Road'. A hoard of Roman bronze coins, dating from AD 135 to 235, concealed in a box, was discovered during the construction of Chamber Mill in 1887 (SMR 110, SD 9135 0343).

The Fairbottom branch canal cuts the west end of the moss north—south. The area has coal-working pits, one pit excavated in the mid-nineteenth century, to the north-east of the moss, on the Coppice estate, was thought by Watkin and Horsfall to have Roman coal 'bell' shafts (SMR 1118, SD 922 038). The place-names Lime Pits and Gravel Hill (First Series Ordnance Survey map) indicate other types of quarrying.

1992—3 field mapping
The whole area is built up except for a small park at the south, which is a slag heap. The park is named after a colliery and there are many mills around. No undisturbed ground survives and no further action is necessary.

NORBURY MOOR

Project Number	33
SMR Number	2597
Parish	STOCKPORT
District	STOCKPORT
NGR	SJ 918 860

Norbury Moor is located south-west of Hazel Grove (called 'Bullock Smithy' on the First Series Ordnance Survey map, Cheshire Sheet 19, 6":1 mile, surveyed 1872, published 1882) and has been crossed by the London—North Western rail line (Stockport and Whaley Bridge Branch). At *c* 80 m OD, the site overlies glacial morainic drift and Coal Measures with mudstones and shales.

The sixteenth-century Norbury Hall (later rebuilt) lay to the east of the moor (SMR 418, SJ 9240 8542). Opposite the hall is the site of an early seventeenth-century chapel (SMR 843, SJ 9263 8545). There is coal mining in the area.

1992—3 field mapping
The moor lay on high ground and is now built over. There are no peaty soils and no sign of a depression in any of the roads. It was was probably blanket peat, now completely removed. No further action required.

OPENSHAW MOSS

Project Number	—
SMR Number	4479
Parish	MANCHESTER
District	MANCHESTER
NGR	SJ 885 975

The original extent of Openshaw Moss is unknown, but it may have been located in the central portion of Openshaw township, at *c* 70 m OD. The area overlies glacial boulder clays and sandstones.

The moss was documented in the 1322 manorial bounds of Manchester, which recorded 100 acres of moor on which tenants in Gorton, Openshaw, Ardwick and the lord of Ancoats had rights of turbary. John de Byron had 'illegally' appropriated 40 acres of the moor (Harland 1861, 351). Thin strips of land on the north side of Openshaw village on the First Series Ordnance Survey map (Lancashire Sheet 104, 6":1 mile, surveyed 1847—8, published 1848) may be the result of moss division into 'rooms'.

1992—3 field mapping
The general area is built up; site not visited.

PILSWORTH MOOR

Project Number	15
SMR Number	5055
Parish	MIDDLETON
District	ROCHDALE
NGR	SD 840 096

Pilsworth Moor lies to the south-west of Heywood, enclosed by two minor roads on the south and west, which cross (at Three Lane Ends) a tributary of the River Irwell draining the moor. The site overlies glacial gravels and clays with aeolian sand, on top of Middle Coal Measures. The place-name indicates an Old English personal name, 'Pils enclosure' (Ekwall 1922, 54). There is little archaeological information available for this area.

1992—3 field mapping
The general area is now an industrial estate, lying in a basin south of Heywood. South of the Pilsworth Road agricultural land rises sharply, with no sign of any surviving peat.

RED CAR MOSS

Project Number	10
SMR Number	4332.1.5
Parish	ECCLES
District	SALFORD
NGR	SD 766 027

Red Carr is a small piece of mossland within the Kearsley Mosses, located between Swinton and Wardley Moss, next to Red Carr Brook. It overlies glacial drift and boulder clays, on top of the Middle Coal Measures.

The general history and 1992—3 mapping is discussed with Kearsley.

ROYTON MOSS (Fig 58)

Project Number	21
SMR Number	6212.1.1
Parish	PRESTWICH WITH OLDHAM
District	OLDHAM
NGR	SD 937 070

Royton and Broadbent Mosses are described in Chapter 5.

SALE MOOR/SALE E'ES

Project Number	41
SMR Number	7926
Parish	ASHTON UPON MERSEY
District	TRAFFORD
NGR	SJ 7991 ?SJ794927

Sale Moor is a wetland site now used as a waterpark, created after extraction of minerals. It is located at *c* 10 m OD, on the south side of the River Mersey, overlying alluvial glacial sands and gravels on top of sandstones. To the west, north of the river, is Stretford, described with Stretford Moss.

The Bridgewater Canal was constructed over the moss, and probably affected some drainage because natural overflow channels into the Mersey were altered (a diagram can be found amongst the muniments of J. E. Bailey, Chethams Library; First Series Ordnance Survey map, Cheshire Sheet 9, 6":1 mile, surveyed 1876, published 1882). The Manchester South Junction railway follows the same line as the Bridgewater Canal.

1992—3 field mapping
The whole area is built over with only a few open spaces remaining. A cemetery at SD 782 911 has sandy soil which is slightly dark with humus at the north-east. A sports centre, at SD 790 908, also has sandy soil, with the groundsman reporting clay at the south. Surrounding houses are mostly of 1930s date; the cemetery is about 1890. No sign of any surviving peat. There was probably only a thin moor here, and the topographical reason for its existence is no longer apparent.

SHADOW MOSS

Project Number	36
SMR Number	2027
Parish	STOCKPORT
District	MANCHESTER
NGR	SJ 8385

Shadow Moss site lies at *c* 40 m OD, on the southern county boundary between Manchester and Cheshire, 1 km north of the River Bollin, bounded on the east side by Styal Road (B5166). Nearby, on the south-west corner, is the extensive complex of Manchester Airport. The area overlies glacial clays and Bunter Sandstone.

About 1 km to the east a flint axe was found in 1975 in a market garden. It is described as grey flint, polished at the distal end, but may have been brought in from a consignment of soil (SMR 1255, SJ 847 847).

Robert Tatton, in his will of 1578, bequeathed 'sufficient turbarie

upon my mosse rowme at Shadowe Mosse, and belo[n]ginge to my mancon house called the Pele' (Crofton 1889, 29). The court leet for Etchells records the 'encroachment by agreement' on to Shadow Moss, which was common pasture for Etchells. In 1700 William Tatton enclosed Bolshaw Outwood, which had been part of Shadow Moss, leaving part for 'ridging clods, sand, clay, gravel' (Groves 1990–1, 37). The area is irregularly drawn on William Stopford's map of Stockport *c* 1800.

1992—3 field mapping
The moss site is now all pasture with slightly dark soil exposed in a few places. No peat appears to survive.

SHAW MOSS (Fig 58)

Project Number	20
SMR Number	6215
Parish	PRESTWICH WITH OLDHAM
District	OLDHAM
NGR	SD 940 085

Shaw Moss is located immediately south of Shaw village, below the end of Sumner Street off Oldham Road (B6194), situated at 190 m OD, on glacial drift overlying Lower Coal Measures. It is a small basin on a plateau at the foot of the western Pennine fault. Shaw is now a very industrial area with numerous mills.

1992—3 field mapping
Shaw Moss lies in a valley hemmed in by a cluster of mills. Part of the site is buried by a tip and the remainder covered by scrub. The wet and marshy site is still probably forming peat. No depths of organic deposits could be determined.

SIDDAL MOOR (Fig 65)

Project Number	14
SMR Number	5076
Parish	MIDDLETON
District	ROCHDALE
NGR	SD 851 088

Siddal Moor lies south of Heywood, now bisected by the M62; another name for the site is 'Doctors moss'. It drains to the south and overlies glacial gravels, clays and aeolian white sands on top of Coal Measures.

The area, at *c* 110 m OD, is shown with peat deposits on the Ordnance Survey Geological Survey, published 1947 (Lancashire Sheet 88).

Lower Whittle Farm, on the south-east side, is of seventeenth century or earlier date, with a timber frame having a rendered stone and brickwork exterior (SMR 2963, SD 8528 0830). Coal mining and sand extraction have occurred on the site (shown on the Geological Survey).

1992-3
The area now lies pasture in a basin. A modern dike showed there was more than 0.3 m peat left and a farmer said that the middle had 1.3—1.6 m of muddy peat lying on sand over clay. Some peaty soils were visible in mole hills in a valley that leads out of the moss at the west (SD 855 087). None of the reported earthworks or ridge and furrow was confirmed, and any remains visible on aerial photographs are likely to be modern.

The site should be ploughed for fieldwalking and sampled with boreholes if threatened.

SLATTOCKS PEAT

Project Number	19
SMR Number	5085
Parish	MIDDLETON
District	ROCHDALE
NGR	SD 894 088

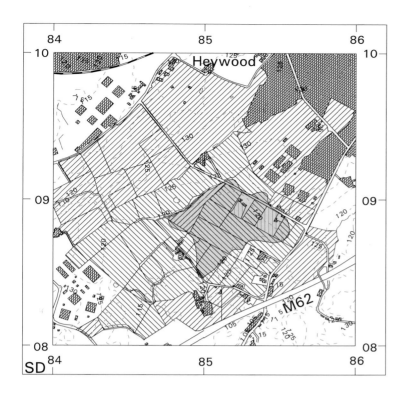

Slattocks Peat is located to the south of Rochdale, east of the main north—south Manchester to Rochdale road, A664. It was a small parcel of peat at *c* 150 m OD, shown on the 1947 Ordnance Survey Geological Survey (Lancashire Sheet 88), overlying glacial sand and gravel with beds of clay, on top of Lower Coal Measures and sandstone. Some Rochdale peats may have developed from recent waterlogging. There is no known associated archaeology.

STOCKPORT LITTLE MOOR
Project Number	29
SMR Number	2598
Parish	STOCKPORT
District	STOCKPORT
NGR	SJ 911 896

Little Moor is located east of Stockport town centre, at the crossroads between Hempshaw Lane and Banks Lane, overlying glacial morainic drift and Coal Measures at *c* 85 m OD. Its extent is not known because the site had been developed before the publication of the Ordnance Survey 1" map (Sheet 81 NW, published 1842).

1992—3 field mapping
The site is completely covered with buildings of 1900 and older. It lies on a hill-top and was probably a blanket peat or moor. No further action needed.

STRETFORD E'ES
Project Number	39
SMR Number	7927
Parish	MANCHESTER
District	TRAFFORD
NGR	SJ 785 935

Stretford E'es is one of the riverside depressions, like Urmston and Chorlton E'es, which seem to have been liable to flooding. It is located at *c* 10 m OD, on an alluvial glacial terrace of sands and gravels, overlying sandstones. There are thin bands of stratified sand and peat deposits (Ordnance Survey Geological Survey, Lancashire Sheet 110, 6":1, published 1910).

It is located on the north side of the River Mersey, cut by a substantial drainage channel of unknown date, called the 'overflow river', that modified a natural tributary to the Mersey (First Series Ordnance Survey map (Lancashire Sheet 110, 6":1 mile, surveyed 1845, published 1848)).

The general history of the site is the same as Stretford Moss.

The Stretford court baron records the seventeenth-century system of leaving the 'eye' free of grazing before the hay crop was harvested (Crofton 1901, 33).

1992—3 field mapping
The area is an east extension of Hennets Moss, now obliterated by a grassed-over rubbish tip. The 1910 Geological Survey records alluvium over peat.

STRETFORD MOSS
Project Number	39
SMR Number	7824.1.0
Parish	MANCHESTER
District	TRAFFORD
NGR	SJ 7997

Stretford Moss is part of a large complex of moss names, making up an extensive area of peat that once covered the area between the River Irwell and the settlement of Stretford. It is situated at *c* 20 m OD, overlying alluvium, peat, and glacial gravels on top of sandstones.

An eighteenth-century Trafford Estate map shows numerous divisions of the moss, indicating that the whole township had common tenancy (reproduced in Crofton 1901).

Taking the moss as a whole, there are several archaeological features in the vicinity. A major Roman road, from Manchester to Chester, cuts north—south through the eastern fringes of the moss, with Stretford at the crossing point over the River Mersey. A section of the road was exposed in 1885, in the vicinity of the 'Great Stone' (north of the Town centre). Described as a boulder road 1 m below the then ground surface, and *c* 1 m lower down was 'a wattle road of brushwood laid upon sand and covered by ling gorse, with a ditch on either side' (Crofton 1899, 4).

A Saxon coin hoard (400 silver coins were in a pot) was discovered in 1774 when digging foundations for a house in Stretford (Harrison 1892, 251).

In the fourteenth century the site of the Mersey crossing was called 'Crossferry' (Harrison 1894, 14—15). A coin hoard discovered in 1903 consisted of a variety of pieces from the reigns of Elizabeth to Charles I (SMR 388, SJ 794 941). Stretford court baron makes frequent mention of mechanical embanking and reinforcing of the Mersey and Irwell as flood defence. Crofton dates one site close to the Stretford overflow weir, as early as 1588. Protection to the adjoining land was given by 'fenders', artificial banks (Crofton 1899, 31, 38). By the mid-nineteenth century most of the area had been reclaimed and cultivated.

1992—3 field mapping
The whole area is built over; site not visited.

SWINTON MOSS

Project Number	10
SMR Number	4332.1.6
Parish	ECCLES
District	SALFORD
NGR	SD 771 030

Swinton Moss is part of Kearsley Moss, located at the far eastern end with relatively good drainage and shallow peat (suggested by Yates' eighteenth-century map which depicts this area of moss in a 'heath-like' way). The site overlies glacial drift of boulder clay and Middle Coal Measures. The River Irwell flows on the northern side.

In 1947, 72 Henry III silver pennies were found in a sand pit at Swinton, the findspot was probably close to the moss (Anon 1947, 224). There is much industrial activity nearby. The 1992—3 field mapping is described in Chapter 4.

TIMPERLEY MOSS

Project Number	42
SMR Number	7928
Parish	BOWDON
District	TRAFFORD
NGR	SJ 780 894

Timperley Moss extent is unknown; it is shown on Burdett's map of 1777, but at the time of the First Series Ordnance Survey map (Cheshire Sheet 18, 6":1 mile, surveyed 1872—6, published 1882), the area had been drained and cultivated. It is located at the junction of Park Road and Moss Lane, and cut on the western side by the Runcorn branch of the Bridgewater Canal, and by the Cheshire Lines railway to Chester. The site, at c 50 m OD, overlies glacial drift and sandstone.

It is less than 0.5 km east of the Roman Manchester—Chester road (A56). On the north side of the moss, a moated site called Ridd Hall of the mid-seventeenth century (demolished 1971) may have been on the site of a pre-Conquest building (SJ SMR 3705, SJ 7791 8968). The place-name *Ridd* means 'clearing'.

The moss and common were divided in 1475 into three holdings belonging to John Arderne of Timperley Hall (south of the moss), William Booth, and William Buckley (Ormerod, I 1882, 546). In 1537—8, there was a dispute over turbary on the moss, when William Arderne complained that Robert Parker of Dunham, George Chaterton, Edward Ryle of Timperley, Richard Hardy husbandman, Robert Vaudrey husbandman and others, under the direction of George and Elizabeth Booth, attacked his employees as they were digging and making turves to sell. Some turves were carried away by the incursors (Stewart-Brown 1916—17, 6—13). In 1584 William Arderne bequeathed 20 loads of turves to be cut yearly on his part of the moss (Pryor 1982, 3).

1992—3 field mapping
The area is built over apart from a small playing field. Surrounding housing dates from c 1910—35 and the sandy soil shows no signs of peat. The moss was probably part of a generally wet Mersey Valley, there being no obvious topographical reason for a basin or valley moss to form.

TRAFFORD MOSS

Project Number	39
SMR Number	7929
Parish	STRETFORD
District	TRAFFORD
NGR	SJ780 975

Trafford Moss is one of the largest parcels of the Stretford Mosses, and the site is now the Trafford Park Industrial Estate (created in 1896), lying between the Bridgewater Canal on the south, and the River Irwell (Manchester Ship Canal) on the north. The area is at c 15 m OD, on alluvial sand and gravel terrace of the Mersey lying on glacial boulder clay overlying Coal Measures.

A bronze axe-hammer of Scottish/Scandinavian type, measuring 16.5 mm long, 8 mm wide, was found at Mode Wheel on the Manchester Ship Canal in 1890 (SMR 543, SJ 7999 9750). From a similar location ('close to the second lock of the Irwell'), a gold pendant was found in 0.3 m of gravel in 1772, originally thought to have been a Roman bulla. The decoration suggests an Irish origin; two concentric bands form an outer border with chevrons in between, and on the reverse were centrally radiating lines with concentric circles (SMR 538, SJ 799 974, Whittaker 1773, 1, 79), Phelps notes a similar piece was found in moss at Trundholm, Denmark, which was dated to c 1000 BC (Phelps 1915, 193).

A hollowed-oak log canoe was discovered in 1889 in the Trafford Hall cutting of the Manchester Ship Canal 'exactly one mile west' of Barton aqueduct. The boat was 3.75 m internal measurement, a maximum of 0.85 m wide and maximum 0.48 m deep. It had a strengthening piece at one end, and projecting 'wooden nose' at the other (possibly a rudder pin, although Bailey supposes it was for a mooring rope (Bailey 1889a, 243)). The boat is now at Manchester Museum, and a date of c cal AD 1007—1231 (920±56 BP; Q-1396) has been determined by radiocarbon (Manchester Museum catalogue, SMR 879, SJ 7721 9778).

Trafford manor was carved out of Stretford township in the Norman period; the area was also known as Whittleswick, an Old English name meaning 'homestead' (Ekwall 1922, 32, 39). Medieval Trafford Park lay on the south bank of the river near the edge of the moss. A sixteenth century manor-house was at Whickleswick Hall, the Old Hall having been abandoned (Crofton 1903, 40). In 1704 the court baron records that Humphrey Trafford was to make a ditch round the park to draw water from Edge Moss.

The construction of the Bridgewater Canal across the moss in 1761 dealt with the difficulty of unstable land at Trafford Moss by using a lining of clay ('clay-puddle') to form an impermeable bank. A new road across Trafford Moss was made in 1781 by John Trafford (Holcroft notebook; Crofton 1906, 25), prior to the division into moss rooms in 1782 (estate survey, Crofton, II 1901). Reclamation of the moss had begun in the seventeenth century, indicated by a deed recording the improvement of 3 acres of moss in Wickleswick in 1632 (Crofton 1899, 43). The greater part of Trafford Moss was drained and cultivated after 1779, as already described in the general account (Chapter 6). A public park was made from 700 acres of the Trafford Estate in 1889 which subsequently developed into the Industrial Estate (Anon 1889, 1, 28).

1992—3 field mapping
The whole area is built up and was not visited.

TURN MOSS

Project Number	38

SMR Number	7924.1.7
Parish	MANCHESTER
District	TRAFFORD
NGR	SJ 804 939

Turn Moss, at *c* 10 m OD, is located on the north side of the River Mersey, and south of the main west—east thoroughfare called Edge Lane (A5145). It lies on alluvial terrace and glacial drift on top of Bunter sandstone. The site is a small basin similar to Urmston E'es, close to the river and therefore probably prone to flooding.

The place-name derives from 'turf', reflecting the rights of turbary. It was first recorded in 1587, then in the occupation of Nicholas Moseley, clothworker of London. A house was associated with the moss in the early seventeenth century (Crofton 1903, 71). The 1839 Tithe Map of Stretford shows part of the area of Turn Moss as a 'brick kiln field', although there are now no visible remains of any structures (Manchester Local History department, Central Library).

1992—3 field mapping
A flat area of riverine alluviated meadow, with two or more small areas of dumping; hemmed in to the north and east by steeply falling slopes. The soil is slightly dark and allotments at the east are sandy. There is a slight dome in the centre of the playing fields, but not much peat visible. An axe find nearby probably came from river dredgings; the soils are suitable for prehistoric settlement.

TYLDESLEY MOSS
Project Number	6
SMR Number	4852
Parish	LEIGH
District	WIGAN
NGR	SD 690 016

Tyldesley Moss lay immediately south of the town, on boulder clay on top of Middle Coal Measures, at *c* 45 m OD. On the First Series Ordnance Survey map (Lancashire Sheet 102, 6":1 mile, surveyed 1845—7, published 1848) there was unenclosed moss, surrounded on the southern side by small farms called Meanley House, Counting House, and Higher Barn. Moss House and Moss Farm on the east suggest the original extent of the moss had been greater. About 1 km to the south was another area of peat at Black Moor.

On the south-east corner of the moss there are two hall sites in close proximity. New Hall moat has a late seventeenth-century hall built at the centre of the moat, with no evidence for an earlier structure (SMR 42, SD 6991 0111), and Dam House, now a hospital, was also of seventeenth-century date, erected by Adam Mort (Farrer and Brownbill 1914, 3, 443).

Tyldesley Moss was destroyed by nineteenth-century coal mining. At Higher Barn Farm, a colliery called Gin Pit was built, which extended northwards, across the moss (Geological Survey, Lancashire Sheet 84 and 85). It was still working in the 1950s, using a mineral railway leading south to the Bridgewater Canal.

In recent years, gardens backing on to the site of the moss have revealed depths of peaty soil (pers comm J Jones).

1992—3 field mapping
The greater part of the site consists of a levelled slag heap lying in a valley. East of Astley Road, parkland and football fields show no sign of peat. Nothing seems to survive.

UNSWORTH MOSS (Fig 66)
Project Number	13
SMR Number	3878
Parish	MIDDLETON
District	BURY
NGR	SD 8307

Unsworth Moss, also referred to as 'Back o' th' mors', is located at the junction of the M62 and M66, east of Whitefield. At *c* 100 m OD, it overlies glacial morainic drift of boulder clay, sands and gravels on top of Middle Coal Measures, shales and mudstones. The Geological Survey shows pockets of peat (Lancashire Sheet 88, 1947).

1992—3 field mapping
All the area is now pasture with sandy soil and a prominent hill at Back o' the Moss Farm. A low basin at the east (centred SD 831 067) was most probably the centre of the moss that spread up on to higher ground. All that is left now is some peaty alluvium in the basin. The farmer from Moss Side, at the north-east, said that when digging drains there are occasional logs and birch twigs left, even on the lower slopes. Underneath is white running sand. The farm was

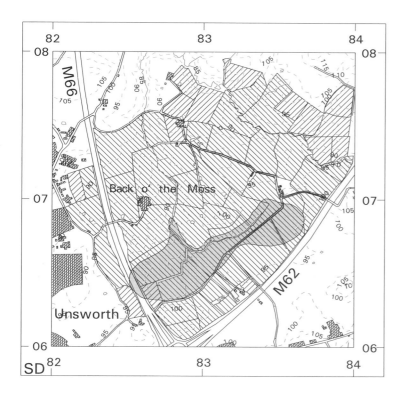

Fig 66 Unsworth Moss: land-use, field-work, and peat extent

Based upon the 1983 Ordnance Survey 1:10,000 map with permission of The Controller of Her Majesty's Stationery Office © Crown Copyright

ploughed before he took it on.

The basin is high up and trapped, but all around (especially to the north) the ground falls away to a much lower level.

The site is threatened by motorway works, and should be ploughed and fieldwalked, as it is an area likely to have prehistoric activity.

URMSTON E'ES (Fig 34)

Project Number	40
SMR Number	7930
Parish	FLIXTON
District	TRAFFORD
NGR	SJ 760 935

Urmston E'es is one of the small basins located at *c* 10 m OD on the north side of the River Mersey, which, because of flooding, have become wetland sites (see Chorlton E'es for the place-name 'e'es'). The area overlies alluvial river terrace deposits, on top of sandstones. No peat is marked on the 1910 Ordnance Survey Geological Survey (Lancashire Sheet 110, 6":1 mile, 1947).

The general history of the area is given with Stretford Moss.

There are two sites of archaeological interest; Urmston Hall, on the north-east side of the basin, probably built in the sixteenth century with a stone base, now demolished (SMR 1215, SJ 7688 9422); and a field called 'Barrow Field' on the 1842 Tithe Map of Urmston, which Langton writes 'is the burial site for the cadavar remains of a battle which took place at the rear of Urmston Hall' (Langton 1898, 21). No evidence for the burial site has been recovered.

1992—3 field mapping
The site is under pasture and lies flat in a basin enclosed on all sides except to the south where it is open to the Old E'es Brook, which was the course of the Mersey. The E'es were part of the river valley. The soil is alluvium, and meadow like, with no features apart from a few drainage channels. In 1910 no peat was observed (SMR, Geological Survey).

URMSTON MOSS

Project Number	—
SMR Number	7931
Parish	FLIXTON
District	TRAFFORD
NGR	SJ 760 950

The extent of this moss is not recorded on any map, presumably having been drained at an early period (see below). The site is located between the Rivers Irwell and Mersey, at *c* 20 m OD, overlying alluvial sands, gravels and clays on sandstone.

The area lies south of 'Sticking island' (SJ 740 960), shown in detail on the Second Series Ordnance Survey map (Lancashire Sheet 103, 6":1 mile, surveyed 1894, published 1896), as a series of loops in the River Irwell, now straightened, which had a number of archaeological discoveries. Charles Roeder found a wooden comb amongst silt and leaf beds, in *c* 1895, *c* 10 mm long, once having teeth on both sides, but one row destroyed (Roeder 1889—90, 204). In 1846, a large polished Neolithic stone axe of Langdale stone was found, 2.4 m deep in a gravel bank at Shaw Hall (now at Manchester Museum, length 32.7 cm, width 8.2 cm). Shaw Hall is a moated site, close to the rail line at Flixton; the latest building before demolition was apparently an early seventeenth century construction (SMR 364, SJ 7532 9388).

An interesting feature is the parish boundary between Barton On Irwell and Flixton parish, called Carr Ditch on the First Series Ordnance Survey map (Lancashire Sheets 110, 103, 6":1 mile, surveyed 1845, published 1848). The line has been claimed as a continuation of the pre-Conquest earthwork called Nico Ditch, circling the south side of Manchester from Ashton to Stretford. Recent excavations, however, have indicated differences between Carr Ditch and Nico Ditch; Carr Ditch may be a later boundary, probably derived from the division of estates or parishes. Another ditch that runs north-south at right angles to Carr Ditch separates the townships of Flixton and Urmston, and was probably constructed in the thirteenth century, when Farrer records the Hulton family estates at Urmston were divided (Arrowsmith and Fletcher 1993). It is probably the ditch that in Stretford court baron of 1626 was ordered to be 'scowred, diched and guttered' by the tenants of the adjacent land (Crofton 1901, 38). Another ditch running at right angles to Carr ditch, dividing the township of Urmston and Stretford, is discussed with Stretford Moss.

Shaw Hall was reconstructed in the early seventeenth century, probably on the site of an older moated house. It has been demolished (SMR 9388, SJ 7532 9388). In 1643 Peter Egerton, then tenant of Shaw Hall, made provision for a schoolhouse, with consideration of 'one daye drawinge of turves with an able p'son...' (Langton 1898, 89). It is not stated whether the rights of turbary were at Urmston. The site had been drained by First Series Ordnance Survey map, but the Tithe Awards for Urmston and Flixton record mossland field-names (1842, Lancashire Record Office).

WALKDEN MOOR

Project Number	10
SMR Number	4332.1.7
Parish	DEANE
District	SALFORD
NGR	SD 746 032

Walkden Moor, part of the Kearsley complex, overlies glacial deposits and boulder clays, with a solid Permo-Triassic geology of sandstones and Coal Measures. The site lies on the north side of the Wigan—Manchester road, between the once scattered settlements of Walkden and Farnworth. The nature of the moor or mossland is not clear, since it was mostly reclaimed or extracted by the end of the eighteenth century (Yates' map of Lancashire, 1786).

Black Leech Brook, springing from Linnyshaw Moss, meanders through Walkden Moor, and forms a boundary (shown on an eighteenth-century map in the Bridgwater Estate archives; Irlam Record Office, M). A boundary stone called the egg stone still exists; it is probably a glacial boulder. Nineteenth-century Peel Hall occupies an older moated site near Walkden Moor on the south side of the A6 Manchester—Wigan Roman road.

On the First Series Ordnance Survey map (Lancashire Sheet 95, 6":1 mile, surveyed 1844—6, published 1850) there is a small settlement called Street Gate on the main road at a crossroads with Spa Lane, leading to Farnworth and skirting the moorland. Roman pottery was found at Peel St Paul's church, Street Gate, described as the neck and shoulder and a handle of a large two-handled, narrow-necked jar of the first century AD military ware (SMR 1906, SD 7233 0368).

1992—3 field mapping
The main central area of the moss is completely built over (SD 733 033). At the north-west open ground contains colliery slag debris only. The original moor was a westwards continuation of Linnyshaw Moss. No further action required.

WARBURTON MOSS (Fig 59)

Project Number	44
SMR Number	7932
Parish	BOWDON
District	TRAFFORD
NGR	SJ 720 895

Warburton Moss is described in Chapter 4.

WARDLEY MOSS

Project Number	10
SMR Number	4332.1.8
Parish	ECCLES
District	SALFORD

NGR	SD 758 036

Wardley Moss was the southern section of the Kearsley Mosses, lying on top of glacial clays over Middle Coal Measures. It is located between Linnyshaw Moss on the west, and Swinton Moss on the east. The general history of the site is given with the Kearsley Mosses.

To the south is the Roman road from Wigan to Manchester (A6).

Wardley Moated Hall lies south of this road; the moat is now dry, but the restored building, in part fourteenth century, still stands with a seventeenth century gatehouse (SMR 511, SD 7572 0228). The area occupied by the hall extended into the moss, and was later bought into the Bridgwater Estate (Irlam Record Office; Bridgwater Estate maps). There is much industrial and mining activity in this vicinity.

The **1992—3 field mapping** is given in Chapter 4; no peat survives.

WHITE MOSS (Fig 60)

Project Number	24
SMR Number	7495
Parish	PRESTWICH WITH OLDHAM
District	OLDHAM
NGR	SD 879 036

White Moss is described in Chapter 4.

WINSTANLEY MOOR

Project Number	1
SMR Number	4908
Parish	WIGAN
District	WIGAN
NGR	SD 5302

The extent and nature of Winstanley Moor is not known, although place-names on the First Series Ordnance Survey map, such as 'Moss House' and 'Moss Cottage', indicate the location (Lancashire Sheet 93, 6":1 mile, surveyed 1845—6, published 1849). This is east of Withington Brook, now partially dammed, but originally a tributary to the River Douglas, and west of the Garswood—Orrell road. It overlies glacial drift and Coal Measures.

The moss area seems to have been exhausted or drained at an early date, with a number of small coal pits shown on the First Series Ordnance Survey map (and more on the east side of the road — Mossey Croft was the Upper Winstanley coal pit (Banks 1939, SMR 3194, SD 5420 0305). Winstanley Hall, less than 0.5 km east of the moss area, is of sixteenth century origin. An inventory made on the death of James Bankes in 1617 lists a £1 worth of turves (Bankes 1942, 89).

1992—3 field mapping
No peat survives anywhere in the area. Soils are variously clays, shale on the high ground with some sandy areas, especially along the brook where there could be lithics. Nearly all ploughed. The soil, rather dark in places, is probably skirtland. There is no local knowledge of a moss here. This moss would have been a blanket moor spreading towards the valley.

WOOLDEN MOSS (LITTLE)

Project Number	46
SMR Number	3033.1.6
Parish	BARTON ON IRWELL
District	SALFORD

Little Woolden Moss lies at the extreme west end of the Chat Moss complex. It is the northern continuation of Great Woolden Moss, both mosses being divided into complementary residential hall estates. The moss overlies sand and gravels of the Mersey alluvial terrace on glacial clays, with an underlying solid geology of the upper Triassic series of sandstones.

The moss history and field mapping is discussed with Chat Moss in

Chapter 3. Little Woolden Hall was built *c* 1830, on an earlier site (SMR 1559, SJ 6848 9424). Little Woolden Moss has a peat extraction and milling hut, complete with milling and bagging machinery, dating from the 1950s (SMR 7190, SJ 6860 9493). There is a small area of moss leased for hand digging. Dried peat bricks are carried on a hand-pushed trolley via tram lines to a milling shed.

WOOLDEN MOSS (GREAT)

Project Number	46
SMR Number	3033.1.4
Parish	BARTON ON IRWELL
District	SALFORD
NGR	SJ 7094

Great Woolden Moss is the southern continuation of Little Woolden Moss. It overlies alluvial sands and gravels, and glacial clays on top of Permo-Triassic sandstones.

The history of the moss is discussed with Chat Moss.

An Iron Age site was excavated in the late 1980s by GMAU. An interim excavation report is available (Nevell 1989b) and a summary has been given in the period discussion, Chapter 6. About 1500 cubic metres of soil were excavated, including topsoil.

Great Woolden Hall, constructed in the early seventeenth century has a brick-built shippon and stable block, and a barn of probable eighteenth century date (SMR 1512, SJ 6940 9350). A Bridgwater Estate plan of *c* 1785 shows the Hall (Irlam Record Office), its garden, the course of Glaze Brook, and a tributary near the Hall dammed to create a water reservoir for a corn mill (also marked on the First Series Ordnance Survey map (Lancashire Sheet 109, 6":1 mile, surveyed 1845—7, published 1849). A field called French Wheat Field shows there was some arable at the time. Other fields are called moss or meadow barn, and brick kiln. Part of the moss was drained.

1992—3 field mapping
Skirt exists near to the Glaze Brook, with clay and sandy clay base. The Iron Age site, GM 1, lies under pasture with nothing visible. Small patches of sand on a ploughed field to the east of the site were carefully searched, but yielded no finds. A possible mill dam lies in a spinney south-east of the farm.

WOODS MOOR

Project Number	30
SMR Number	2582
Parish	STOCKPORT
District	STOCKPORT
NGR	SJ 905 880

Woods Moor is one of a group of small wetland areas scattered to the south of the River Mersey. It lies at *c* 85 m on an area of glacial moraine overlying Coal Measures. The site had been developed before publication of the Ordnance Survey 1" map (Lancashire Sheet 81 NW published in 1842)

1992—3 field mapping
The area is built up with houses of *c* 1900. A levelled school playing-field is the only open ground and was inaccessible. Gardens nearby have slightly dark sandy soil; no peat left. There is no obvious topographical reason why a moor should be here.
No further action needed.

WORSLEY MOSS

Project Number	46
SMR Number	3033.1.3
Parish	BARTON ON IRWELL
District	SALFORD
NGR	SJ 715 975

Worsley Moss lies at the north-eastern extreme of Chat Moss, and has been partly reclaimed. It overlies alluvial sand and gravel, and glacial clay on Manchester marls and sandstone. Botany Bay Wood is a late eighteenth-century plantation (Anon 1876a).

The history of the moss and previous archaeological finds are discussed under Chat Moss, Chapter 3.

A former large house in the locality — Whitehead Hall recorded in 1683 — was in possession of Richard Whitehead of Astley. Owned in 1909 by Lord Ellesmere, it is now razed and covered by a slag heap (Farrer and Brownbill 1914, 3, 448). A branch of the Bridgewater Canal, built *c* 1759, extended into the moss and served to drain it and to supply water for the Bridgewater Canal. Waste material from mines was dumped on the moss (Hadfield and Biddle 1970, 12). A description of the botany of the moss was made in 1876 recording pansies, roses, water burr, and wild hops, and woodland of elder, silver birch, and Scotch fir (Anon 1876b, 32).

Appendix 2:   Greater Manchester radiocarbon dates

Site	Context	Lab Code	^{14}C date BP	2 σ cal BC	2 σ cal BP	1 σ cal BC	1 σ cal BP
Nook Farm 1	25-40	GU-5271	4590±70	3599-3047	5548-4996	3493-3142	5442-5091
Nook Farm 2	38-48	GU-5272	3710±60	2300-1940	4249-3885	2199-2034	4148-3983
Nook Farm 3	30-40	GU-5273	4670±60	3629-3207	5578-5156	3598-336	5547-5311
Nook Farm 3	*Betula*	GU-5280	4570±50	3497-3100	5446-5049	3369-3142	5318-5091
Nook Farm	Carb wood	GU-5325	3930±80	2853-2149	4802-4098	2569-2333	4518-4282
Nook Farm 1	5-10	GU-5356	2170±50	380-100	2329-2049	362-1672	2311-2116
Nook Farm 3	5-10	GU-5357	4020±50	2861-2460	4810-4409	2603-2480	4552-4429
Worsley Farm	33-42	GU-5359	3280±60	1735-1430	3684-3379	3684-3379	3619-3463
Worsley Farm	64-66	GU-5360	4050±70	2878-2460	4827-4409	2860-2491	4809-4440
Worsley Farm	70-77	GU-5361	4320±50	3082-2785	5031-4734	3024-2908	4973-4857
Worsley Farm	110-13	GU-5362	4950±60	3945-3640	5894-5589	3787-3694	5736-5643
Worsley Farm	150-60	GU-5363	5270±50	4240-3990	6189-5939	4226-4002	6175-5951
Worsley Farm	300-20	GU-5364	7980±80	7132-6597		7033-6654	
Worsley Farm	410-30	GU-5365	9140±70	8345-8025		8327-8065	
Barton Moss	20-30	GU-5366	3280±50	1690-1440	3639-3389	1629-1516	3578-3465
Barton Moss	70-80	GU-5367	4300±60	3080-2705	5029-4654	3018-2891	4967-4840
Barton Moss	105-115	GU-5368	4870±60	3780-3519	5729-5468	3774-3547	5723-5496
Barton Moss	152-162	GU-5369	6020±60	5199-4780	7148-6729	5040-4848	6989-6797
Barton Moss	215-25	GU-5370	6850±60	5840-5583	7789-7532	5749-5640	7698-7589
Barton Moss	330-60	GU-5371	8480±50	7571-7430		7539-7485	
Barton Moss	380-400	GU-5372	7750±60	6700-6440		6673-6476	
Red Moss 1		GU-5373	2260±50	400-200	2349-2149	394-251	2343-2200
Red Moss 2		GU-5374	1260±50	520-263	2469-2212	677-700	1273-1160
Red Moss 3		GU-5375	2330±50	520-263	2469-2212	405-387	2354-2336

Appendix 3: Aerial photographic interpretation in Greater Manchester

Chris Cox

This survey was undertaken as part of the North West Wetlands Survey investigation of the archaeology and palaeoenvironment of the Metropolitan County of Greater Manchester. The aims and methods of the aerial photographic survey were previously established and employed during survey of Lancashire (Cox 1991) and Merseyside (Cox 1992).

Aerial archaeology within the NWWS

The NWWS uses aerial photographic data to locate buried archaeological features, and to locate and define the limits of both extant and former wetland areas.

The aerial viewpoint allows the archaeologist to see an elevated overview of the countryside, and elucidates pattern and relationships of features which are not apparent or meaningful to the ground-based observer. Archaeological features are mainly revealed from the air via the media of differential crop growth (Riley 1946, Riley 1979), varying soil tones and colours (Wilson 1982, 39), and shadows cast by earthworks or built features (Wilson 1982, 27). Within its limitations, aerial photography is a very economical and effective archaeological survey tool. Aerial photographic interpretation forms an integral part of multi-disciplinary archaeological investigations such as the NWWS, where communication between field and aerial archaeologists allows for cross checking and verification of both aerial and field evidence.

Specialist oblique-angled photographs, usually taken by an airborne archaeologist from the window of a light aircraft were interpreted alongside the 'blanket coverage' provided by the vertical surveys which are usually commissioned by county councils for administrative, archival and planning purposes on a regular basis. Vertical photographs are not taken for archaeological purposes, and their fixed viewpoint and often unsuitable 'timing' (*eg* when crops are unripe, or earthworks are masked by vegetation), may be problematic for the archaeological interpreter. However, verticals are often the only data source, particularly in counties such as Greater Manchester (*see below*). Interpretation of verticals requires use of a stereoscope to provide a three dimensional and enlarged view of the photographic data.

After interpretation, archaeological and environmental data are transferred to translucent overlays to the 1:10,000 OS maps, either as defined areas of archaeological potential, or as accurate plans of the buried features, depending upon the nature of the recorded features and the validity of the photographic interpretation. An accurate cartographic representation of archaeological and environmental features is achieved via use of aerial photographic rectification software (Haigh 1989), which provides compatibility with GIS and allows integration with the main NWWS archive. Sketched additions to the maps can also be added to the GIS databank via digitisation. The graphical record is supported by a database which can be interrogated via location, site name or site type.

The survey area

The survey area was covered by the following 1:10,000 OS quarter sheets:

SJ68 NW NE SW SE
SJ69 NW NE SW SE
SJ78 NW NE SW SE
SJ79 NW NE SW SE
SJ88 NW NE SW SE
SJ99 NW
SD60 SW SE
SD70 NW NE SW SE
SD80 NW NE SW SE
SD90 NW SW

The area was thus defined in order to examine the wetlands and adjacent areas in as significant and broad a landscape context as time would allow. In this case, the division of survey areas by modern administrative boundary has been adhered to, but it should be noted that archaeological features seen on aerial photographs covering the adjacent counties of Cheshire and Merseyside must be considered as integral to a landscape study of the lowlands of the North West, and this report stands simply as an interim to a final synthesis, which will be possible when aerial photographic interpretation has been carried out over Northern Cheshire, and the data integrated with those from Merseyside, Lancashire and Greater Manchester. Two sites of interest in this case are noted here, at Winwick and Moss Hall, but are not included in this survey, as they are in Cheshire.

The geology and hydrology of the study area have been described by Nevell (1992a).

Previous aerial photographic survey and analysis

The county has been surveyed regularly from the air since the 1970s by Jones and Higham (Jones 1979), and by the RCHME APU. In both cases, oblique photographs were taken for specialist archaeological purposes. Restrictions imposed upon light aircraft operation in the county by the proximity of Manchester International Airport effectively limit aerial photographic flights to a narrow, low-level, corridor on the West side of the county. The extent of urbanisation in the county is also great, in comparison to the proportion of agricultural land. Industrial areas, ranging from coal mining areas to industrial estates also abound, and contribute to the high level of disturbance to the natural landscape and the limitation to the arable. These factors greatly reduce the chances of discovery of buried sites by observation of crop or soil marks, which require arable agricultural regimes and regular reconnaissance flights (Nevell 1992a, 28—9). The number of oblique photos available for consultation was therefore smaller than that in neighbouring counties. However, Greater Manchester is comprehensively covered by vertical aerial photographs, which have been taken regularly by commercial survey companies for non archaeological purposes.

Data from both oblique and vertical photographs have been used extensively throughout the county during analysis of the archaeology of the Mersey Basin, notably by Nevell (1985 [not consulted], 1987, 1988, 1989, 1992a), Higham (1988) and Philpott (1991) and of Tameside (Nevell 1992b).

Photographic sources
This survey is based on aerial photographs from the following sources:

Obliques
1   The National Library of Air Photographs (NLAP), RCHME, Swindon.

2   The Greater Manchester Countryside Unit, Ashton-under-Lyne. Colour obliques covering the Moss lands of Greater Manchester.

3   The Greater Manchester Archaeological Unit and Greater Manchester SMR. Colour slides depicting known sites

4   Greater Manchester Archaeological Unit. Mainly uncatalogued collection of negatives and contact prints.

Verticals
1   Greater Manchester Archaeological Unit. Prints at 1:2500 black and white prints covering Northern Cheshire. Selected 1:10,000 black and white prints.

2   Greater Manchester Geological Survey. Vertical photographs dating from 1945—8, 1961, 1971—2, 1979, 1983—4, 1988—9 (predominantly monochrome prints at 1:10,000 scale) were consulted. The 1988—9 coverage was taken in colour.

Comment on the available database
Oblique photographs were examined prior to the verticals, and areas of potential interest were marked on 1:10,000 base maps. Some areas of possible archaeological interest were located on the available oblique photo sources, but no sites of possible prehistoric or Romano-British date were identified on the available obliques which have been accessioned into the NLAP. However, this does not mean that there is no evidence — photographs covering the published site at Great Woolden Hall (McNeil and Nevell 1989) were not available for consultation in the NLAP, not having been accessioned. This site is not included in the survey, having been mapped from aerial sources, excavated and published. Several other major photo sources which have been cited in publication or research documents were not easily available for public consultation, presumably due to their being prepared for accessioning to the NLAP. Vertical photographic coverage was comprehensive and easily accessible.

Survey method
The survey concentrated on all areas of arable and pastoral land-use within the county, taking into consideration, but not interpreting and mapping, sites lying outside the county boundary. The strengths of a landscape approach are here acknowledged, and it is intended that this approach be adopted in the final synthesis of the material as suggested above.

All sites located within the lowlands of Greater Manchester by Nevell, (gazetteers Nevell 1989, 1992a) were then checked to their photo sources, as his research brief covered periods up to c AD 400, and this survey covered all periods to the post-Medieval. Some discrepancies were found between his interpretations and my own, and the author would advocate a more conservative approach to the identification of ditched sites showing as possible cropmarks on vertical photographs, especially, as is the case in many of these examples, vertical photographs taken for non-archaeological purposes are the only source of evidence (Scollar 1990, 26—7). It is important to note that the available research data kindly provided by Nevell was a preliminary version of his finally submitted thesis, which was substantially revised at a later date. The addition of reliability ratings to Nevell's data (Nevell 1992a, 324) provides a much more conservative and balanced view of these data in the light of my above comments. It is also important to acknowledge that vertical photographs are taken from one viewpoint only, and may not take advantage of lighting conditions to emphasise the shadows cast by earthworks. Therefore interpretations made from vertical sources alone cannot be regarded as archaeologically 'definitive' in any way.

Results

The results of this work were tabulated as a working gazetteer, which was used as a basis for field investigation of possible archaeological sites. There were no additions to be made to the sites recorded by Nevell's and Higham's research within Greater Manchester, although some sites (particularly dating to the medieval and post-Medieval periods) were noted in northern Cheshire, and will be recorded during synthesis of the evidence from that county. I have not re-synthesised published work within the county covering sites already known from aerial photographic sources and cited above.

The resulting list of possible sites was passed to David Hall and Adele Mayer at GMAU for field checking, and the results of interpretation of vertical photos were also re-checked by David Hall after the field survey had been carried out.

The gazetteer of possible sites was concorded with Nevell's then unpublished research data which has been extensively revised since this contribution was completed. It records the site location, Nevell's revised site number, and the NGR. These items are shown in bold typeface. The photographic source has been consulted by the author where cited, and noted where not consulted due to unavailability at the time of survey. This information is followed by the author's own short commentary on features. The commentary is followed in each case by the results of field survey undertaken and authored by David Hall over each listed site.

Winwick
SJ 592 934   Site no: 400
Cheshire CC verticals 197709, Run 28 9664
RCHME Obliques SJ 5993/1—3

A complex of enclosures and boundaries shows as a clear crop mark at this location, encompassing Nevell's sites Winwick 1—111, Nos. 400—2. These sites lie within the county of Cheshire.

Field Survey Data: *The chief cropmark is a rectangular enclosure with rounded corners straddling a Roman road marked on the OS 1:10,000 map. The cropmark has had previous fieldwork and a section has been excavated yielding second-century Roman pottery.*

A field visit showed that a division across the photograph dividing into dark and light parts is a geological boundary. The light part is sandy high ground and the dark area lies lower, with a cover of colluvium on sand that would yield no finds. The field had very little modern nightsoil rubbish,

and was sown with winter corn just emergent. Careful searching revealed no dark area of occupation, and only a single Roman pottery sherd of local rustic fabric was discovered, lying outside the enclosure and associated with burnt stones. The only finds within the enclosure were a few 'archaeologically' burnt stones. These stones are important markers for sites in the region, rather like fire-cracked flints ('pot boilers') on prehistoric sites in lowland England.

Thelwall
SJ 653 867   Site no: 405
A large single ditched oval enclosure. No photo source was available, but the site is published and known (Higham 1988, noted by Nevell 1992).

Hole Mill Farm
SJ 681 944   Site nos: 411
GMC BW Vertical 1984 Run 12 2384 030
Amorphous crop marks indicating possible buried ditches and a slightly raised, light toned, area was seen at SJ 682 948. The Glaze Brook Valley has high archaeological potential due to its situation and proximity to the excavated site at Great Woolden Hall, but these features were not conclusive enough to map with confidence.

Field Survey Data: *This field has sandy soil and was newly planted when visited; no finds were made. To the north were soilmarks of wide medieval 'ridge and furrow'. A subsequent viewing of the original photograph, taken in early June, revealed only vague geological marks and no likely archaeological cropmark.*

Moss Hall
SJ 687 914   Site no:412
GMC Colour verticals 1988—9, Run 13, 2989 004—5

Sub rectangular double ditched enclosure showing as a crop mark in cereal. This site is visible on vertical photographs, but the internal features are very tenuous. The site is in Cheshire and will be mapped during survey of that county.

Field Survey Data: *The sandy-clay field was covered with a dense corn crop when first visited. There were no finds apart from a few burnt stones to the south. The aerial photograph showed curved strips of medieval form south of the cropmark. A subsequent field visit in autumn 1993 revealed only modern debris over the cropmark, which seems to be related to the coalmine and dump site to the west (pers comm M Leah 1994).*

Mount Pleasant
SJ 697 916   Site no: 419
GMC BW vertical run 13 2384 050
A ditched feature is likely to be old field boundaries, which often show as crop marks on aerial photographs in exactly the same manner as more ancient buried ditches.

Field Survey Data: *Field under grass with no finds on the land immediately next to the west. Sandy clay soils.*

Glazebrook
SJ 691 921—SJ 699 927   Site nos: 421 and 422
Cropmarked features, described by Nevell, which are probably relict streams. No features of archaeological significance were seen by the author on the cited photo sources. Nevell notes field walking finds from this field.

Field Survey Data: *Field under thin winter corn in good condition. No finds. A slightly wet furrow lying along a local watershed forms one side of the reported enclosure. The south-west of the proposed enclosure is an old hedge and ditch line and the north-west side is a crop division.*

Caldwell Brook
SJ 730 893   Site nos: 426, 428, and 429
No photo source located, cited as Nevell 1989, 32, Fig. 5, NC10; Nevell 1992, 331

Possibly crop marks showing enclosures — nothing seen on the verticals. Photographed by Dr N Higham.

Field Survey Data: *Peterhouse Farm. Sites not visited because they were left uncultivated as 'set-aside'.*

Barton Grange/ Eccles
SJ 731 959   Site no: 427
GMC BW Vertical 1961 Run 12 9655 135
A possible double ditched enclosure was identified on the fringes of Chat Moss. Nothing seen at this location, but some drainage ditches seen at SJ 752 985.

Field Survey Data: *At the site of the cropmark is a high mound of boulder clay and modern rubbish; a recent dump.*

Redhouse Farm
SJ 734 898   Site no: 430
A possible single ditched enclosure, noted by Nevell, but photos not consulted for this survey.

Field Survey Data: *Sinderland House Farm. Sandy field under thin winter corn. No finds, but burnt stone lies in the cropmark area, with none on the remainder of the field. Lot of nightsoil background. Site GM 15*

Dunham
SJ 738 943   Site no: 431
A large curvilinear enclosure, photographed by Higham and examined by resistivity survey (Nevell 1985, 133, Fig 9.2).

Rhodes Green Farm
SD 840 053   Site no: 438
GMAU oblique colour slide. A series of possible trackways, field boundaries and enclosures. Inconclusive grass marks in a pasture field. From these photos alone, and from examination of vertical photos, conclusive pattern cannot be discerned with confidence, therefore the site has not been mapped for this survey, but is of potential interest.

Ashton Moss
SJ 923 998, SJ 925 998   Site nos: 441, 442
Sites recorded around Ashton Moss on both GMC verticals and on photographs held by the Countryside Unit at Ashton were interpreted by the author as non-archaeological.

Medieval and post-medieval landscape
This survey also recorded medieval and post-medieval ridge and furrow ploughing, both as upstanding and plough-eroded features. The majority of ridge and furrow and 'narrow rig' was recorded in at the interface of Greater Manchester and Northern Cheshire, in the areas adjacent to Manchester International Airport, from vertical photographic sources.

Conclusions
The consultation and interpretation of aerial photographs within Greater Manchester has facilitated the NWWS field survey, but has not yielded any new or extensive archaeological evidence beyond that already published or contained within the Greater Manchester SMR. Most of the targets presented to the field surveyors as 'potential areas' have been discounted upon ground examination. The difficulties of the aerial survey method and its application within the county have been outlined above, and it is wholly unfruitful to compare the evidence from this region to that from rural areas such as Wessex or the East Anglian Fenland edges, where buried archaeological sites are very easily located by the aerial photographic interpreter. Counties such as Greater Manchester reveal little of their past landscape history to the aerial observer, due in large part to the very heavy urbanisation and industrialisation of the last two centuries. However, the county is only one part of a regional survey, and it will be in the regional context that the seemingly meagre evidence from Greater Manchester can be compared with that from other counties, and results from all survey sources combined to help synthesise the regional archaeological perspective. Aerial photographic sources have revealed the major settlement site at Great Woolden Hall, and further sites are known to the west in Cheshire (at Winwick), in Merseyside, and in Northern Cheshire to the South of this study area. Unless sites are very well defined, such as those in evidence at Winwick, aerial photographic sources alone are not suitable for the analysis of urbanised areas such as Greater Manchester, and must always be used in conjunction with field survey and documentary research such as that employed by the NWWS, and studied in a regional landscape context.

Appendix 4:   Wetland sites in Greater Manchester

List of sites associated with Greater Manchester mosses discovered and investigated during fieldwork in 1992/3. The numbers refer to the page of the text where the site is referred to.

Chat Moss
GM 1	SJ 6900 9350	Iron Age site revealed by a cropmark.Excavated 1986-8 by GMAU (27, 120-1)
GM 2	SJ 6838 9469	Early ditch below peat at Little Woolden (26-7)
GM 3	SJ 6888 9927	Late medieval building stone near Morley's Hall (28)
GM 4	SJ 6970 9874	Flints from a sandy knoll south of Astley (26-8)
GM 5	SJ 7907 9797	Nook Farm Mesolithic site (26, 50-62)
GM 6	SJ 7390 9905	Worsley Moss canal, SMR 1903 (29)
GM 7	SJ 7474 9940	Burnt stone area west of Worsley Grange (29).
GM 8	SJ 7475 9900	Dark area of post-medieval pottery and other rubbish near Worsley Grange (29).
GM 9	SJ 7490 9875	Burnt stone area south of Worsley Grange (29).
GM10	SJ 7376 9659	Burnt stone area near Boysnope, Barton (29).

Warburton Moss
GM 11	SJ 7344 8974	Burnt stone area near Sinderland House Farm (112,158 (under heading Redhouse Farm)).

Highfield Moss
GM 12	SJ 6092 9565	Burnt stone area north of Highfield Moss (106).
GM 13	SJ 6167 9583	Burnt stone area north of Highfield Moss (106).

Appendix 5: Artefacts associated with Manchester wetlands

Artefacts preserved in Museums. Numbers in brackets are the Museum reference/catalogue number, where known.

Manchester Museum

1   Human skull from Worsley, SMR 1487 (Plate 5)
2   Canoe from Barton upon Irwell, SMR 1220
3   Perforated axe-hammer from Heaton Chapel, SMR 848
4   Stone axe from Withington (0.5042), SMR 392
5   Bronze Age axe from Northenden (0.5840), SMR 43
6   Flat Bronze Age axe from Wythenshaw (1987.2217)
7   Stone axe fron Bickenshaw Hall, Abram, Wigan (25931)
8   Polished Neolithic stone axe from Shawe Hall, Flixton (25916), SMR 364
9   Basalt axe from Moston, (25919)
10   Chert axe from Fallowfield (31502), SMR 1238?
11   Adze from Timperley (25935)
12   Axe from Timperley (26216)

Bolton Museum

13   Roman quernstone from Red Moss, SMR 4389, Plate 16
14   Antler pick from Red Moss, SMR 442, Plate 16

Cambridge Museum of Archaeology and Anthropology

15   Human skull from Ashton Moss, SMR 8237

Warrington Museum

16   Axe head from Brinnington, SMR 3665
17   Perforated hammer-stone from the Manchester Ship Canal at Irlam, SMR 1762 (Plate 3)
18   Bronze axe from Breightmet (Bolton, near Darcy Lever Moss), SMR 552
19   Finds from the Winwick barrow excavation, Cheshire SMR 4719

Wigan Museum

20   Spearhead from Ince in Makerfield (acc no 29A1977)

Stockport Museum

21   Perforated stone from Cheadle Heath, SMR 799
22   Flint knife from Heaton Moor, Stockport, SMR 784.

Merseyside Museum

23   Winwick flint dagger (acc no 65.98), Plate 17

Artefacts recorded in the literature, but now lost, and artefacts still remaining near their original location. The first six had no SMR numbers in 1993, the remainder have SMR numbers and the SMR should be consulted for more information if the item is not discussed in the text.

1   Stone 'adze' from Chat Moss, SMR 234; *Trans Lancashire Cheshire Antiq Soc* **9** (1868-9) 204-5. *Same as no 17 below*
2   Quernstone from Sidall House (Bramhall, probably from Kitts Moss; Heginbotham 1892, 316)
3   Wickerwork beehive from Chatmoss,;Whitaker 1773, 316
4   Wickerwork beehive from Linnyshaw Moss (Kearsley complex; Whitaker 1773, 316)
5   Antlers from near Warburton
6   Saddle quern from Red Moss, kept in the garden of Gibb Farm, SMR 4382 (SD 6320 1055)
7   Alexandra Park; two axes and a perforated pebble, SMR 80 and 79
8   Alexandra Park; perforated oval pebble of sandstone, SMR 1240
9   Flint axe from Stockport, SMR 1255
10   Bronze palstave from Cockey Moor, SMR 3633
11   Wooden comb, bones, and a 'weapon' from the Manchester Ship Canal near Sticking Island, SMR 387
12   Gold bulla or pendant from the River Irwell, SMR 538
13   Hoard of 30 Roman coins from Stretford, SMR 110
14   Roman coins from the River Mersey, SMR 790
15   'Axe' from Cheadle, SMR 2822
16   Roman pottery vessel from Peel church, SMR 1906
17   Adze from Irlam, SMR 234. *Same as no 1 above*
18   Three 'coins and urns' from Droylsden Moss, SMR 612
19   'Axe' from Droylsden, SMR 613
20   Bronze tanged spearhead from Boggart Hole Clough, SMR 2011
21   Roman coins from Oldham, SMR 110
22   Human skull from Red Moss, SMR 442
23   Two Roman coins from Boggart Clough, SMR 1395.1.0
24   Roman coin hoard from Boothstown, SMR 514 (Anon 1947, 224)
25   Bronze flanged axe from Radcliffe (Black Moss), SMR 77
26   Hoard of 20 Roman coins from Denton Moor, SMR 8236
27   Saxon coin hoard from Stretford, SMR 368 (Harrison 1892, 251)

See also the list of finds under Crumpsall Moor in Appendix 1.

Bibliography

Yates W, 1786 Lancashire 1"

Geological Survey, Lancashire Sheet 85, 1970

Geological Survey, Lancashire Sheet 95NW, 1:10,560. Surveyed 1844, revised 1915/16

OS 1st edn 6", Lancashire Sheet , 1848

OS 1st edn 6", Lancashire Sheet 88, 1851

OS 1st edn 6", Lancashire Sheet 96, 1848

OS 1st edn 6", Cheshire Sheets 8, 9, 1882

OS 1st edn 6", Lancashire Sheet 86, 1849

OS rev edn 6", Lancashire Sheet 86, 1930

OS rev edn 6", Lancashire Sheet 86, 1909

Anon, 1860 Excursion to Botany Bay Wood and the grounds of Worsley Hall, *Trans Manchester Field Naturalists' Hist Archaeol Soc*, 27

Anon, 1875 Excursion to Flixton: the grasses, *Trans Manchester Field Naturalists' Hist Archaeol Soc*, 22

Anon, 1875a Manchester Guardian, *Trans Manchester Field Naturalists' Hist Archaeol Soc*, ,

Anon, 1876a Visit to Botany Bay Wood: trees in towns: meteorology and biology, *Trans Manchester Field Naturalists' Hist Archaeol Soc*, 33

Anon, 1876b Worsley, *Trans Manchester Field Naturalists' Hist Archaeol Soc*, 33

Anon, 1889 Alexander Park, *Manchester Faces and Places*, **1**, 27

Anon, 1892 Carrington Moss, *Trans Manchester Field Naturalists' Hist Archaeol Soc*,

Anon, 1894 The Ship Canal, Barton Catholic Church, and Trafford Park, *Trans Manchester Field Naturalists' Hist Archaeol Soc*, 23

Anon, 1894a Worsley, *Trans Manchester Field Naturalists' Hist Archaeol Soc*, 43

Anon, 1897—8 Trafford Park, *Manchester Faces and Places*, **9**, 17

Anon, 1905 Barton Moss, *Trans Manchester Field Naturalists' Hist Archaeol Soc*, **12**

Anon, 1907 Boggart Hole Clough, May 16th, *Trans Manchester Field Naturalists' Hist Archaeol Soc*, 17

Anon, 1907 Worsley, *Trans Manchester Field Naturalists' Hist Archaeol Soc*, 19

Anon, 1947 Antiquarian notes, *Trans Lancashire Cheshire Archaeol Soc*, **59**, 224—7

Anon, 1988 Mystery of the vanished commons, *Stockport Heritage*, **5**, 3—12

Anon, 1989 *Mosslands strategy*, City of Salford MBC and Wigan MBC, unpubl

Anon, 1991 Blake Lake gibbet chains, *Stockport Heritage*, **6**, 2, 6

Aikin, J A, 1795 *Description of the country for 30 miles around Manchester*, London

Albert, D, 1842 Process of carbonizing turf, *Mem Phil Soc Manchester*, 2 ser, **6**, 399—408

Andersen, S T, 1988 Pollen spectra from the double passage grave, Klekkendehøj, on Møn: evidence of swidden cultivation in the Neolithic of Denmark, *Journ Danish Archaeol*, **7**, 77—92

Andersen, S T, 1993 Early and middle Neolithic agriculture in Denmark: pollen spectra from soils in burial mounds of the Funnel Beaker culture, *Journ European Archaeol*, **1(1)**, 153—80

Arrowsmith, P and Fletcher, M, 1993 Nico ditch and Carr ditch, a case of mistaken identity?, *Archaeol North West*, **5**, 25—31

Ashmore, O, 1958 Household inventories of the Lancashire gentry 1550—1700, *Hist Soc Lancashire Cheshire*, **110**, 59—105

Ashmore, O, 1969 *The industrial archaeology of Lancashire*, Newton Abbot

Ashmore, O, 1982 *The industrial archaeology of North West England*, Manchester

Axon, E, 1906 William Crabtree's plan of the Booth Hall Estate, *Trans Lancashire Cheshire Antiq Soc*, **23**, 32—64

Bailey, J E, nd *Field names in Flixton*, mss unpublished

Bailey, J E, nd unpubl mss coll, Chetham's Library; Field-names in Flixton

Bailey, W H, 1889a On an old canoe recently found in the Irwell valley near Barton, *Mem Phil Soc Manchester*, 4 ser, **2**, 243—51

Bailey, W H, 1889b Prehistoric Chat Moss, and a new chapter in the history of the Manchester and Liverpool railway, *Trans Manchester Geol Soc*, **5**, 119—27

Baines, E, 1836 *The history of the County Palatine and duchy of Lancaster*, **1—4**, London

Baines, T, 1867 *Lancashire and Cheshire past and present*, London

Baker, A R H, and Butlin, R A, 1973 *Studies of field systems in the British Isles*, Cambridge

Banks, J H M, 1939 Records of mining in Winstanley and Orrell, near Wigan, *Trans Lancashire Cheshire Antiq Soc*, **54**, 31—65 Bankes, J H M, 1942 James Bankes and the manor of Winstanley 1594—1617, *Hist Soc Lancashire Cheshire*, **94**, 56—94

Bankes, J, and Kerridge, E, 1973 *The early records of the Bankes family at Winstanley*, Chetham Soc, 3 ser, **21**

Barber, K E, 1981 *Peat stratigraphy and climatic change: a palaeoecological test of the theory of cyclic peat bog regeneration*, Rotterdam

Barber, K E, Cambers, F, and Maddy, D, 1994 The climate of northern Cumbria since the Neolithic, *NWWS Ann Rep 1994* (eds R Middleton and R Newman), Lancaster, 21—8

Barnes, B, 1982 *Man and the changing landscape*, Liverpool

Barton, B T, 1874 *History of the borough of Bury and neighbourhood, in the county of Lancaster*, Manchester

Barton, B T, 1887 *History of Farnworth and Kearsley*, Bolton

Behre, K E, 1981 The interpretation of anthropogenic indicators in pollen diagrams, *Pollen et Spores*, **23** (2), 225—45

Behre, K E, and Kučan, 1986 Die Reflektion archäologisch bekannter Siedlungen in Pollendiagrammen verschiedener Enternung-Beispiele aus der Siedlungskammer Flögeln, Norwestdeutschland, in *Anthropogenic indicators in pollen diagrams* (ed K E Behre), Rotterdam, 95—115

Birks, H J B, 1964—5 Chat Moss, Lancashire, *Mem Phil Soc Soc Manchester*, **106**, 24—43

Bonney, A P, 1972 A method for determining absolute pollen frequencies in lake sediments, *New Phytol*, **71**, 393—405

Booker, J, 1854 *A history of the ancient chapel of Blackley in Manchester parish*, Manchester

Booker, J, 1857 *A history of the ancient chapels of Didsbury and Chorlton*, Chetham Soc , **42**

Bowman, S, 1990 *Radiocarbon dating*, British Museum Publications, London

Boyd-Dawkins, W, 1911 Note on a find by Mr T R Morrow in the alluvium of the Mersey at Irlam, *Trans Lancashire Cheshire Antiq Soc*, **29**, 101—3

Briggs, D J, and Courtenay, F M 1985 *Agriculture and environment*, London

Brisbane, C, 1987 Roman dig at Prestwich, *Brit Archaeol*, **1**, 10—14

Brockbank, W, 1865—6 Notes on a section of Chat Moss near Astley station, *Mem Phil Soc Soc Manchester*, **5**, 91—5

Bronk-Ramsey C R, 1994

Bu'Lock, J D, 1958 The Pikestones: a chambered long cairn of Neolithic type on Anglezarke Moor, Lancashire, *Trans Lancashire Cheshire Antiq Soc*, **68**, 143—6

Bu'Lock, J D, 1961 The Bronze Age in the North-west, *Trans Lancashire Cheshire Antiq Soc*, **71**, 1—42

Burdett, P P, 1777, *A survey of the County Palatine of Chester*, Hist Soc Lancashire Cheshire, occ ser, vol 1 (1974 facsimile reprint, introduction by J B Harley and P Laxton)

Burton, R G O, and Hodgson, J M, 1987 *Lowland peat in England and Wales, Soil Survey Special Survey*, **15**, Harpenden

Cameron, K, 1967—8 *The place-names of Cheshire*, **2**, English Place-Name Society, **45**

Carter, C F (ed), 1962 *Manchester and its region, a survey prepared for the British Association*, Manchester

Chambers, F, 1991 Peat humification: proxy climatic record or indicator of landuse history?, in *Soils and human settlement*, (ed S Limbrey), Welsh Soils Discussion Group, Bangor, **26**, 27—44

Chandler, G, 1953 *William Roscoe of Liverpool*, London

Chistjakov, V I, Kuprijanov, A I, and Gorshkov, V V, 1983 Measures for fire prevention in peat deposits, in *The role of fire in northern circumpolar ecosystems* (eds R W Wein and D A Maclean), Chichester, 259—72

Coles, B, and Coles, J M, 1989 *People of the wetlands*, London

Coles, J M, Hibbert, F A, and Orme, B J, 1973 Prehistoric roads and tracks in Somerset: 3, The Sweet Track, *Proc Prehist Soc*, **39**, 256—93

Cooke, G A, nd, *c*1861 *Topographical and statistical description of the county of Lancaster*, London

Cowell, R W, 1991 The prehistory of Merseyside, *J Merseyside Archaeol Soc*, **7**, 21—61

Cowell, R W, and Innes, J B, 1994 *The wetlands of Merseyside*, NWWS 1, Lancaster Imprints 2, Lancaster

Cox, C, 1991 Aerial archaeology in the North West Wetlands: Lancashire 1990—91, *NWWS Ann Rep 1991*, Lancaster, 1—4

Cox, C, 1992 Aerial photography in the North West Wetlands: Merseyside 1992, *NWWS Ann Rep 1992*, Lancaster, 37—41

Crofton, H T, 1889 Lancashire and Cheshire coalmining records, *Trans Lancashire Cheshire Antiq Soc*, **7**, 26—74

Crofton, H T, 1899—1903 *A history of the ancient chapel of Stretford*, **1**—3, Chetham Soc , n ser, **42**, 45—52

Crofton, H T, 1902 How Chat Moss broke out in 1526, *Trans Lancashire Cheshire Antiq Soc*, **20**, 139—45

Crofton, H T, 1905a Agrimensorial remains round Manchester, *Trans Lancashire Cheshire Antiq Soc*, **23**, 112—71

Crofton, H T, 1905b *A history of Newton chapelry in the ancient parish of Manchester, Failsworth section*, **2** (2), Chetham Soc, n ser, **54** 303—8

Crofton, H T, 1906 Dumplington and the Holcrofts, *Trans Lancashire Cheshire Antiq Soc*, **24**, 21—46

Crofton, H, T, 1907 Moston and White Moss, *Trans Lancashire Cheshire Antiq Soc*, **25**, 32—65

Cundill, P R, 1981 The history of vegetation and land use of two peat mosses in south-west Lancashire, *The Manchester Geographer*, n ser, **2** (2), 35—44

Dean, E B, 1991 Bygone Bramhall, *Stockport Heritage*, **2** (2), 20

Dodds, J, and Middleton, R, forthcoming GIS in the North West Wetlands Survey, in *GIS Applications in Archaeology* (eds R van der Noort and D Donaghue), Hull

Dodgson, J McN, 1970 *The place-names of Cheshire*, **1**, **2**, English Place-Name Society, **44, 45**

Drake, E S, 1861 *Commercial directory of Bolton, Bury, Wigan, Chorley, Darwen, Leigh, Radcliffe, Ramsbottom (and adjoining townships)*, Sheffield

Edwards, K J, and Ralston, I, 1984 Post-glacial hunter-gatherers and vegetational history in Scotland, *Proc Soc Antiq Scotl*, **114**, 15—34

Ekwall, E, 1922 *The place-names of Lancashire*, Chetham Soc, n ser, **81**

Ekwall, E, 1960 *Dictionary of English place-names*, 4 edn, Oxford

Erdtman, G, 1928 Studies in the post-arctic history of the forests of north-western Europe: **1,** Investigations in the British Isles, *Geol Foren Stock Forh*, **50**, 123—92

Evans, A T, and Moore, P D, 1985 Surface pollen studies of *Calluna vulgaris* (L) Hull and their relevance to the interpretation of bog and moorland pollen diagrams, *Circaea*, **3** (3), 173—8

Faegri, K, and Iversen, J, 1989 *Textbook of pollen analysis*, Chichester, 4 edn

Farrer, W (ed), 1900 *The chartulary of Cockersand Abbey*, **2** (2), Chetham Soc, n ser, **43**

Farrer, W (ed), 1902 *Lancashire pipe rolls and early Lancashire charters*, London

Farrer, W (ed), 1903 *Lancashire inquests, extents and feudal aids AD 1205—1307*, Lancashire Record Soc, **48**

Farrer, W (ed), 1907 *Lancashire inquests, extents and feudal aids AD 1310—1333*, Lancashire Record Soc, **54**

Farrer, W, and Brownbill, J (eds), 1914 *Victoria County History of Lancashire*, London

Faul, M L, and Moorhouse, S A, 1981 *West Yorkshire: An archaeological survey to AD 1500*, Wakefield

Faulkner, H, 1989 *Warburton, the history of a village*, unpublished mss, Trafford Archives

Fielding, nd, *Hist Gleanings*, 204

Fishwick, H, 1889 *History of the parish of Rochdale*, Rochdale

Fishwick, H, 1913 *Rochdale Survey 1626*, Chetham Soc, n ser, **71**

Fletcher, T W, 1982 The agrarian revolution in arable Lancashire, *Trans Lancashire Cheshire Antiq Soc*, **72**, 93—122

Freke, D J, Holgate, R, and Thacker, A T 1980 Excavations at Winwick, Cheshire, *Journ Chester Archaeol Soc*, **70**, 9—39

French, G J, 1894 The stone circles on Chetham's close, *Trans Lancashire Cheshire Antiq Soc*, **12**, 42—52

Garland, N, 1987 The skull on the moss, *Medioscope*, **65** (3), 32—3

Garton, D, 1991 *North Peak ESA, Tintwistle Moor survey, interim report*, Trent and Peak Archaeol Trust, and Peak Park

Gaskell, A, 1964 *The history and traditions of Clifton*, Salford

Godwin, H, 1975 *History of the British Flora*, Cambridge, 2 edn

Godwin, H, and Switsur, V R, 1966 Cambridge University natural radiocarbon measurements, **VIII**, *Radiocarbon*, **8**, 390—400

Goodier, J, 1971 *Chat Moss: its reclamation, its pioneers*, Lectures 70—1, Eccles and District Historical Society

Gowlett, J A J, Hedges, R E M, and Law, I A, 1989 Radiocarbon accelerator dating of Lindow Man, *Antiquity*, **63**, 71—9

Granlund, E, 1932 De Svenska hogmossarnas geologi, *Sver Geol Unders Afh* Ser C, **26**, No. 373, 1—19

Grayling, C, 1983 *The Bridgewater heritage*, Manchester

GMAU, 1990 *North Western Relief Route: archaeological assessment*, unpubl

GMAU, 1992a *Dumplington, Trafford Park: archaeological assessment*, unpubl

GMAU, 1992b *The old farmhouse Villa Farm, Warburton: archaeological survey*, unpubl

GMAU, 1993 *Red Moss, Bolton: archaeological survey*, unpubl

Grindon, L, 1867 The Manchester peat bogs, locally 'mosses', *Trans Manchester Field Naturalists' Hist Archaeol Soc*, 27—35

Groves, J, 1990—91, The memorandum book of a Cheshire yeoman, John Ryle of High Greaves, Etchells 1649—1721, *MRHR*, **4** (2), 37—9

Hadfield, C, and Biddle, G, 1970 *The canals of North West England*, vol 1, Newton Abbot

Haigh, J G B, 1989 Rectification of aerial photographs by means of desk-top systems, in S Rahtz and J Richards (eds), *Computer applications and quantitative methods in archaeology 1989, BAR Int Ser,* **548**, 111—19

Hall, D N, 1982 *Medieval fields,* Aylesbury

Hall, D N, 1995 *The open fields of Northamptonshire,* Northampton

Hampson, T, 1882 *History of Blackrod,* Manchester

Hardman, D B, 1961 *The reclamation and agricultural development of the north Cheshire and south Lancashire mossland area,* unpubl thesis, Uni Manchester

Harland, J, 1861 *Mamcestre: being chapters from the early recorded history of the barony; the lordship or manor; the vill, borough, or town, of Manchester,* **1, 2,** Chetham Soc, **53, 56**

Harley, J B (ed), 1968 A map of the county of Lancashire, 1786, *Hist Soc Lancashire Cheshire,*

Harley, J B, and Laxton, A, 1974 *A survey of the County Palatine of Chester* (by P P Burdett), Hist Soc Lancashire Cheshire Occas Series **1,**

Harris, B (ed), 1979 *The Victoria History of the County of Chester,* **2,** Oxford

Harrison, W, 1888 Commons inclosures in Lancashire and Cheshire in the eighteenth century, *Trans Lancashire Cheshire Antiq Soc,* **6,** 112—27

Harrison, W, 1892 Archaeological finds in Lancashire, *Trans Lancashire Cheshire Antiq Soc,* **10,** 249—55

Harrison, W, 1894 Ancient fords, ferries and bridges in Lancashire, *Trans Lancashire Cheshire Antiq Soc,* **12,** 1—30

Hart, C R, 1984 *The North Derbyshire Archaeological Survey,* Sheffield

Haslam, C J, 1987 *Late Holocene peat stratigraphy and climatic change: a macrofossil investigation from the raised mires of North Western Europe,* unpubl thesis, Uni Southampton

Hausding, A, (trans H Ryan), 1921 *A handbook on the winning and utilisation of peat,* Department of Scientific and Industrial Research, London

Hayes, G, *c* 1986 *Collieries in the Manchester coalfields,* De Archaeologische Pers Netherlands

Hedges, R E M, Housley, R A, Bronk-Ramsey, C R, and Klinken, G J van, 1993 Radiocarbon dates from the Oxford Radiocarbon Accelerator System, Archaeometry Datelist 16, *Archaeometry,* **35,** 147—67

Heginbotham, H, 1882 *Stockport, ancient and modern,* **1, 2,** London

Hewitt, H J, 1972 *The building of the railways in Cheshire down to 1860,* Manchester

Hibbert, F A, Switsur, V R, and West, R G, 1971 Radiocarbon dating of Flandrian pollen zones at Red Moss, Lancashire, *Proc Roy Soc London B,* **177,** 161—76

Higham, N J, 1988 The Cheshire burhs and the Mercian frontier to 924, *Lancashire Cheshire Antiq Soc,* **85,** 193—222

Higham, N J, 1993 *The origins of Cheshire,* Manchester

Higson J, 1859 *History and descriptive notices of Droylsden,* Droylsden

Hindle, B P, 1988 *Maps for local history,* London

Holt, J, 1795 *General view of the agriculture of the county of Lancaster: with observations on the means of its improvement,* London, reprinted 1969, Newton Abbott

Howard-Davis, C, Stocks, C, and Innes, J B, 1988 *Peat and the past,* Lancaster

Huckerby, E, and Wells, C E, 1993 Recent work at Solway Moss, Cumbria, *NWWS Ann Rep 1993* (ed R Middleton), 37—42

Huckerby, E, Wells, C E, and Middleton, R, 1992 Recent palaeoecological and archaeological work in Over Wyre, Lancashire, *NWWS Ann Rep 1992* (ed R Middleton), 9—18

Hulton, W A (ed), 1847 *The coucher book or chartulary of Whalley Abbey,* **2,** Chetham Soc, o ser, **11**

Innes, J B, and Tomlinson, P R, 1983 An approach to palaeobotany and survey archaeology in Merseyside, *Circaea,* **1,** 83—93

Ivanov, K E, 1981 *Water movement in mires,* London

Jacobi, R, Tallis, J, and Mellars, P, 1976 The southern Pennine Mesolithic and the ecological record, *Journ Archaeol Sci,* **3,** 307—20

Johnson, R H, 1985 *The geomorphology of North West England,* Manchester

Johnston, F R, 1967 *Eccles: the growth of a Lancashire town,* Eccles

Jones, G D B, 1979 The future of aerial photography in the North, in *The changing past,* (ed N J Higham), Manchester, 75—87

Jones, G J, and Price, J, 1985 Excavation at the Wiend, Wigan 1982—84, *Greater Manchester Archaeol Journ,* **1,** 25—33

Jones, O T, 1923—24, The origin of the Manchester plain, *Trans Manchester Geol Soc,* **39—40,** 89—123

Jones, R C B, Tonks, L H, and Wright, W B, 1938 *Wigan District Memoirs Geological Survey of Great Britain,* **84,** London

Just, J, 1842 An essay on the Roman road in the vicinity of Bury, Lancashire, *Mem Phil Soc Soc Manchester,* 2 ser, **6,** 409—25

Kaaland, P E, 1986 The origin and management of Norwegian coastal heaths as reflected by pollen analysis, in *Anthropogenic indicators in pollen diagrams* (ed K E Behre), Rotterdam, 19—36

Kenyon, D, 1983—4, Addenda and corrigenda to Ekwall, *The place-names of Lancashire, Engl Place-Name Soc Ann Rep,* **17,** 20—106

Kenyon, D, 1991 *The origins of Lancashire,* Manchester

Kidd, A, 1993 *Manchester*, Keele

Komarek, E V, 1973 Ancient fires, *Proc 12th annual tall timbers fire ecology conference*, 219—40

Kromer, B, and Becker, B, 1993 German oak and pine ^{14}C calibration, 7200—9400 BC, *Radiocarbon*, **28**, 125—36

Kromer, B, Rhein, M Bruns, M, Schoch-Fisher, H, Münnich, K O, Stuiver, M, and Becker, B, 1986 Radiocarbon calibration data for the sixth to the eighth millenia BC, *Radiocarbon*, **28**, 954—60

Lamb, H H, 1972 ∑*Climate, past, present and future, 2: climatic history and the future*, London

Langton, D H, 1898 *A history of the parish of Flixton*, Trafford

Leavitt, T H, 1867 *Facts about peat as an article of fuel*, Boston, 3 edn

Lindsay, R A, 1989 Bogs, in *Nature Conservancy Council, guidelines for the selection of biological SSSIs*, Peterborough

Linnick, T W, Suess, H E, and Becker, B 1985 La Jolla measurements of radiocarbon in south German oak tree-ring chronologies, *Radiocarbon*, **27**, 20—32

Lobban, R D, 1973 *Farming*, London

Lowther, N, and Raines, M, 1987—8 Ancient lowland peat mosses, *Countryside Wildlife*, **1** (winter), 12—13

Lumb, A E, 1958 *A study in town development*, unpubl thesis, Uni Manchester

Lunn, J, 1968 *A short history of the township of Astley*, Manchester

Lunn, J, 1971 *History of Atherton*, Atherton

Malet, H, 1961 *The Canal Duke: a biography of Francis, 3rd Duke of Bridgewater*, Dawlish

May, T, 1903 Notes on a Bronze Age barrow, *Trans Lancashire Cheshire Antiq Soc*, **21**, 120—7

McGrail, S, 1978 Dating ancient wooden boats, *Dendrochronology in Europe*, Brit Archaeol Rep Int ser, **51**, 239—50

McNeil, R, and Nevell, M D, 1989 Great Woolden Hall: a settlement on the fringes of the wetlands, *CBA Group 5 Newsletter*, **57**, 5—6

Mellars, P A, 1976 Fire ecology, animal populations and man: a study of some ecological relationships in prehistory, *Proc Prehist Soc*, **42**, 15—45

Middleton Civic Society, c 1990 *Mineral exploitation at Hopwood, 1450—1850*, unpubl mss, Middleton Archives/Middleton Civic Society

Middleton, J, 1895 *A brief account of the enclosure of Hollinwood Common*, Manchester

Middleton, R (ed), 1990 *NWWS Annual Report 1990*, Lancaster

Middleton, R (ed), 1991 *NWWS Annual Report 1991*, Lancaster

Middleton, R (ed), 1992 *NWWS Annual Report 1992*, Lancaster

Middleton, R (ed), 1993 *NWWS Annual Report 1993*, Lancaster

Middleton, R, and Newman, R M, (eds) 1994 *NWWS Annual Report 1994*, Lancaster

Middleton, R, and Wells, C E, 1990 A research design for an archaeological survey of the wetlands of North West England, *NWWS Ann Rep 1990* (ed R Middleton), 1—6

Middleton, R, Wells, C E, and Huckerby, E, 1995 *The wetlands of North Lancashire*, NWWS **3**, Lancaster Imprints **4**, Lancaster

Mills, D, 1976 *The place names of Lancashire*, London

Moore, P D, 1975 Origin of blanket mires, *Nature*, **256**, 267—9

Moore, P D, 1984 The classification of mires: an introduction, in *European Mires*, (ed P D Moore), London

Moore, P D, 1986 Hydrological changes in mires, in *Handbook of Holocene palaeoecology and palaeohydrology*, (ed B E Berglund), Chichester

Moore, P D, and Bellamy, D, 1974 *Peatlands*, London

Moore, P D, and Willmot, A, 1976 Prehistoric forest clearance and the development of peatlands in the uplands and lowlands of Britain, *Proc 5th Int Peat Congr, Poznan*, Poland

Moore, P D, Evans, A T, and Chater, M, 1986 Palynological and stratigraphic evidence for hydrological change in mires associated with human activity, in *Anthropogenic indicators in pollen diagrams*, (ed K E Behre), Rotterdam, 209—20

Moore, P D, Webb, J A, and Collinson, M E, 1991 *Pollen analysis*, Oxford, 2 edn

Morris, M, 1983 *Medieval Manchester: the archaeology of Greater Manchester v1*, Manchester

Morse, E, 1969 *The development of Irlam*, unpubl dissertation, Padgate Coll Education

Muirburn Working Party, 1977 *A guide to good muirburn practice*, Edinburgh

Mullineaux, F, 1959 *The Duke of Bridgwater's canal*, Salford

Nevell, M D, 1985 *Late prehistoric and Romano-British settlement in North East Cheshire: an aerial survey*, thesis, University of Manchester

Nevell, M D, 1987 Legh Oaks Farm: an investigation of two crop mark sites 1985—6, *Manchester Archaeol Bull*, **1**, 24—29

Nevell, M D, 1988 Arthill Heath Farm, Northern Cheshire: a prehistoric settlement, *Manchester Archaeol Bull*, **3**, 4—13

171

Nevell, M D, 1989a An aerial survey of Southern Trafford and Northern Cheshire, *Greater Manchester Archaeol J*, **3**, 27—35

Nevell, M D, 1989b Great Woolden Hall Farm, excavations on a late prehistoric/Romano-British native site, *Greater Manchester Archaeol J*, **3**, 35—45

Nevell, M D, 1991a *Tameside 1066—1700*, Manchester

Nevell, M D, 1991b The exploitation of Ashton Moss: a brief review of the archaeological and historical evidence, *NWWS Ann Rep 1991*, 21—5

Nevell, M D, 1992a *Settlement and society in the Mersey Basin c 2000 BC to c AD 400: a landscape study*, unpubl thesis, Uni Manchester

Nevell, M D, 1992b *Tameside before 1066*, Manchester

Odgaard, B V, 1988 Heathland history in western Jutland, Denmark in *The cultural landscape, past, present, and future* (ed H H Birks, H J B Birks, P E Kaaland, and D Moe), Cambridge, 311—19

Ormerod, G, 1882 *History of the County Palatine and city of Chester*, London, 2 edn

Paget-Tomlinson, E, 1993 *Canal and river navigation*, Sheffield

Parker, J (ed), 1904 *Lancashire Assize rolls*, 1, Lancashire Record Soc, **47**

Pearson, G W, 1987 How to cope with calibration, *Antiquity*, **61**, 98—103

Pearson, G W, and Stuiver, M, 1986 High-precision calibration of the radiocarbon time scale, 500—2500 BC, *Radiocarbon*, **28**, 839—62

Pearson, G W, Pilcher, J R, Baillie, M G, Corbett, D M, and Qua, F, 1986 High-precision ^{14}C measurement of Irish oaks to show the natural ^{14}C variations from AD 1840—5210 BC, *Radiocarbon*, **28**, 911—34

Peglar, S M, 1993 The mid Holocene *Ulmus* decline at Diss Mere, Norfolk, UK: a year by year pollen stratigraphy from annual laminations, *Holocene*, **3** (1), 1—13

Pennington, W, 1975 The effect of Neolithic man on the environment in North West England : the use of absolute pollen diagrams, in *The effect of man on the landscape: the Highland zone* (eds J G Evans, S Limbrey, and H Cleere), *CBA Res Rep*, **11**, 74—86

Phelps, J J, 1915 A gold pendant of early origin, *Trans Lancashire Cheshire Antiq Soc*, **33**, 192—201

Phillips, A D, 1980a Mossland reclamation in the Manchester area in the nineteenth century, *Ind Archaeol Rev*, **4** (3), 227—32

Phillips, A D, 1980b Mossland reclamation in the nineteenth century, *Hist Soc Lancashire Cheshire*, **129**, 93—107

Phillips, A, 1988 Stockport Castle — final proof?, *Stockport Heritage*, **5** (3), 13

Philpott, R A, 1991 Merseyside in the Roman period, *Journ Merseyside Archaeol Soc*, **7**, 61—74

Piccope, G J (ed), 1857 *Lancashire and Cheshire wills and inventories from the ecclesiastical court, Chester*, Chetham Soc, o ser, **33**

Plant, J, 1870—71 On some logs of oak found in the Irwell valley gravels, *Mem Phil Soc Soc Manchester*, **10**, 169—70

Platt, C, 1978 *The English medieval town*, London

Preece, G, and Ellis, P, 1981 *Coalmining*, Salford

Pryor, F M M, 1984 *Excavations at Fengate, Peterborough, England, the fourth report*, Northants Archaeol Soc Mono **2**, Peterborough

Pryor, H, 1982 *Looking back at Timperley*,

Rackham, O, 1986 *The history of the countryside*, London

Radley, J, 1965 Significance of major moorland fires, *Nature*, **205**, 1254—9

Raines, F R, 1856 *Examynatyons towcheynge cockeye more, temp Hen VIII*, Chetham Miscellanies, **2**, Chetham Soc, **37**

Reilly, J, 1861 *History of Manchester*, Manchester

Riley, D N, 1946 The technique of air-archaeology, *Archaeol Journ*, **101**, 1—16

Riley, D N, 1979 Factors in the development of crop marks, *Aerial Archaeol*, **4**, 28—32

Robinson, W, 1880 *Clifton valley painting*, Manchester City Art Gallery

Roeder, C, 1888—90 Some notes on the Barton section of the Manchester Ship Canal, *Trans Manchester Geol Soc*, **20**, 285—95

Roeder, C, 1889—90 A new archaeological discovery on the Ship Canal at Sticking Island, *Trans Manchester Geol Soc*, **21**, 204—11

Roscoe, H, 1833 *Life of William Roscoe*, **1**, 2, Liverpool

Rothwell, W, 1850 *Report of the agriculture of the county of Lancaster*, London

Rymer, T S, 1992 *Problems, sources, interpretations in historical geography: a study of the formation of the Wigan flashes, Ince in Makerfield 1849—1947*, unpubl mss, Salford Archives

Scola, R, 1992 *Feeding the Victorian city: the food supply of Manchester, 1770—1870*, Manchester

Scollar, I, 1990 *Archaeological prospecting and remote sensing*, Cambridge

Shimwell, D W, 1985 The distribution and origins of the lowland mosslands, in *The Geomorphology of North West England* (ed R H Johnson), Manchester, 299—312

Shimwell, D W, and Robinson, M E, 1993 *Biostratigraphy and pollen analysis of a peat profile from Ashton Moss, Tameside,* unpubl, Greater Manchester Archaeological Unit

Shirt, J L, 1978 *A list of material relating to Irlam, Cadishead, and Chat Moss,* unpubl mss, Salford Archives

Shone, W, 1911 *Prehistoric man in Cheshire,* Chester

Sibson, E, 1846 An account of a Roman public way from Manchester to Wigan, *Mem Phil Soc Soc Manchester,* 2 ser, **7**, 526—58

Simmons, I G, and Tooley, M J (eds), 1981 *The environment in British prehistory,* London

Slater, F G, 1910 The story of Ince in the eighteenth century: extracted from the parish records and other sources, *Journ Chester Archaeol Soc,* n ser, **43**, 55—80

Smiles, S, 1857 *The story of George Stephenson,* London

Smith, A G, 1970 The influence of Mesolithic and Neolithic man on British vegetation: a discussion, in *Studies in the vegetational history of the British Isles* (eds D Walker and R G West), London, 81—96

Smith, A H, 1961 *The place-names of the West Riding of Yorkshire,* vol 2, English Place-Name Society, **31**

Smith, B M, 1985 *A palaeoecological study of the raised mires in the Humberhead levels,* unpubl PhD thesis, Uni College Cardiff

Smith, D, 1988 *Maps and plans for the local historian and collector,* London

Smith, M D, 1988 *About Horwich,* Chorley

Stead, I M, Bourke, J B, and Brothwell, D (eds), 1986 *Lindow man: the body in the bog,* London

Steele, A, 1826 *The natural and agricultural history of peat moss or turf bog,* Edinburgh

Stewart-Brown, R, (ed) 1916—17 Star Court Chamber proceedings, *Lancashire Record Soc,* **71**,

Stockmarr, J, 1972 Tablets with spores used in absolute pollen analysis, *Pollen et Spores,* **13** (4), 615—21

Stuiver, M, and Pearson, G W, 1986 High-precision calibration of the radiocarbon time scale, AD 1950—500 BC, *Radiocarbon,* **28**, 805—38

Stuiver, M, and Reimer, P J, 1993 Extended ^{14}C database and revised CALIB v3.0 C14 calibration program, *Radiocarbon,* **35**, 215—30

Stuiver, M, Kromer, B, Becker, B, and Ferguson, C W, 1986 Radiocarbon age calibration back to 13,300 years BP and the ^{14}C age-matching of the German oak and US bristlecone pine chronologies, *Radiocarbon,* **28**, 969—79

Switsur, V R, 1986 884 BP and all that!, *Antiquity,* **61**, 214—16

Tait, J, 1904 *Medieval Manchester,* Manchester

Tait, J, 1916 *The Domesday Survey of Cheshire,* Chetham Soc, n ser, **75**

Tait, J, 1917 *Lancashire Quarter Session Records,* **1**, Chetham Soc, n ser, **77**

Tate, W E, 1978 *A Domesday of English Enclosure Acts and Awards,* Reading

Taylor, J A, 1983 The peatlands of Great Britain and Ireland, in *Ecosystems of the world,* **4b**: *mires, swamp, bog, fen, and moor, regional studies* (ed A J P Gore), Amsterdam, 1—46

Thompson, F H, 1967 The Roman fort at Castleshaw, Yorkshire (West Riding): excavations 1957—64, *Trans Lancashire Cheshire Antiq Soc,* **77**, 1—19

Thomson, W, 1889 On leaves found in the cutting for the Manchester Ship Canal, 21 feet under the surface, and on green colouring matter contained therein, *Mem Phil Soc Soc Manchester,* 4 ser, **2**, 216—19

Thorpe, D, 1984 *Railways of the Manchester Ship Canal,* Manchester

Tillotson and Son Ltd, 1899—1901 *Post Office Bolton Directory,* Bolton

Tillotson and Son Ltd, 1902—1904 *Post Office Bolton Directory,* Bolton

Tillotson and Son Ltd, 1906 *Post Office Bolton Directory,* Bolton

Tillotson and Son Ltd, 1907 *Post Office Bolton Directory,* Bolton

Tillotson and Son Ltd, 1916 *Post Office Bolton Directory,* Bolton

Tillotson and Son Ltd, 1922 *Post Office Bolton Directory,* Bolton

Tillotson and Son Ltd, 1932 *Post Office Bolton Directory,* Bolton

Tindall, A, 1985 Wigan: the development of the town, *Greater Manchester Archaeol Journ,* **1**, 19—23

Tindall, A, 1986 Some aspects of the history and archaeology of Bolton, *Greater Manchester Archaeol Journ,* **2**, 83—9

Tomlinson, V I, 1956 Salford activities connected with the Bridgewater Canal, *Trans Lancashire Cheshire Antiq Soc,* **46**, 51—87

Tonks, L H, Jones, R C B, Lloyd, W, and Sherlock, L, 1931 *The geology of the country around Manchester and the South East Lancashire coalfield,* Mem Geol Surv Great Britain, London

Tooley, M J, 1978 *Sea-level changes in North West England during the Flandrian stage,* Oxford

Toulemin-Smith, L (ed) 1909 *The itinerary of John Leland in or about the years 1535—1543 pt VII andVII ,* London

Troels-Smith, J, 1955 Characterization of unconsolidated sediments, *Danmarks Geol Unders,* 4 ser, **3** (10)

Turner, J, 1962 The *Tilia* decline: an anthropogenic factor in vegetation history, *New Phytol*, **61**, 328—41

Turner, J, 1975 The evidence for land use by prehistoric farming communities: the use of three-dimensional pollen diagrams, in *The effect of man on the landscape: the Highland zone* (eds J G Evans, S Limbrey, and H Cleere), CBA Res Rep, **11**, 86—96

Twigger, S N, 1988 *Late Holocene palaeoecology and environmental archaeology of six lowland lakes and bogs in North Shropshire*, unpubl thesis, Uni Southampton

Walker, F, 1939 *Historical geography of south-west Lancashire before the Industrial Revolution*, Chetham Soc, n ser, **103**, 1—151

Walker, J (ed), 1986 *Roman Manchester, a frontier settlement*, Manchester

Walker, J (ed), 1989 *Castleshaw; The archaeology of a Roman fortlet*, Manchester

Walker, J, and Tindall, A (eds), 1985 *Country houses of Greater Manchester*, Manchester

Warburton, N, 1970 *Warburton*, Warburton

Ward, G K, and Wilson, S R, 1978 Procedures for the comparing and and combining of radiocarbon age determinations: a critique, *Archaeometry*, **20** (1), 19—31

Watkin, W T, 1883 *Roman Lancashire: a description of Roman remains in the County Palatine of Lancaster*, Liverpool

Watts, W A, and Winter, T C, 1966 Plant Macrofossils from Kirchner Marsh, Minnesota: a palaeoecological Study, *Geol Soc Amer Bull*, 77, 1339—60

Weber, C A, 1900 Über die Moore, mit besonderer Berucksichtigung der zwischen Unterweser und Unterelbe liegenden, *Jahres-Bericht der Männer von Morgernstern*, **3**, 3—23, Geestemunde

Wells, C E, and Huckerby, E, 1991 Macrofossil and pollen investigations at Fenton Cottage, Out Rawcliffe, Over Wyre, Lancashire, *NWWS Ann Rep 1991* (ed R Middleton), 21—5

West, R G, 1968 *Pleistocene biology and geology*, London

Wheaton, C, nd *c* 1987 *Barton and Irlam Moss*, Irlam and Cadishead Local History Society

Wheeler, B D, 1984 British fens: a review, in *European mires* (ed P D Moore), London, 237—81

Whitaker, J, 1773, *The history of Manchester*, **1**, **2**, London, 2 edn

White, F, and Co, 1860 *History, gazetteer, and directory of Cheshire*, Sheffield

Wilson, D R, 1982 *Air photo interpretation for archaeologists*, London

Wimble, G T, 1986 *The palaeoecology of the lowland coastal raised mires of South Cumbria*, unpubl thesis, Uni College Cardiff

Worsley H, 1988 *The dwindling furrows of Lowton*, Wigan

Yates, G, 1895 Bronze implements of Lancashire and Cheshire, *Trans Lancashire Cheshire Antiq Soc*, **13**, 124—42

Index

Airbrushing
with Vince Goodeve

Vince Goodeve

C000119336

Published by:
Wolfgang Publications Inc.
P.O. Box 223
Stillwater, MN 55082
www.wolfpub.com

Legals

Airbrush How To with Vince Goodeve

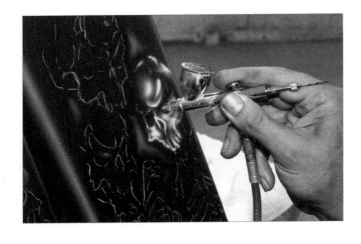

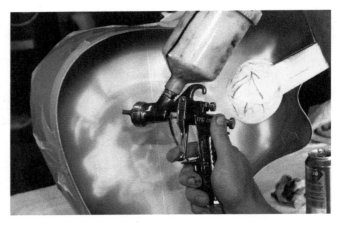

3

From the Publisher

I've known Vince for over fifteen years. And like most working artists, Vince's skills have evolved and improved during those years. In Vince's case, however, not only have his skills improved, but also his understanding of those skills and thus his ability to explain those skills has improved as well.

Like the earlier book we did with Vince, this one is made up primarily of photo sequences designed to show not only the obvious progress of an image from raw sketch to finished art - but the subtle changes that take place as Vince sprays a fine mist of blue on one side of the painting to "cool it down."

Hanging out in an artist's studio taking photos and notes is the fun part of this job. Putting it all together when I get home is the work part of the deal. But when it's all done, I like to think that through teamwork Vince and I have created a book that provides both hard-core, how-to-paint information - and the inspiration that makes both experienced painters, and first-timers head out to the studio or garage and paint and paint and paint some more.

Timothy Remus

Introduction

It was way back in 2005 that I did the *Pro Airbrush Techniques* book with Wolfgang Publications and Timothy Remus. I did nearly all the photos and captions for that first book. In the case of this new book, I wrote most of the captions, and Tim took at least half of the photos. I think this new combination works; good photos and my (sometimes lengthy) captions. With this book I've tried to do more than just give the minimal amount of information in each caption box. I did a lot of these with voice-to-text technology, which meant I could truly express what I was feeling as I was painting, or immediately afterwards.

I like variety, and I paint all types of images on all types of media, on all types of vehicles. The art itself changes by the week. I think all this variety is a good thing. If you want to make a living as an artist, you have learn how to paint all kinds of subjects in all types of situations. I always remind myself that the goal at the end of the job is good art - and a happy customer.

I've tried to explain not only the mechanics of applying paint with an airbrush, but the importance of planning, of always knowing where the light is coming from, and the importance of using subtle applications of paint until I have exactly the color that I want.

Chapter One

Skull Fairing

Hairy, Scary Skull

In this chapter our subject is going to be a skull painted on a fairing for a bagger. It's a fairly stock H-D with a lot of chrome, but the paint work will encompass the entire bike and should really make it stand out and express my client's individuality. For our purposes we will only be concentrating on the single piece, which is the faring. As the rest of the bike is similar, the same basic approach and techniques are used.

Once again, I was grateful to my client who

A skull with a difference. Sometimes you have to reach way back in your head to create something different. It's good to stretch your skills.

gave me basically free reign on this motorcycle, except for the fact he wanted it dark and mysterious, and with skulls of course. Over time, even though this client has given me cart blanche, I still like to get a feel for what he likes and dislikes - whether through the work of other artists he likes, or just by looking at his lifestyle. It may seem that I'm over zealous with this rule, but really, you don't want to give a steak to a vegetarian, and the success of the project really depends on the type of style or feeling your artwork will give. As life goes on, every artist evolves and is influenced by his or her surroundings or different goals.

We want to achieve goals and express our interpretation of the way we see light and color. A project like this is a great time to let loose the different things bouncing around in my creative head. On this bike I wanted to achieve a certain color palette as well as create a lot of depth with abstract shapes. I find it helps me to set out a couple of specific goals with each piece of artwork and then let it happen, not forcing it, or stressing out that it has to be "the best ever."

My advise is to concentrate on what you want to achieve most; whether it's outstanding color, or dramatic lighting or combinations of affects. Don't try to use every trick in your bag on one illustration, chances are it will end up over worked and stiff. That's just my opinion. Everyone has a process that works for them.

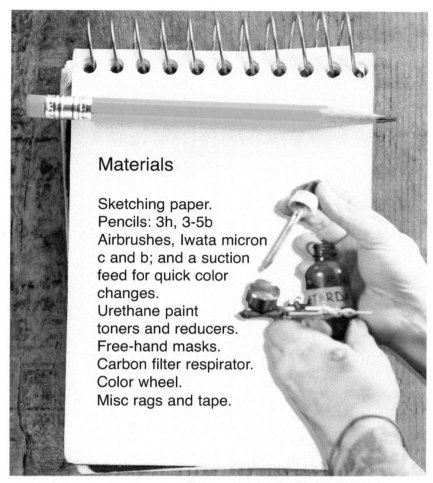

Materials

Sketching paper.
Pencils: 3h, 3-5b
Airbrushes, Iwata micron c and b; and a suction feed for quick color changes.
Urethane paint toners and reducers.
Free-hand masks.
Carbon filter respirator.
Color wheel.
Misc rags and tape.

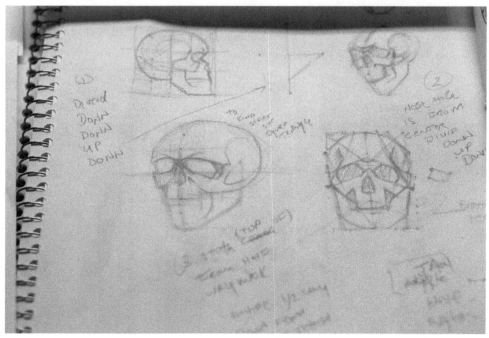

This is a picture of my personal sketchbook. I always sketch certain ideas or concepts before I begin working on a project, it helps me refine my ideas and get into my groove.

Sketches seen here illustrate areas where I want to pay special attention.

Thia particular sketch really shows the light and dark areas in my design.

In this sketch I'm trying to concentrate on the step down above the nose on the brow - reminding myself what will be important in the final images.

This is a basic color wheel. It's a three-primary wheel, a contemporary one would do just as well depending on what your color theory is.

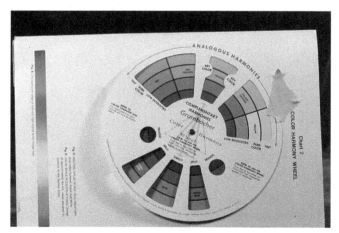

Here, I'm deciding on which colors and what type of harmony I want in this particular painting. In this case it's a split triadic, complementary color scheme.

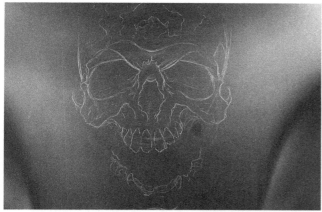

1. For a base I painted it black with a small amount of bronze flip-flop pearl - hard to see without the clear. As usual I sketch out the basic idea with white chalk or you can use a small Conte crayon.

4. I move forward slowly, teaching myself as I go how I like the things in front to look, the way they curve and overlap. I'm just going with the flow here, just painting and paying attention to my light source.

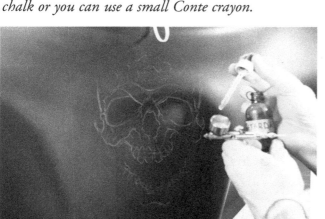

2. I adjust viscosity and check evaporation time so the brush sprays without spitting. Even when set up properly you still need to clean the tip regularly, wiping it with a soft cloth, be careful not to bend the needle.

5. Note the hard edges created with an ellipse template. Then I work freehand to embellish the shapes, softening edges a bit and adding light as I go.

3. Keeping my light source in mind I start underpainting in white. If I'm in the groove I start thinking as if I'm painting with light. This sets up a great bright underpainting for me to apply transparent colors over.

Note from the Pub: The photo sequences used on our 6-panel pages run as shown here.

6. At the mid right of this photo you can see on the curved object the shadow - which suggests a fin-like shape on the top. I often like to show a form by the shadows that they cast giving a nice 3-dimensional look.

Here's a close-up of more of my shaping. I tell students to be sure to leave lights and darks with smooth transitions between them. This will really add to the 3-dimensional look of the things you're portraying in later stages. I also tell them to overlap as much as they can, it's a good way to create opportunities to cast shadows to further the depth of the design.

I finish up one side with my white then I start deepening the shadows using a thin mix of black. You can use straight black or create black by mixing three of the darker primary colors. This way you can "lean" your custom black to a warm side or a cool feel by mixing different amounts of each of the three primaries: blue, yellow, and red respectively.

I continued the exact same process on the other side, happily creating shapes and form. I'm still not sure what they are, but I like them and it has a very interesting flow. If you look to the top left over the eye you can see what I mean by adding drop shadows for depth.

I continue to add details and pump up some of the light areas that I know I want to really dominate with color in the next stages. I can't emphasize enough the importance of a good solid white underpainting - if you're going to use the glazing or transparent system of layering that I use. Without it your colors and detail will be dull and uninteresting, as well as dramatically reduced in saturation.

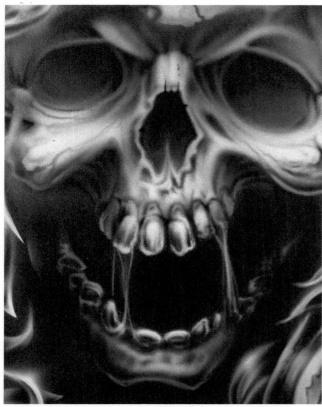

For me this is the fun part. By using colors that are transparent in nature, at this stage you can slowly layer each color smoothly and get a nice bright undertone. I generally work from light to dark. I find you get a sharper image that way.

Here you can see I've added a nice light wash, or thin layer, sprayed smoothly over areas that I want to be blue or cooler in color. Build this up slowly, and leave some areas lighter and darker for a more dramatic affect.

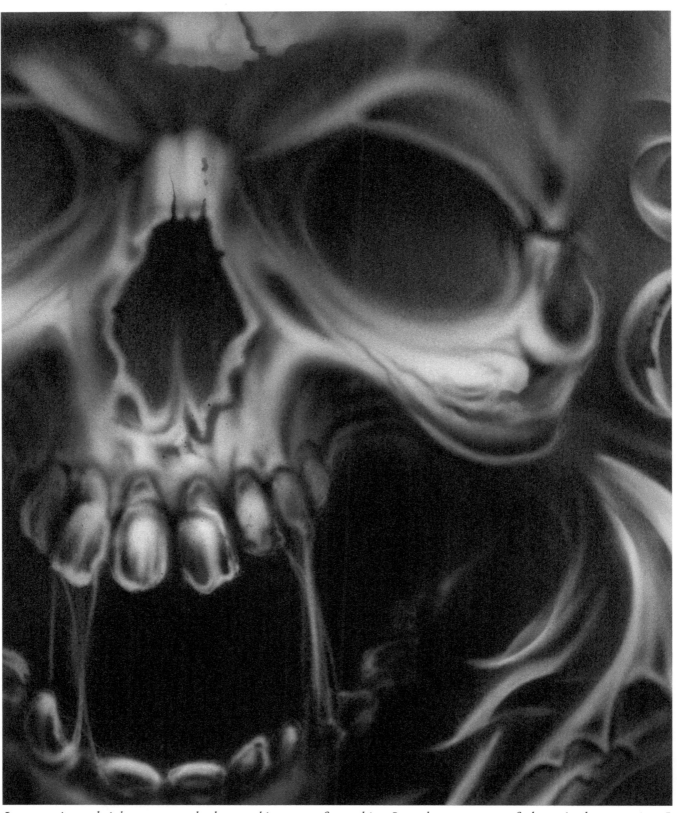

I now go in and tighten up my shadows, taking care of any thing I see that seems out of place. At the same time I add small details and take care of any little spits or spatters that might've been created during my thrashing in the initial stages of creating. It's important to clean up your painting so you end up with a nice clean, crisp image. It's easy to skip the step, but it doesn't take long and will take your work to that professional level.

Now I step back and look at my painting from a distance. I walk away for a while and come back. That's usually when I remember my goals at the start, and see things that I want to adjust. In this case, I wanted to add a sense of depth to the brow line and really get a feel for the way the nose is recessed into the that area.

As I did earlier I take a short break and come back to it with a fresh set of eyes before making my final touch ups. Sometimes I look at the subject backwards in a mirror. This gives me a different perspective so I can find little problems that were hidden.

Chapter Two

Goalie Helmet

Spirit of Radio Tribute Goalie Mask

This project was for a friend of mine who is a DJ and program director for a big local rock station. We have done several projects together over the years, from guitars to promotional posters - and goalie masks of course. Working with Jay is always a lot of fun and we connect at a level which makes the process very entertaining. We can talk for hours without realizing how time is flying by. He has even gone so far as to commission me to do a sleeve tattoo for him that should be quite a blast. The idea behind the design on this mask was to incorporate the forties and fifties

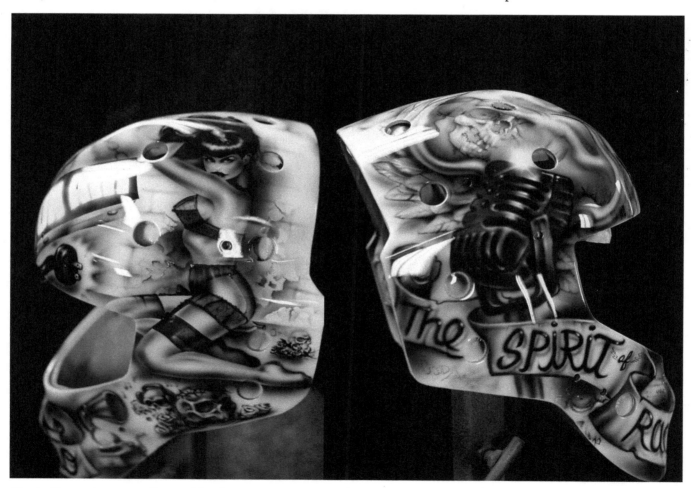

Bettie Page, vintage radio, and hockey... not an easy combination to put in the blender in hopes that the end product will be a cohesive design. I think though that it all came together in the end - all done in black and white, which adds to the vintage feel of the art.

imagery with the nostalgia of radio. Jay also likes Bettie Page and I felt the old microphone really provided a link to that era.

The combination of light contrasts against darker tones, plus soft edges versus hard edges, gave me a dynamic design. I've done quite a few goalie masks and I have to admit they're not my favorite thing to work on - because I'm lazy and masks are a challenge as there are so many holes and curves to deal with. Doing a strong design and making it still look cohesive is tough. But I will continue to paint them because it looks so cool when they're done, and you see the finished product out on the ice.

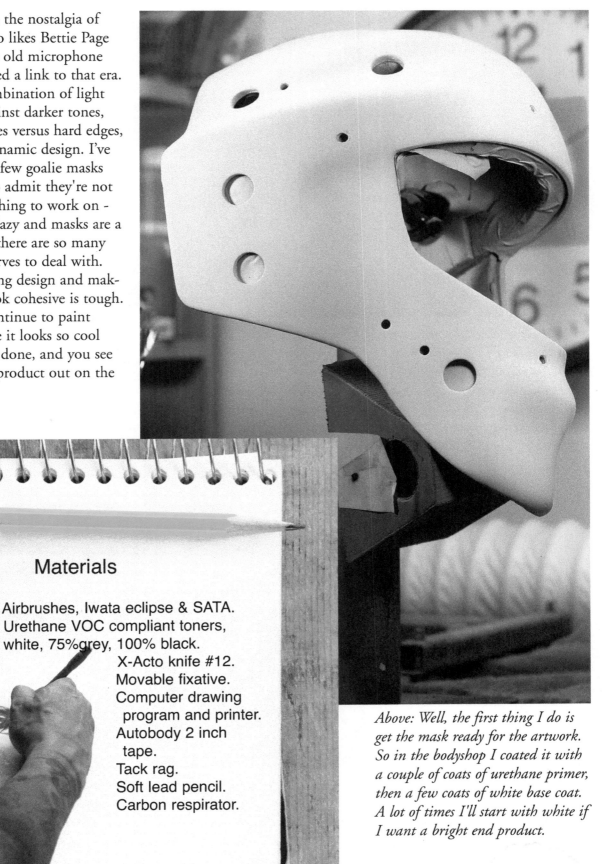

Materials

Airbrushes, Iwata eclipse & SATA.
Urethane VOC compliant toners,
white, 75%grey, 100% black.
X-Acto knife #12.
Movable fixative.
Computer drawing
 program and printer.
Autobody 2 inch
 tape.
Tack rag.
Soft lead pencil.
Carbon respirator.

Above: Well, the first thing I do is get the mask ready for the artwork. So in the bodyshop I coated it with a couple of coats of urethane primer, then a few coats of white base coat. A lot of times I'll start with white if I want a bright end product.

15

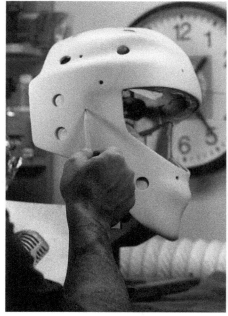

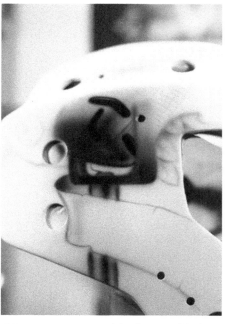

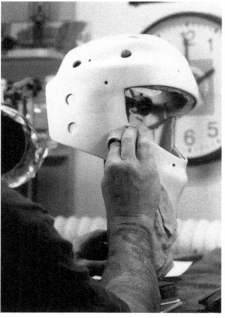

Here, using a soft lead pencil, I'm starting to design the shape of the iconic old microphone. I'm just aiming to put in the general shape and determine placement of the light and dark patterns that make up the repetitive grill pattern.

If I make a mistake I simply take a rag and some final wash, which is a very mild reducer, and wipe the pencil lead off and start over. If you don't make mistakes you can skip this step.

Using my 100%, black I establish my darkest dark, which is in the shadow of the microphone. All I'm doing is looking for general shapes in light and dark, I'll worry about the details later.

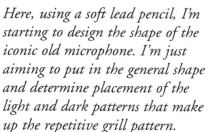

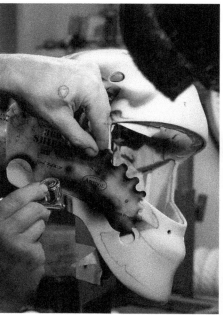

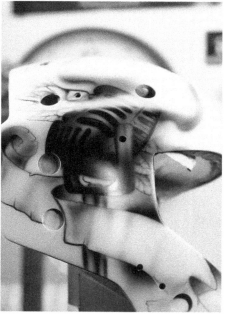

Using the straight black, I continue with dark areas. As I do this I keep flicking my eyes back and forth between the painting and a reference photo of a mic. In this way I can quickly see any differences.

Here you can see me using a loose mask to get the sharp edges I want on the banner that flows behind the microphone. It's a good way to get a nice separation of light and dark without having to do a lot of static masking.

Now you can see the resulting hard edge and how it separates the nice clean edge from the darkness. This also illustrates how some simple shading in the half tones and embellishments can give the banner a three-dimensional appeal.

16

I start to suggest feathers in the wings that frame the mic. I do this freehand, thinking of my light source, trying to create soft shadows that make it 3-dimensional.

On the other wing I work on pushing the banner forward with a dark background. I also add a light source to the mic, with a cool white to show the forward planes of the object.

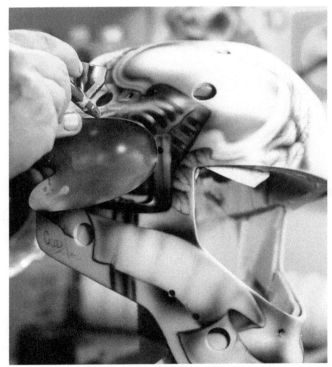

If I need a sharp edge on a highlight, I use my trusty mask template that I cut out from a piece of acetate. These loose templates have saved me hours and hours of hard taping with masking tape, or cutting things out of masking film.

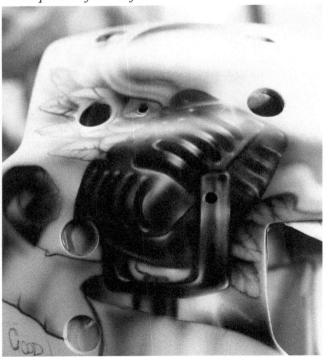

Here you can see the results of the loose mask on the highlight areas of the grill on the microphone - nice clean edges without a lot of time spent. Just the way I like it!

I airbrush the lettering in freehand with a relaxed font and do a final once over checking for any adjustments. OK, this side of the mask is done.

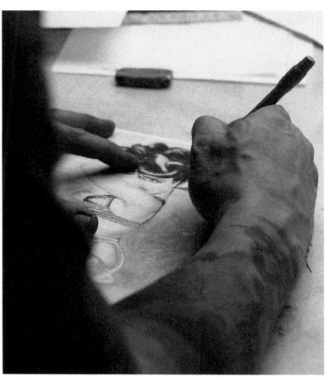

I prepare for the other side of the mask by printing out, on a regular sheet of paper, the figure that I want to airbrush. Then I use the X-Acto knife and cut out the silhouette of her figure.

Using a workable fixative, I spray the back side of the figure I cut out, let it dry 5 minutes and then apply the template to the helmet in a position I find appealing.

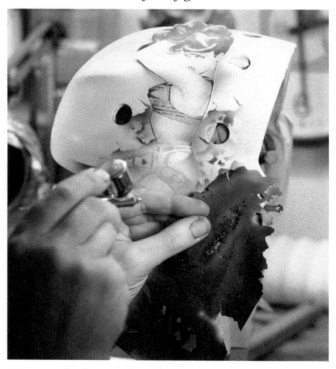

Using a funky template that I made I create an interesting pattern behind Bettie, making sure it's dark enough to leave negative space behind after I pull away the paper template I glued down in the first steps.

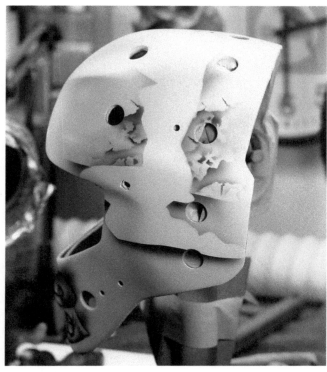

I carefully remove the paper template I placed there in the initial steps and save it. At this point I'm happy with the results - the figure is plain to see and is differentiated from the darker background.

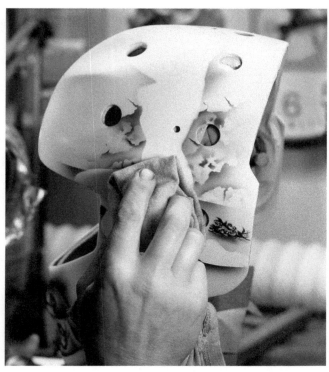

Here I use a soft cloth and some gentle primer wash to remove any glue or residue that might have been left behind from our template. If you skip this step you'll see an ugly pattern anywhere there's overspray.

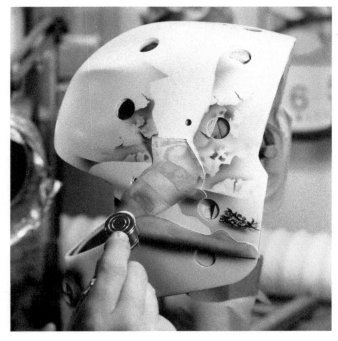

Using the paper template we made at the beginning, and an X-Acto knife, I carefully and strategically cut out parts and use them as edge masks to separate the different forms of the human figure.

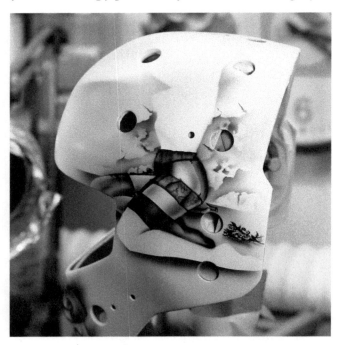

Remove the desired area you wish to render. Slowly build up pigment to modulate the light from the shadows. Remove paper template which created the edge you just sprayed, exposing the negative shape.

19

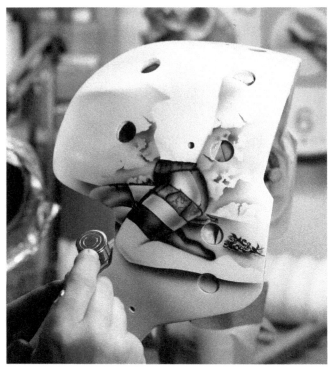

Here you can see on the lower leg where I've removed the template and sprayed the dark shadow edges.

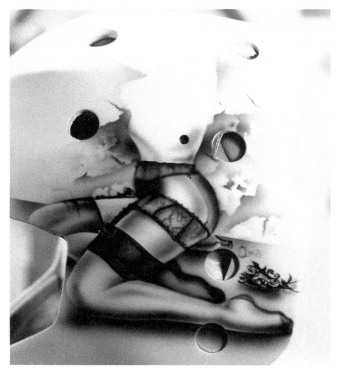

You have to pick your light source - the shading techniques and half tones will create 3-D forms from the shapes you make. You can take this to an extreme, it all depends on your enthusiasm and the budget.

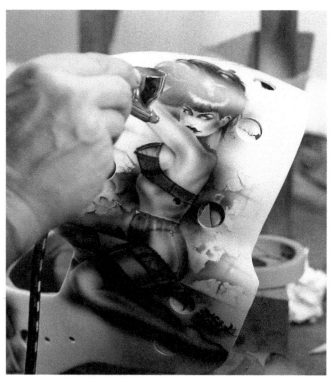

I carry-on, finishing things as I go. Usually it doesnt take very long to complete a figure like this. I find doing the face freehand, rather than cutting out little noses and lips, is much more convincing.

Everybody has their own style, here you can see my freehand face and eyes. I like to give them a sultry look. I also feel that minimal shading in the face is much cleaner and looks more feminine.

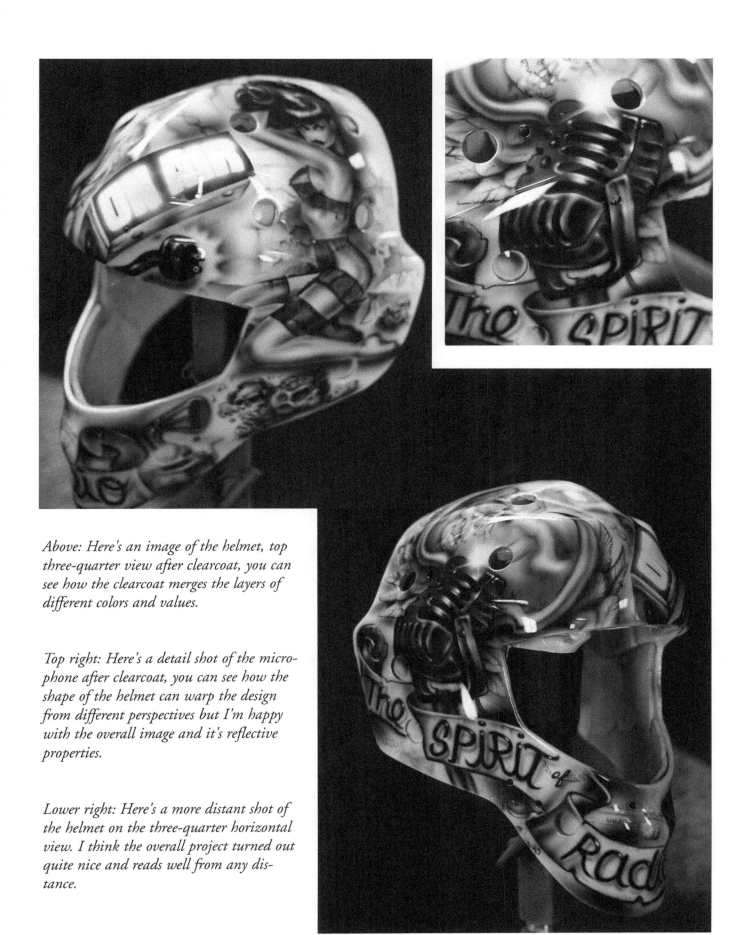

Above: Here's an image of the helmet, top three-quarter view after clearcoat, you can see how the clearcoat merges the layers of different colors and values.

Top right: Here's a detail shot of the microphone after clearcoat, you can see how the shape of the helmet can warp the design from different perspectives but I'm happy with the overall image and it's reflective properties.

Lower right: Here's a more distant shot of the helmet on the three-quarter horizontal view. I think the overall project turned out quite nice and reads well from any distance.

Chapter Three

Leering Wolf

Creating a Big BAD Wolf

Some projects are just meant to be created, they are fun and seemingly effortless and turn out well. They make both the customer and me happy, they don't happen very often, but once in a while it's like I wasn't even there! The project just happens by itself and I'm on autopilot.

The client approached me at a car show and invited me to see his ride. I was quite impressed, the build was done extremely well and you could eat off the Italian leather seats. It was that sweet. I was a little bit anxious to take the project on because the car was so clean and I didn't want to

The small 3 dimensional evil wolf's head was the inspiration for this illustration. I obviously took the wolf's basic profile as the starting point, and added details and lots of color. Note the light source on this one, always something to consider and keep consistent.

make it vulgar by adding something that didn't work with the colors and the lines of the car. But after some thinking and a bit of coaxing I decided to take his idea to the design point. Using a napkin and a pen, he and I came up with a few ideas in the beer tent. He really liked the direction it was heading so I decided to go ahead and come up with more detailed drawings for his approval.

Over time I've found three things that are most important to the success of any project. The first one is to listen to the client. And the other two really don't matter as long as you listen. I've also learned there are two kinds of clients. One can visualize what you're talking about with just rough sketches and small concept designs. The other type of client is the one who has a hard time picturing things until they are more clearly defined, in which case I make sure that I do a detailed rendering for the client's approval to avoid any blank stares or problems later. It also helps me greatly to come up with things on the fly, on paper, rather than having to solve problems when I'm actually doing the finished artwork.

Materials

Standard paint palette consisting of blue, red, green, yellow, violet, orange, white, and black. Drawing materials: pencil, eraser, paper etc. Airbrushes, Iwata & SATA

The customer supplied me with a hot rod wolf head. I'm assuming it was the gearshift knob, obviously this is the type of style he would like to see in the finished art. So my brain filled up with ideas from our conversation. Once in the studio I came up with several different ideas which I presented to him at our next meeting.

After prepping the car and taping her up, I was ready to start sketching the design straight onto the trunk lid.

Using a mock black - by combining the three primary colors - I begin carving out the dark shadows. Mixing your own black lets you push it in a cool or warm direction, depending on your overall color harmony.

Here you can see I'm plotting out part of the sketch with pencil. As you can see I sprayed a nice fade of white base coat over the back portion of the trunk lid to act as my underpainting.

Here you can see me sketching in some of the background ideas that I come up with, in this case it's a taxicab, viewed from the backseat out.

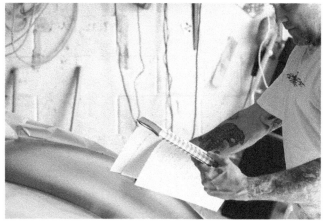

Here's a better view of the white basecoat area that I sprayed onto the hood. (It's not a prerequisite or even necessary to make a grimacing face while sketching, heck I guess I'm just concentrating.)

After blocking in the background with a warm grey of a lesser value I start to build the warm light side of the face.

The next part is kind of bouncing back and forth representing the shadows side with cool colors and the light side with warm colors always keeping my light source in mind. In this case we're going to have two light sources, the first being the cool outside diffused light from the night sky (blue) and secondly the warm direct light from the glowing cigar.

The next job I have is to spray in the atmosphere around my figure with blue, which will help me combine the bluish color into the final purple violet color of the car. I should end up with a seamless blend.

I now zoom in with my mind and focus on the details of the piece, saturating colors where they need to be richer and smoothing out any edges or contour lines that just don't look right. I'll usually step back now and then, about 10 feet away, throughout the process of painting, to keep things in perspective.

I continue to add details. Moving around the piece looking for little bits of eye candy that I can add, like the numbers on the fare counter and his buttons and ring.

Here's a good close up to see different elements I've added - textures, reflections and shadows help make this cartoonish figure appear more realistic. Note the bloodshot right eye.

Checkout the cool little creature crawling out of his left eye, the tiny shadows on his whiskers and the wetness of his mouth and gums. They aren't big changes in themselves, but they all add up in the end.

Here's the finished trunk lid. It blends nicely with the rest of the car. Of course there will always be the debate, whether a car should have murals and artwork, graphics - or just plain clean color. But in the end . . .

. . . it's the customer who needs to be happy, that's the bottom-line.

Chapter Four

Murphy's Big Twuck

Pros & Cons of working on a Large Scale

I received a phone call from my client, who expressed to me a yearning for some artwork on his truck. He had been in the trucking industry for quite some time, let's just say he was a veteran. All the reality television shows don't have anything on them that Murray hasn't done twice, in

,my opinion of coarse. As we got deeper into our conversation it became clear to me that Murray had owned several trucks through the years and this one was in all intents and purposes his last and his favorite. Later, in confidence, his wife jokingly told me, "that's what he says every time.",

For this project I chose to go with a double-split complementary color design, with blue-purple complemented by red-yellow. The other thing I did differently is to have more than one focal point which kind of breaks the rules of composition. But I think they're far enough apart to hold their own. I used larger equipment to make bolder strokes, and applied color more rapidly. You just can't do this with a conventional airbrush.

Here's the truck all wet sanded with 800 paper, taped up and covered with masking paper and poly. I use a system of rolling scaffolds and tables to get around and make it easier to be comfortable and work at different levels. I don't do a lot of these large vehicle jobs, frankly they're very labor-intensive with the up and down all the time. Yet I sure love how they look when you're finished and the freedom it gives me to add detail on such a large canvas.

When I arrived at his location I realized this was no small operation. There were several large transport trucks, tractor-trailers everywhere and a lot of heavy equipment. Here I met Murray, a fairly large and definitely gregarious man. He shook my hand and most of my arm. I began to have this funny feeling that this truck was not small pick up. As we rounded the corner of his large garage /hanger he pointed with much pride in the direction of his RIDE. Before me was beautiful and sparkling black tractor-trailer that I could live in full-time anytime (although I don't think my beautiful wife would be cool with that idea).

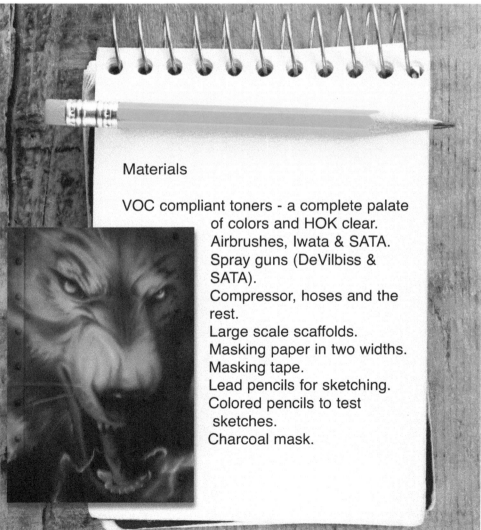

Materials

VOC compliant toners - a complete palate
    of colors and HOK clear.
Airbrushes, Iwata & SATA.
Spray guns (DeVilbiss &
SATA).
Compressor, hoses and the
rest.
Large scale scaffolds.
Masking paper in two widths.
Masking tape.
Lead pencils for sketching.
Colored pencils to test
 sketches.
Charcoal mask.

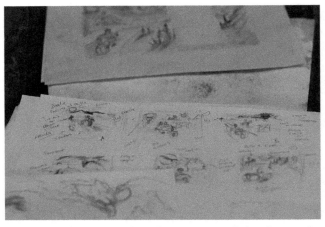

On this job I remember doing a ton of sketches and workups to get to the final design. It makes me feel confident before I begin and I left the customer secure that I have a firm grip on what I'm doing.

A lot of these drawings are just thumbnail sketches and small color compositions. I don't spend a ton of time until I find the one he likes, and then I expand upon that one to provide a more finished example.

Using my medium-sized Iwata touchup gun, I use white base coat to emulate a foggy background by varying the distance and speed of my gun.

Using a lead pencil, I sketch out the proportions of the subject. You can also see I kind of lightened up some of the areas that I know are going to be brighter - I just do that instinctively.

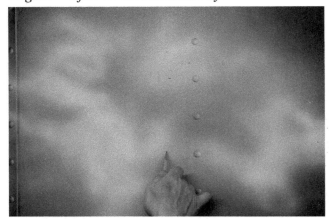

Here I think you can see what I'm trying to accomplish, just the main shapes that I can use as references, nothing too detailed because I'll probably change some of this as I go.

I dive right into it with my Eclipse gravity feed airbrush. Using a mixture of red, a little green and a little blue I come up with a reddish brown tone, kind of neutral to the overall design.

I continue the process, and as I explained in other parts of this book, I move in close for detail and back out for softer spray effects. It's a personal preference, but I tend to start with the eyes on anything that has eyes. I like to capture the expression before I move on. I know this is completely opposite of the way some other guys paint - goes to show you there are many ways to approach things.

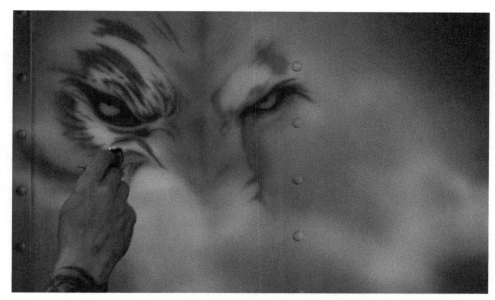

Here's a good example of how I approach things that are misted-in or fade to nothing. I know I could just spray white paint over the structure that I already painted, but I like to indicate shapes in the shadow or mist vaguely. To me they look more convincing that way.

I've decided to put a very dramatic Shadow-side to this wolf's face so I can play up the forms on the shadow side created by the secondary light source, (you won't see this effect until very near the end of this segment). I think you need to have dramatic contrast in your shadow areas with this type of artwork, to grab the viewers attention and anchor the design.

The very dark areas look kind of clunky right now, but it will make sense, trust me. I laid in some white underpainting of mist areas where I think I'd like to have dark shapes show up in the distance, to give my foreground images something to relate to.

You see where forethought pays off - the dark area on the side of the face becomes a wicked shape indicating fur. The dual light source gives a full look, capturing the entire spectrum from warm to cool. Highlights on the right illuminate dark areas without a lot of detail.

Here I'm using white with a few drops of blue to cool it down and suggest mountains in the background, as well as storm clouds and lightning. I also added some purple pearl with a larger spraygun over the white.

Here's the mid ground and the fore ground. By keeping the value lighter in the background it gives the illusion of distance. I wanted reds and purples in the foreground, it gives an autumn feel and ties the colors together.

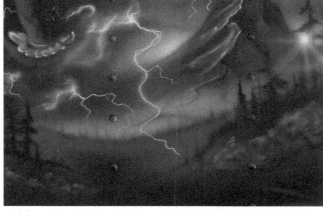

Although sometimes overused, I think the lightning gave me an opportunity to add depth and energy. As well as the glints of light from behind the mountain - some things just work and are timeless.

RIGHT: A close up of the color patterns and the actual brush strokes. Even though the airbrush never actually touches the surface the character of the lines and strokes you create is very similar to that of a traditional paintbrush depending on your personal technique.

Here it would appear that I'm having a conversation with the wolf or perhaps trying to climb up beside the vehicle, But in reality I'm tacking off (a term used in the industry for using a sticky cloth and wiping lightly over the surface to remove any dust and lint).

Here's the finished face of the wolf. You will notice that I've adjusted saturation of the colors. By that I mean richening some of them by applying more layers of color. I fiddle with the details, but all in all I like the results of this project and I had a happy customer who paid me right away - always a good sign.

Chapter Five

Skull Tank

Some Artists do Flames, and some do SKULLS!

In this chapter the canvas is a motorcycle and I will be concentrating on the left side of the tank, it may look similar to the chapter with the skull faring and that's exactly the point. The client's son purchased a motorcycle and liked his father's paint job so much he asked me to do something similar - but not exactly the same. I decided to paint his tank with a different procedure; mainly in the way I approached the lighting. This time I went from dark to light. More

The finished fuel tank. I try to make sure that the image fits the area on the bike geometrically. Basically I'm putting round objects in round spaces, while trying to flow the other shapes through the longer portions of the tank.

like an oil painting; starting with the dark tones and building on top of them, only applying a very small amount of gray to give me some reference for the imagery. This will be a great example to show how to approach the subject from two different perspectives and in the end there are some definite differences that you can see. It's all a matter of personal choice. That's the beauty of being an artist, you have so many techniques and ways to express a concept or idea, the choice is yours.

Some techniques work more successfully with specific media. This is something you need to discover on your own - what works best for you. What was nice about this project is that I had quite a bit of freedom to try some things I've been thinking about doing, as I'm often influenced by objects in my environment.

The job and the parameters set by the client dictate how much pre-planning I do before I start. Since I didn't have strict design elements for this job, I like to kind of learn as I go. More specifically, I can put my brain in a free flowing consciousness where I leave it up to my instincts to let my hands move when my eyes tell me how something should look.

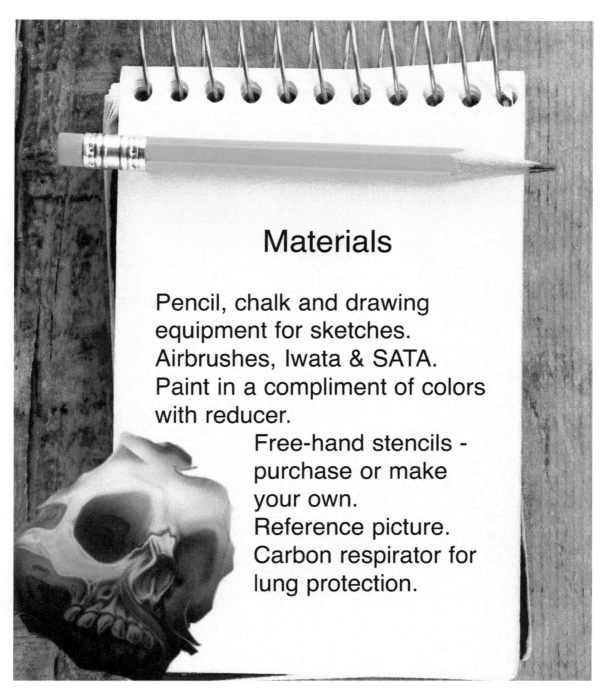

Materials

Pencil, chalk and drawing equipment for sketches.
Airbrushes, Iwata & SATA.
Paint in a compliment of colors with reducer.
Free-hand stencils - purchase or make your own.
Reference picture.
Carbon respirator for lung protection.

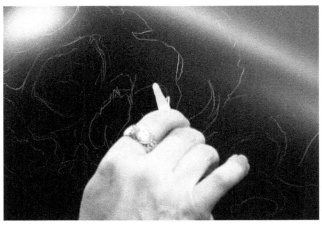

I like to start with a soft chalk crayon and lightly sketch out my design using only a minimal amount of lines.

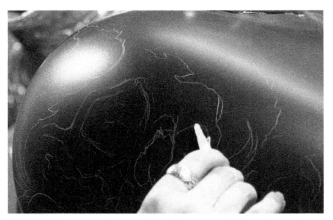

I then move to the smaller shapes and details within those which I have already created.

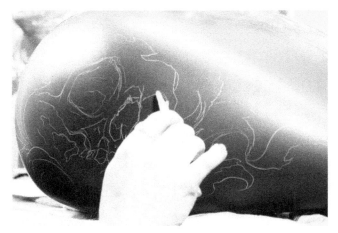

I finish off by lightly drawing in the rest of the design. How elaborate you make these sketches is up to you.

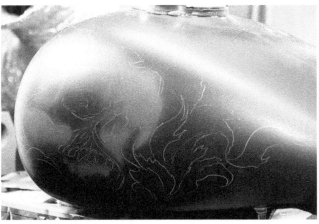

I mixed up a warm, neutral gray using white and a little purple and a little transparent yellow or orange to knock back the chroma and make it a more natural earth tone.

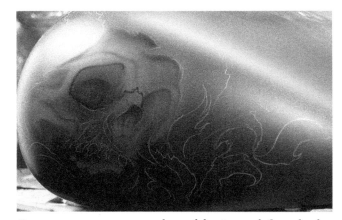

I use a transparent purple and begin to define shadow areas, keeping it fairly loose. I want to see how much saturation I can get in dark shadow areas without adding any more dark color to my mixture.

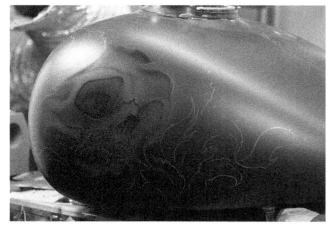

It's hard to see, but I sprayed a small thin layer of purple pearl into my shadow areas and over some of the background - very lightly so it's barely visible, but the clear will make it pop later.

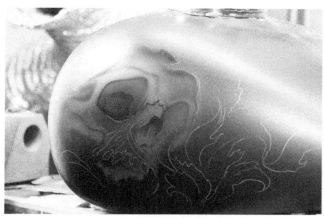

I mix up a halftone, a color slightly lighter than the purple base, and begin to build up my light areas. As I begin I notice that it isn't light enough - no problem - I can adjust it.

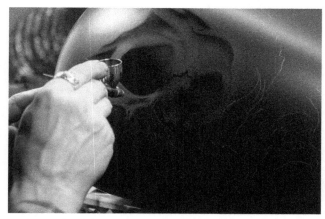

I slowly build up my smaller shapes, it takes a little while because the mixture is thin, with a slow reducer. Using this slower drying mixture allows me to avoid a lot of spitting and tip-dry on the airbrush.

Here you can see me feverishly adding white and a little bit more purple to brighten things up.

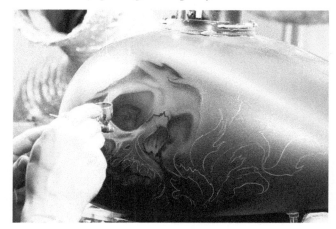

I continue this process until I reach the desired texture forms and under painting.

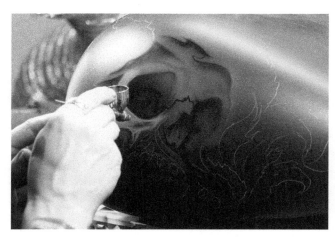

Much better! I now begin to define my light area mid tones with the warm purplish mixture, always keeping in mind the direction of my light source.

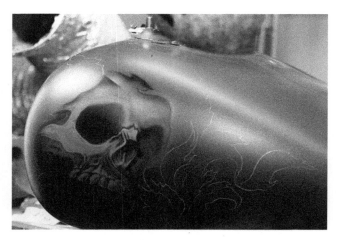

This shows my under painting completed for that section, and you can see where I've been experimenting with some of my re-darkening color.

I mixed up a transparent purple and a few drops of transparent red and began to build upon and enhance the darker areas, working slowly with the same type of slow reducer used earlier.

I switch back and forth between my light shadow tones. This is where you use your personal judgment as to the best way to express the shape and form of your subject.

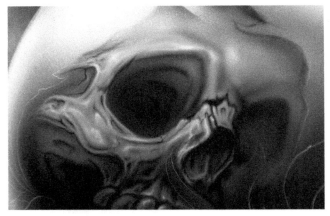

I lightened that mixture even more, with more red than purple, and go back in and work off the edges of the dark shadows. I'm creating a transition into the light, kind of a step from darkest to lighter.

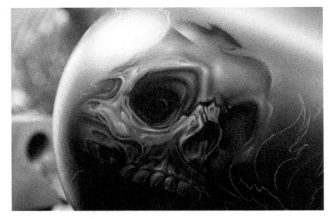

Here my light side and shadow side are pretty much finished except for tiny details. Note that the colors I used early are only suggestions, you can add any color you see fit, that's what makes us all unique.

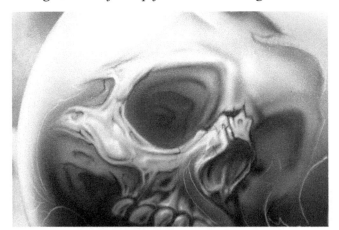

Now I go back in with white mixed with a few drops of orange. My goal is to increase the illusion of height in my shapes and form, again keeping in mind the direction from which your object is lit.

Here I'm using a white base with a little blue to cool it down. I wanted some of the leaf-like objects to pass in front of the skull - and create additional depth and shadow.

I used a free-hand template to create the hard edges on the leaves. The biggest thing to remember is the play of light and shadow. I also wanted to have the leaves surrounding the skull create a cool cast in contrast with the warmth of the skull face.

The last thing I did was to put a slightly warm, secondary light source under the leaves for a little variety. Now's the time to step back, take a break, and see if the painting needs any adjustments. Then I'm off to the clearcoat, booth. I think the clear will really make that purple pearl pop, and help it complement the other colors in the design.

Chapter Six

Betty Boop

An Icon Comes to Life

This project really illustrates how diverse you must be in your airbrush skills to be able to carve out a living in this business full time. Some may disagree and say you should specialize in one particular genre, but I'm of the school of thought that believes that being diverse keeps your doors open and your belly full year after year. I rarely turn down commissions because the challenge of designing and rendering *anything* well is still challenging and fulfilling to me.

This is my final artwork, the version the client really liked - finally.

The folks I did this paint job for are very sure of what they like and don't mind letting you know if the end result is not exactly what they had in mind. To be completely honest with you, I did this job three times before it was the one she really loved. Thinking back, the reason this happened was my fault; for not sitting down and actually talking directly to the end client. As I've said before, this is something I always do. And on those rare occasions when I stray from my protocol it almost always turns around and bites me in the bum. The real lesson of the story is not the fact that I did it three times, but the fact that I took the time to make the job right, and parted ways with a happy customer and I was a happy artist. A year later the same customer commissioned me to do a large mural in their incredible garage. That would not have happened if I took offense and walked away. On my first attempt, That probably would not have been something I would've done 20 years ago, but I now understand that my job is to create what the client wants not what I want. I still am completely in charge of composition and color choices as well as the craftsmanship, I just try to make sure the idea in their minds stays true and comes through. So, As far as my thoughts

go when designing this Betty Boop scene I wanted to keep everything fairly lighthearted. even the skulls aren't sinister but kind of happy. There were certain details on the Betty that were personal to the young lady, for example the color of the dress and the horns were part of her criteria. I picked my colors to give the feeling of an early morning sunrise in a large city, New York perhaps. That way I can have the cool blues and purples of the shadow areas and the nice fiery red and orange of the morning sunrise raking across the landscape.

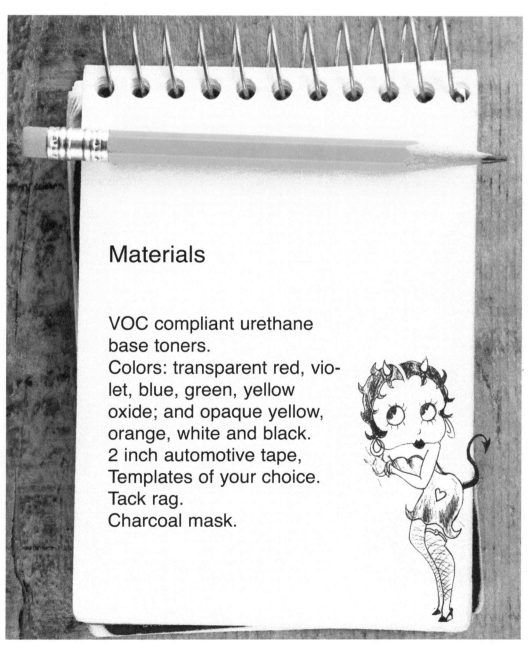

Materials

VOC compliant urethane base toners.
Colors: transparent red, violet, blue, green, yellow oxide; and opaque yellow, orange, white and black.
2 inch automotive tape,
Templates of your choice.
Tack rag.
Charcoal mask.

Here's the sketch that I submitted to the client after several false starts on the project.

I use a white pencil, which is kind of like chalk, to sketch out the main elements, which gives me kind of a road map of the project.

Sometimes I use simple shapes to visualize the bulk of the subject I'm going to draw. This gives me an idea of how much space my subject will occupy.

I proceed to draw rough details in the shape so I can be sure of balance, and get a feel for what it's going to look like when I start to apply the paint.

I use a tack rag to gently wipe the surface, removing the dust - I don't want that dust to remain and mess up my clearcoat later.

I decide where I want to make my focal point, which will be the point of highest contrast, and bam, hit it with straight white paint.

I carry on with white, planning where I want contrast with dark shapes. The most important rule in painting is - you need dark to show light and light to show dark.

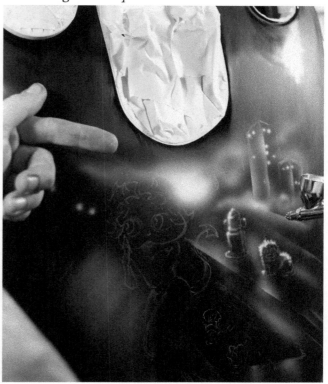

This shows how I save time by creating shadows, by leaving out the areas I want to remain dark, like the fire hydrant and the small puppy.

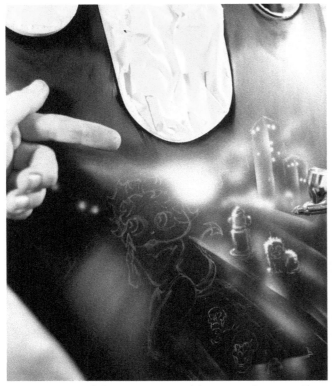

I continue modulating areas and begin working on Betty. Be sure to have your brights intense enough to make the transparent colors that are going to follow really pop.

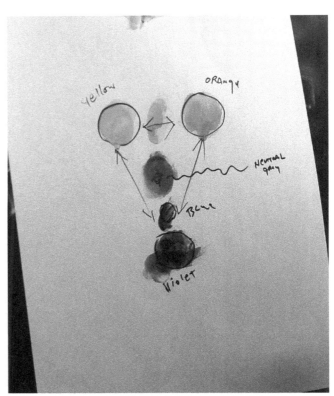

I do a quick test with my pigments to show that it's going to work. I use a split complement type of harmony.

I start working on the background, experimenting with different color combinations in the warm sunset.

Satisfied with the colors I've chosen, I darken the outside of the design with magentas and transparent reds.

I start testing the waters with my shadow colors in the apartment building. I decided that I like their vibration against the warm oranges.

More moving around the design adding blues where shadows will look nice, taking advantage of places where I know it will vibrate against the red and orange.

Using opaque yellow and transparent magenta, I add some intensity to the sunset and halftone light areas.

Using a very dark mixture of purple and black I create the darkest dark in my painting. The plan is to make this my area of highest contrast or focal point.

I use the same dark color that I used in the hair to suggest the buildings and background in the distance.

Here I bounce around the design working the cooler areas, for example the pantyhose and maybe the shadows on the little puppy. Awwww a cute lil' bastard.

Using magentas and purples I begin to work the shape of the body and start detailing the lips and tail.

This is where it gets tight. I thin my mixture a little more and switch to my Iwata micron. With my dark color I work in details like eyelashes and stockings.

The finished piece without any clearcoat. Here are a few things I think were successful in this design: The way the cool and hot colors blend together giving it a very full look. I also like the way her face and skin looks so bright and clean, almost bathed in moonlight. And I like the way the bright red tail sings against the large dark area I left in the background. There are also things that I would change, but it's best to walk away and let the piece be what it is rather than give it that really labored look.

Chapter Seven

Skull Bags

Simplicity is the Key

I consider myself a commercial artist and as a commercial artist I get paid to create and paint things on a deadline within certain parameters. The customer expects a visually predetermined "look" or content, by contract.

I know that sounds pretty straightforward ,and obvious, but let me explain a little further. As an artist with a passion to create and express myself more dynamically, I think it's quite natural, as I progress through my career, to want to become more of a fine artist and utilize my full knowledge of color, perspective, and composi-

The finished design is simple rather than complex with just enough contrast to make it pop. And the two bags are not identical, which is why they call it art. If people want absolute perfection and two mirror-images, let 'em buy vinyl.

tion. I think the same thing happens to most of us, and that's great! But there comes a point, at least for me, when the amount of effort and time needed to create these complicated pieces outgrows the budget, as determined by my contract for a particular project - the amount of money my customer and I agreed upon. For this job, I decided to give my head a shake and remind myself that it isn't what *I want* painted on this bike that's important,

That's when I realized it was the *look* that was important. What I mean by that is, the style of art work that I've done in the past, art that was seen and responded to by the customer. Work I've done that turned them on, that's what they desire. It isn't always the most super detailed, wicked fine art, sometimes it's the more simplistic styles they like. So for this project I decided to save my finer art studies for personal projects, or projects where the customer has an understanding of the time necessary, and stick with a strong, simple powerful design.

Here you can see me starting to sketch out the first little skull with my white conte pencil.

I continue sketching, the bandanna is optional as it was extremely hot that day.

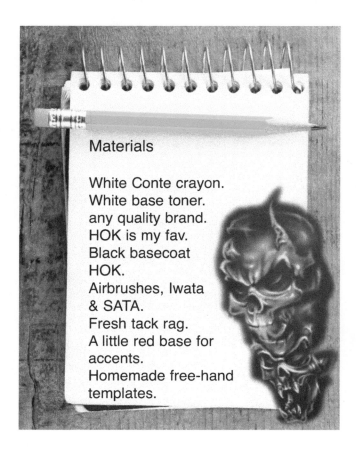

Materials

White Conte crayon.
White base toner.
any quality brand.
HOK is my fav.
Black basecoat
HOK.
Airbrushes, Iwata
& SATA.
Fresh tack rag.
A little red base for
accents.
Homemade free-hand
templates.

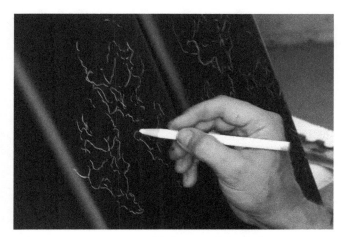

Here you can see a small cluster of skulls beginning to evolve. I'm keeping the two sides similar but not symmetrical.

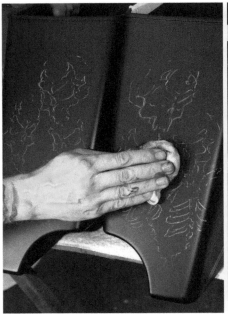

This is where the tack rag comes in handy. It will remove any excess chalk dust and grunge you'd rather not airbrush over.

Using my suction feed airbrush, I apply a layer of red basecoat fading it off at the edges to insinuate glowing areas of doom and intrigue.

It's hard to see in the photo, but I just hit the centers with the red basecoat a little harder and it seems to make it appear to be glowing centrally.

This frame illustrates how close I get to spray the tight areas. Notice my hands, how they support the airbrush sturdily.

I'm using very aggressive white, one that's not too slow, so that I can create bright contrast quickly - i.e. the white is mixed with a medium reducer. This may cause the airbrush to spit, but by cleaning the tip intermittently I reduce that spitting, and I can cover a spit with darker colors later.

Using white I begin to interpret the shapes where I think the light would hit the different forms of the skull. I'm thinking of the direction of my light source all the time.

Here you can see I moved back a little bit to get a wider spray pattern and do the larger faded white areas.

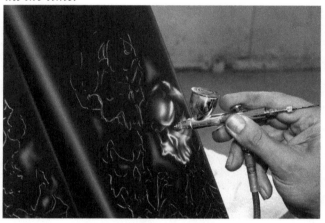

Again, I go back in close to get a tighter spray pattern in additional detailed areas.

All the time when it comes to the highest detail areas, like small teeth, I steady my hand with my other hand to give me ultimate control of my equipment.

I build the white up in layers, this way I can control the density and opaqueness. Basically, the more layers the brighter it gets.

Once again two-handed grip, more detail, building layers in the lower jaw area.

I continue to build the lower teeth using the fabled dagger stroke - it's actually just thicker on the origin end and thin on the completion end. The teeth are quite small.

Continuing on the body, I'm not using any type of anatomical chart, really just making things up as I go along. This a loose, almost graffiti style.

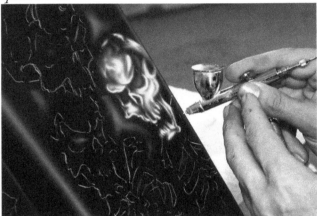

I continue on the lower jaw, basically just suggesting it with a couple of strokes.

Notice here how I move back again for a wider, looser type of spray pattern?

Continuing on the body, the lower rib portion, just doing the underpainting in white which will serve as base for the color layers.

  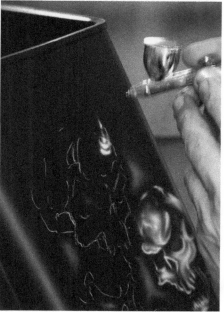

Even though the skulls are exaggerated, I still add a degree of realism and detail to the bony structures.

Here the white underpainting is completed on one skeleton - it may look a little bright now, but when we overpaint with layers of color it will be just right.

Moving to the next ghoul, I begin the sequence again. Remembering my light source, I start building layers using two hands for tight detail.

 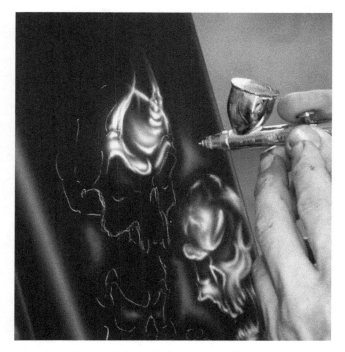

Confidence in your strokes is important. The resulting nice, clean lines gives you something to build off of, and helps everything look more polished and professional.

Moving along, although the two skulls are separated by a few inches I try to keep their brightness, or value, consistent.

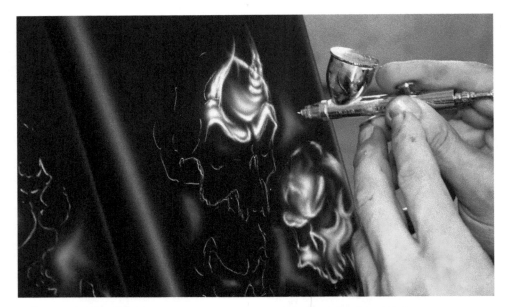

The advantage of not having your lines too defined before you start painting is that you continue drawing as you go, correcting or improving the look of your original sketch as you create the final look.

Sometimes the best design game is to simplify your drawing, leaving out sections that don't do anything to improve the overall look of your art.

Often I'll use my baby finger on my support hand to steady my movements. Be sure your hands are very clean - a funny statement I know by looking at my hand, but I mean free of grease and oil.

For example, as I go along I'm adding a line here or moving a shape there - editing each section to my advantage.

I really like to control the overspray on the small details, it looks much cleaner and more professional. Do this by backing off on the pressure, getting in tight, and keeping the tip immaculately clean.

With this particular style I like a clean steady line on which to build my light sources. Here you can see the left side of the cheek, the line is nice and steady no squiggles or stop-start sections. It's a good skill to practice all the time.

Here I move in close for the kill, and make the line for the nose hole.

I do the same thing for the side of the skull. See how close I am? Actually the tip looks a little dirty, but I think I get away with it here. Nope, there's a speck of white. See? Gotta keep the tip clean.

I'm adding more definition to some shapes on the side of the head, as if the edges are hit by the light source.

Now I start to move back, and you can see my line softening up.

You can really see how far I've moved back. You can also see the difference in the softness of the lines or material I'm adding to my subject.

Next, I move in close again to lighten up by adding more layers like we talked about before. Making bright contrasts for our color application later.

Top guy underpainting finished. On to the forehead of his brother beneath him. This is a good chance to just improvise and morph the jaw into the forehead of the lower skull.

Another example of not having to follow the line you put down in your reference exactly, and just creating and improving as you progress.

Here's a good shot of how I hold my airbrush. I am very careful not to shake and I tend to move my entire wrist. My hands and wrist move as one.

Now to the smaller skull below the top skull. I'm getting fairly tight here if you compare the airbrush to the nose structure of the bottom skull.
You need to remain steady; the smaller you get the more amplified your mistakes become.

I build the next object the same way I did the top one. The only difference is that the strokes are smaller, the action and movements remain the same, just in miniature.

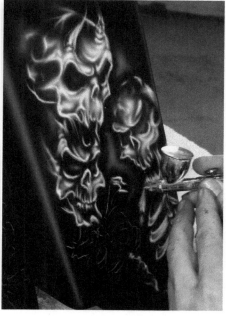

Let's move back a bit, with all those close shots we were losing the overall design. My skull is a three-point design, one of several types of compositions that are used commonly because they work so well.

From back farther, I really can tell that I like how the red is working with the white. It's the simplicity that I'm attracted to, so I enhance that by making some of the images very suggestive - like the bottom jaw of the skull I just finished.

In some cases, I believe having less detail in some of the surrounding subjects enhances the detail in the main focal point. Like when the lead guitarist is doing a riff and the rest of the band quiets down.

I continue with the white under-painting, keeping an eye on my light source.

I like to have things that are going into shadow just simply fade away to nothing, making the back-ground the atmosphere.

That's the beauty of working light into dark, you can fade things off to nothingness and still have a very convincing effect.

59

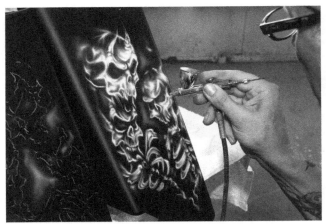

Continuing with the white base coat, I step back a ways and look at what I've got going on and just adjust things as I see fit until it looks right to me. Sometimes it doesn't hurt to take a break and come back later and look at your picture with fresh eyes - you will see things that need adjustment.

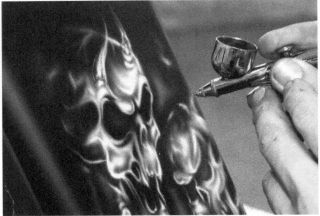

Fade to black here, I'm using the black base-coat reduced a little more than usual. I use it to carve out shadow areas where the light is not reaching, and also to add some small corrections and coverups.

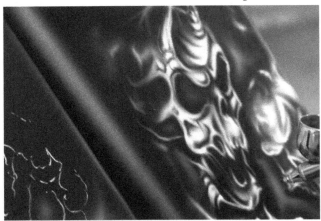

It's the same principle applied again with the white - come in tight for details and back off for a looser, softer spray for shadows and soft edges.

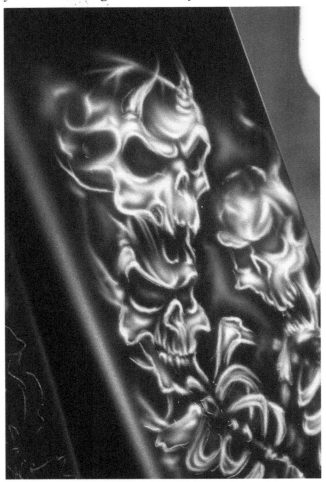

This is what it looks like ready for the next step. There are things that need to be covered up; like small spit marks from the airbrush and maybe a few drawing corrections, so carry on I must.

You can see how much contrast you can develop using the black against the white underpainting. If the underpainting had been subdued we would not have such dramatic contrast.

Now that everything is laid down I can adjust the contrast for more of a subdued look. If I'm careful about it I can get back about 12 inches and spray carefully, giving it a fine wash to darken the entire area.

Here's another shot of how bright it was before I sprayed the wash of grey, just to give you a comparison.

Here's a shot of the whole gang grated down and subdued, but looking rather dastardly, wouldn't you say?

The last thing is to go full circle, using white with a little touch of blue green to complement the red. I add (ever so slightly) some highlights to key areas, being careful to not overdo it or you might as well have skipped the gray wash tint altogether.

Well thats it, without spending a crazy amount of time we still end up with a saleable image and a very happy customer.

Chapter Eight

Harley-Davidson Tank

Vinyl Cutter to the Rescue

A lot of the paint jobs we do come about because somebody gets into a small accident, comes by to get an estimate of the damage, and decides to let the insurance company pay for the repair. But they also want more than just a stock paint job so they throw in a few extra bucks to get a sexy paint job. This is cool because everybody wins.

The owner of this bike came armed with all the things he wanted in the paint job. He knew

The final paint job honors the customer's wish-list. From the color, to the Celtic flavor and the overall simplicity of the design.

the color, and he stressed the fact that it should be a simple, clean design with a definite Celtic flair. So, basically I took his ideas and made them fit. Yet, even after 24 years at this job I still find it hard to walk away from a simple design. I have to consciously restrain myself from adding more - and remind myself that what's important is what my client wants, not what I want.

Here's the fuel tank coated with hi-build two-part primer, It's a urethane primer and has great adhesion to bare metal and seals in any bodywork underneath. I sand it with 800 grit wet sandpaper - making it a great substrate to work on.

Materials

Silver base coat mixed to a custom color.
Pearl green basecoat mixed to desired color.
Gold pearl mix.
Black and green base for shadowing.
1/8th, 1/4 and 2 inch tape.
Vinyl paint mask.
Charcoal mask.
Airbrushes, Iwata & SATA.

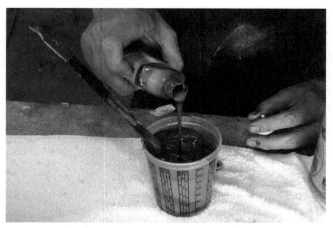

Here's the silver base coat I use, but I'm adding Goldpearl to give it a warm feel. That will go nicely with the green pearl which is warm as well.

During the wet sanding process I found another dent that we missed, so I repaired it, recoated the area with primer, then dry sanded this time with 600 grit dry, so as not to make such a mess.

63

I'm checking the consistency of the paint as I reduce it. Sometimes I'll just go by feel to see if it's what I'm looking for rather than go by the formula. But unless you're fairly adept at mixing I would follow the manufactures suggestions closely.

With a large format spraygun, I apply basecoat until it's uniform and the metallics are straight and even.

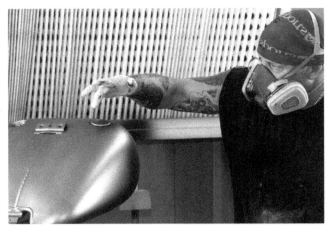

After 6 to 8 minutes, depending on the air temperature, I check to see if the basecoat has flashed off enough to continue. Then I apply another coat of basecoat and let it dry for an hour before I start my

Using fine 1/8th inch tape I establish a centerline on the fuel tank which allows me to keep the upcoming graphics as symmetrical as possible.

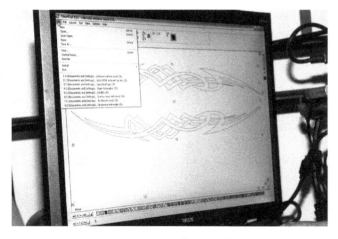

Using a cutting program and a vinyl cutter I create a simple graphic to use as a template on the tank, I could get into more detail here but that would require a whole book in itself.

Here's my finished mask being cut by the computer-driven cutter. If you're a purist, well I guess you could get your razor blade out and do it by hand, but make sure you pack your lunch.

Here you see the vinyl design we're using, with the parts I don't need taken off. The application tape applied on top of the mask holds the design in place so I can lay it down on the tank.

Here I'm moving pieces around so they flow with the shape of the tank. Some artists know how to use the computer program so that the graphics actually conform to the tank's shape. I'm not that smart so I do it this way.

65

I carefully remove the application tape, making sure I don't tear or stretch the mask out of shape. Remember you're probably going to have to move and adjust parts of this design so don't press it down completely.

Be sure to press down anywhere the vinyl wrinkled or you'll end up with paint leaking underneath and have a big mess to deal with. Take time here and you will save much time later.

If you're careful you can cut and remove sections or edit the design as long as you have repositionable vinyl or mask. Sometimes I use this system with regular masking tape to create different types of graphics.

This shows more of the same, notice how the back tails of the graphics are not secured down - they still have to be positioned aesthetically.

Here I'm laying down 1/8th inch fine line tape to create the section that I want to mask off for the two-tone part of this paint job The second line of tape will actually create a blast off pinstripe when I'm done. Quick, easy, cool.

I'm laying down more of the repositionable vinyl with some hooks that were left over, to create three-dimensionality into the second color that I'm going to spray. I think it'll be a good way to have the two parts inter-act with each other.

After a few minutes of mixing and tinting, I come up with the color that is what I like to spray for the green secondary portion of this bike tank. I use the customer's approval as a guideline for this color.

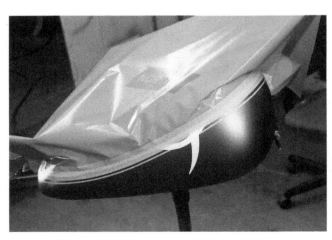

Here's the tank on the stand taped up and ready to go. So much time taping and thinking for such a small amount of time actually spraying.

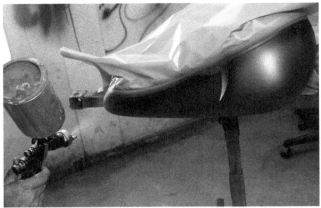

Using a large-format spray gun and a dark blue-black color, I spray a small edge along the two-tone and a fade-to-nothing about 2 inches wide. This will give me a darker green along the edge of the two colors when everything is said and done.

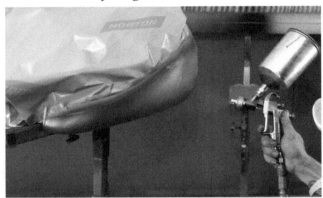

Using the same equipment, I spray the green base coat being careful not to build up too much of an edge on the two-tone. I'm careful not to soak the tank, using several thin coats not one wet coat, to reduce the chances of bleed-throughs on the tape edge.

Here's the green base coat. You can see that the metallics are straight and uniform. It's important to have a good fixture for your object so you're able to reach all angles and lay down a uniform finish.

I remove the masking system, being sure to pull the tape away at a forty five degree angle toward the edge that was just sprayed, I don't know why I just do it that way it seems to come off with a cleaner edge. I'm sure there's a more scientific answer, but let's leave that to the squints to explain.

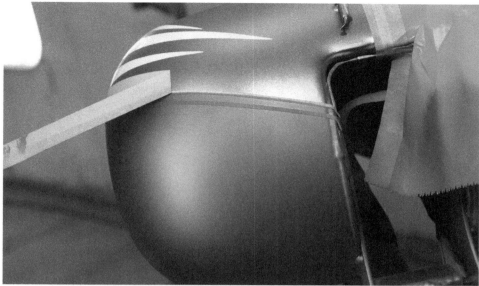

Here I'm spraying a purple pearl, I'm thinking I'm going to like this - it will go along with the green since they're close together on the color wheel. This color will create an interesting element and add to the final overall look.

Next I spray a transparent blue along the bottom of the two-tone just to deepen the green, and enhance the shadow and create more contrast.

I take the transparent blue and mist it over areas of the vinyl to give it a deeper, cool appearance. It also ties the silver color with the green on the bottom.

I've decided to add some more purple pearl along the top edge just because I can, and it'll look sweet!

Using a thin black, I darken the bottom edges of the graphic at the bottom of the vinyl. I also left a little of the black mist on the top edge, not too much, you want it to look heavier on the bottom side.

Note the overall shading and colors that I've added on top of the vinyl mask. Be sure to leave it dark enough so that it's easy to see the negative shapes that are left behind when you remove the vinyl.

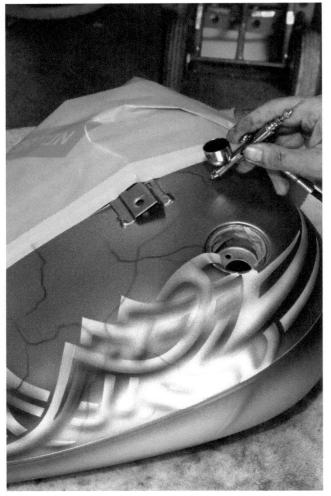

To occupy some of the space that's around the graphic I decided to create a cracked appearance. If I add it behind the graphic it'll give the graphic more depth.

I continue to add the cracked look underneath the graphic and think about ways to make it interesting with unique shapes - not just the same repetitive patterns.

After assessing my progress I decide it needs a bit more contrast on the top sides of my vinyl mask. So I adjust it accordingly with my darker purple and black until I think it looks right.

Using a template that I cut out of Mylar, (or acetate) with my stencil burner, I create some interesting holes, nooks and crannies with the black paint I used to enhance the shadows of the graphic.

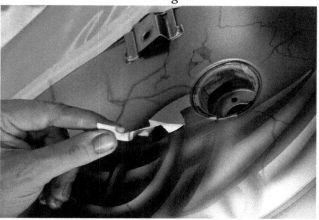

I carefully remove the vinyl mask from the basecoat revealing the negative space below, yeah. And check out the thumb, you think red suits me?

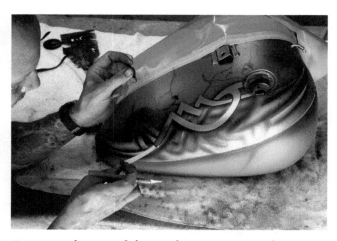

I remove the rest of the masking system. I take my time here, I don't need to damage the base coat or leave scratch marks from the tip of the razor blade.

71

In this type of work it's a good thing to be constantly asking yourself - what do I want it to look like? Is this where I want it to be? You should be constantly adjusting things and cleaning up edges until you're satisfied.

Here you can see the blue showing up nicely from the earlier stages of our spraying. It's subtle, but enough to give the graphic a little bit more life and also tie the two colors together.

Time to peel off that fine line tape we put down in the early stages of this project. You can see how, by the sequence we laid out, the masking system adds depth by overlapping the fine linear stripes with the hook shape from the main graphic.

When it comes to obstacles that block the flow of your graphic you have two choices: one is to tape around it so that it looks like it's part of the graphic, the other is to tape through it like it's not even there.

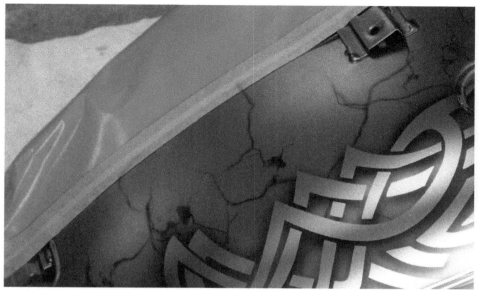

Take time to pick the objects you want to appear on top of other objects by carefully shading them appropriately. Note how I put a shadow on each side of the hook so it looks like it's on top of the linear stripes.

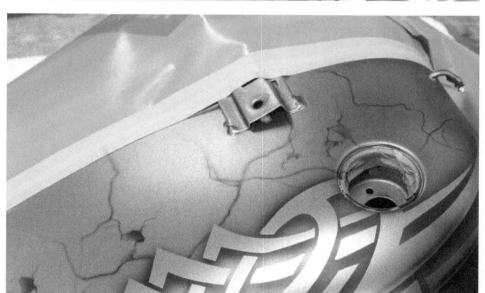

Using a thin black, I start to add more shape to the graphic by adding a shadow on the bottom edge to give it a three-dimensional appearance.

I methodically work along each graphic and shade the bottom trying to maintain a uniform thickness and width to my edges. Always pick the light source and keep that in mind as you work, I usually pick top right light.

I find that by keeping my movements quick and fluid I get a better result with my shadows. I really try to keep them light - not too dark and overpowering. I also work to make them smooth and not jerky looking.

Here's a shot of the completed shadows on the graphic. Notice that they're fairly uniform and smooth, as well as not too dark. The light source is consistent too. Shadowing is an important part of a paint job.

Using a nice slow-drying, medium thickness white, I begin to lay in the highlights on the graphics. The goal is the same as with the shadows - keep in mind the light source and keep things moving.

It's very important not to overdo this - if you make them too wide and overpowering it destroys the integrity of the paint job. Be sure to thin the white to a consistency that allows good control, with a slow enough drying speed to avoid tip-dry.

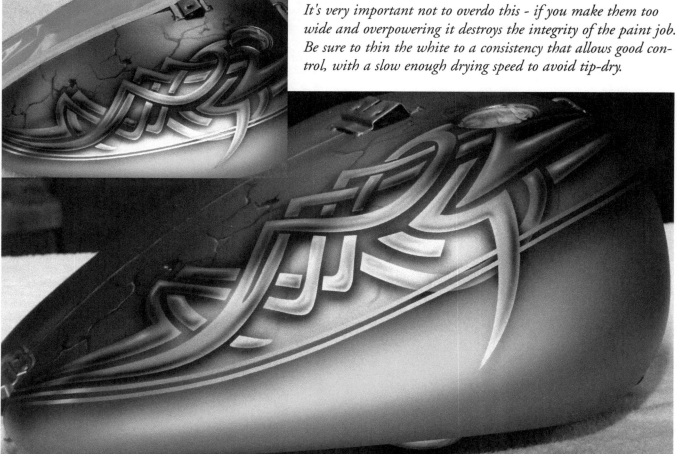

The finished graphic ready for clearcoat. You can see I highlighted the edges of the cracks as well, keeping the light source consistent with that of the graphic. The clearcoat will blend everything together nicely and also activate the pearls and metallics underneath so they sparkle. Speaking of sparkle, I added a bit of rainbow sparkle to the clearcoat to give it a little extra pop.

Chapter Nine

Guitar

Paint: It's All About the Preparation

It's not often that things roll into my shop without wheels, but once in awhile we have projects a little out of the ordinary. Like this acoustic guitar. Although we don't see them in the shop that frequently I'm no stranger to the various pitfalls that come with painting guitars. I would guess

that I've painted at least 50 guitars over the course of my career. So here are a few things that might help things go smoothly. The first thing is to determine the condition of the guitar when it comes through the door. Is it bare wood that's never been sealed? Or is it wood that's been rubbed with wax

Here's the guitar after art work, in base coat ready for a clear protective finish. The clearcoat will also merge the various layers of paint and help to highlight the pearls.

or coated with sealants? Is it new with a painted surface, is the paint oil based or urethane? Was the surface finish painted years ago, and now bears a variety of scars and chips? These are all things you must consider before attempting a paint job.

The reason it's so important to figure out what you're starting with is that each one requires a different approach in terms of preparation if the painting is to be successful. For example, some time ago a client brought me a guitar which he built it himself explaining that he just brought it straight from the wood shop. He failed, however, to tell me that he had rubbed a wax sealant into the wood and since it was clear I had no idea the surface was essentially contaminated. Even though I had adhered to my standard prepping procedures, and had cleaned it thoroughly with a wax degreaser, I had incredible problems when I attempted to put on the clear coat finish. You could imagine the fish eyes and contamination that took a lot of sanding, and some old tricks, to get it finished to where I was happy to give it back to the client. So be sure to dig in to the history of the guitar so you know which direction you want to take the preparation.

There's not enough room in these paragraphs to explain all the various ways to prepare these different surface conditions. A good book on paint preparation would be a great investment or you could take it to somebody who has thorough knowledge to help you out.

Another important thing is to strip the guitar as much as possible of all its hardware. This is especially important on electric guitars because of the various electronic equipment on the face of the guitar. Not only could you damage the delicate components, you might well encounter poor paint adhesion because it's difficult to sand into all the small nooks and crannies. Another good tip is to be sure to use an adhesion promoter on the scratch plate of the electric guitar if it's plastic. Also, if you can remove the neck of the guitar on an electric it makes it much easier to work with and negates the possibility of damaging the nice finish of the hardwood. So fortunately for this guitar it came to me already stripped and with a nice smooth factory finish to work on. All I had to do was sand the finish with 600 grit dry paper, mask and tape the areas I want to, stay clean, and get to work.

My client and I discussed the various elements he wanted to include in the design, and since he worked at a nuclear plant I think it's fairly obvious what the whole thing is about! The main thing is that my client gets it.

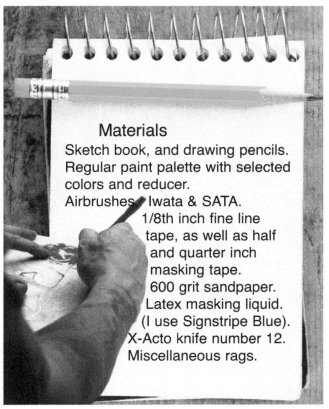

Materials
Sketch book, and drawing pencils.
Regular paint palette with selected colors and reducer.
Airbrushes, Iwata & SATA.
1/8th inch fine line tape, as well as half and quarter inch masking tape.
600 grit sandpaper.
Latex masking liquid. (I use Signstripe Blue).
X-Acto knife number 12.
Miscellaneous rags.

Using 1/8th inch green, fine line body shop tape I carefully mask off everything I want to protect from overspray.

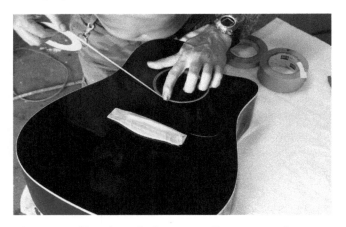

Starting off with such thin tape allows you to be very accurate as to the position of the tape's edge. It also allows you to navigate around corners that would be virtually impossible to follow with a thicker tape.

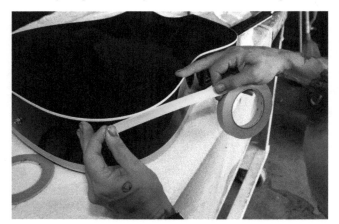

Taping: I lay down the 1/8th inch tape first, then apply the half-inch tape so it covers half the 1/8th inch tape. By rubbing it firmly I can remove all wrinkles.

I continually clean the surface using my wax and grease remover, even though my hands are clean. Be sure to use 2 rags; one that is saturated with the cleaning product, one that is dry. Work on one small area at a time, never letting the product dry on the surface without wiping the dry rag over it.

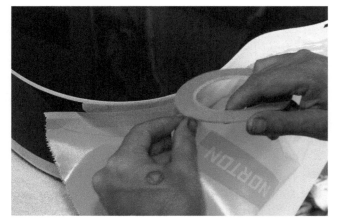

For a really nice clean edge, use 1/8th inch, fine line body shop tape, slightly overlap your first tape you laid down and this will allow you to remove this tape before the clearcoat hardens completely - this way it has a chance to lay down as it flows out.

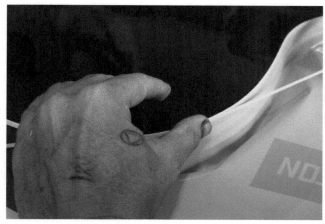

This picture will give you an idea of how to lay that 1/8th inch fine line body shop tape over your original 1/2 inch tape line.

I use 600 grit dry sandpaper to scuff the surface of the factory painted guitar. This will give my base coats something to adhere to. Take your time, this is important.

There are many different products to help paint stick to plastic, for convenience sake I'm using this rattle can product. It works great, though it's just a little more expensive buying it in small amounts.

Here I'm using a plastic preparation product designed to clean off any kind of contaminants. It ensurers that there's nothing between me and the adhesion promoter that I'm going to apply.

Wearing the proper equipment, specifically a respirator, I spray a medium coat, let it sit for a minute and spray a second coat. I let this dry for 5 minutes or whatever the manufacturer suggests,

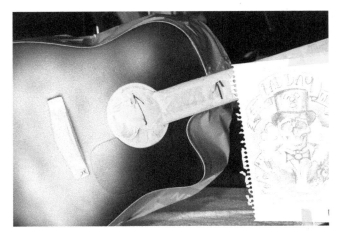

Placing my sketch nearby I set things up to be comfortable, and have decent lighting before beginning.

As a side note, you can see that I've marked the upside of the guitar while he's playing. Using a tack rag I remove any dust or crud that may be on the surface. You don't have to press hard, just be thorough and turn the cloth frequently..

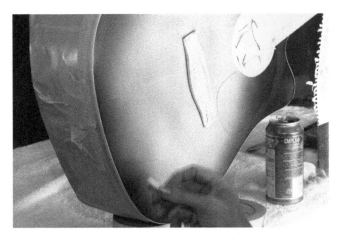

I usually use a piece of chalk in the dark areas to sketch out my basic ideas, as I move into the white I switch to a darker piece of chalk or lead pencil.

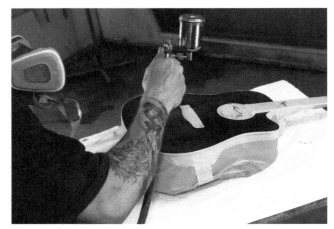

With everything taped up tight I use a detail gun to shoot a fine mist of white base. I make sure to use a slow enough reducer to avoid a dry spray. All I want here is a light mist.

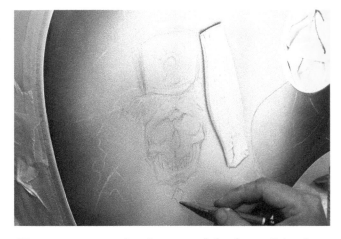

Here you can see that I use a soft lead pencil in the light areas where the white chalk won't show up. The idea is to lay out my design in the two different areas of light and dark.

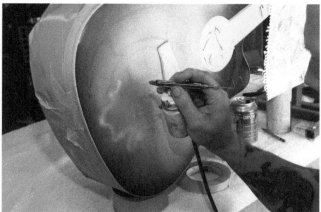

I begin to separate the light from the dark here. The thin white base makes it easier to apply more white base, or not, depending on the color of the topcoat.

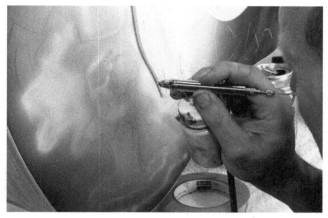

Adding a little purple, red and white mixture, I want to create something slightly warmer from the background. This color will be the under-painting for the skull face.

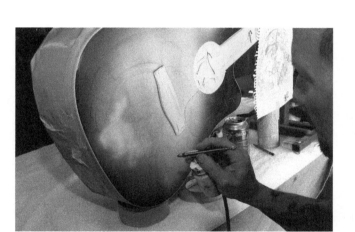

Using variable distance and trigger control, I begin to lay out the cloud - the aftermath of a nuclear melt-down. All I really want is areas of light and dark - just like I might interpret falling water or clouds.

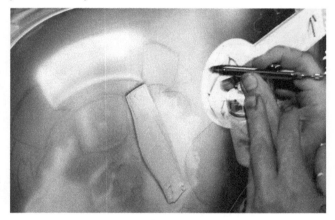

Switching back to my pure white, which in my opinion is a cool color. I start to loosely model the light effects on the banner which will contain the text of the design.

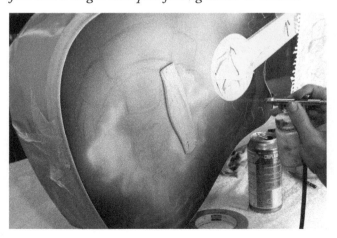

I continue to work using the same procedure and technique on the other side, trying to get a feel for the movement of expanding gas.

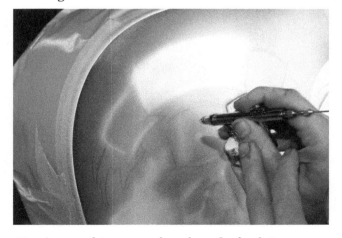

Here's a good image to show how far back I am to produce an edge. It also shows the position of my hands and how I lock down my wrists to avoid any wobbly movements.

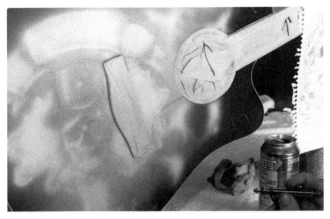

All the white under-painting completed. Though it looks extremely bright you need a bright under-painting to make your artwork pop when using transparent colors on top of your under painting.

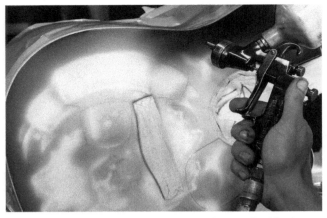

With a large format spray gun and straight black (which is the color of the guitar overall) I begin to cut in the background, which will blend with the rest of the guitar's natural color.

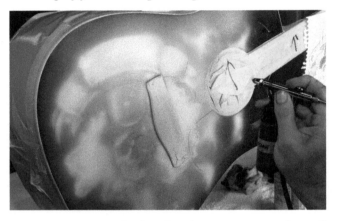

Using my suction feed airbrush I mist a nice glow of purple pearl around the outside of my design and in the dark areas. This will really show up later when the clearcoat is applied.

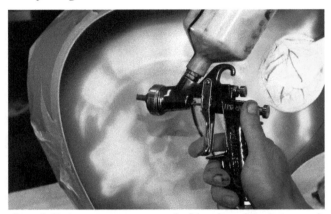

Carefully I continue to gently blending black in to the purple pearl I sprayed earlier.

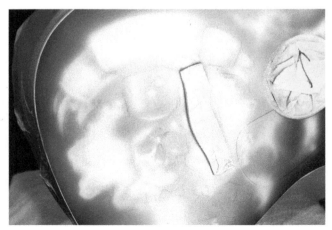

I then go back in with my white and enhance any areas that I want to really come forward. This is just a white spray, enough to give me a good base for the colors.

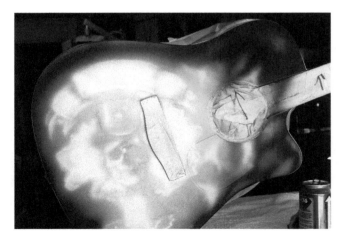

You can see the basic under-painting. The idea is to merge the background color with the foreground. To create a seamless transition from my colors to the original factory paint.

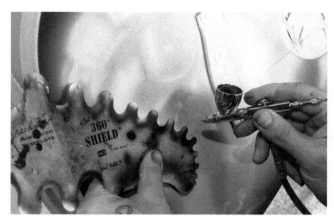

Using a free-hand shield and black toner, I decide where I want to have sharp edges between light and dark, in the areas of highest contrast. It's hard to create crisp edges with the airbrush without a template.

Here the edge of the atomic cloud is almost finished, in some areas I blend it with the background. Notice the distance of airbrush from the surface. I want to keep the strokes wide and the shapes very soft.

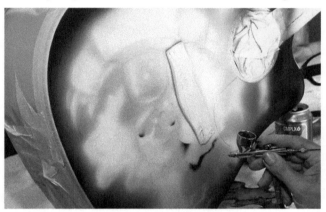

Using a straight black toner (or a custom black that you mix up from your 3 primaries as I explained in other chapters) I continue to shape my shadow areas.

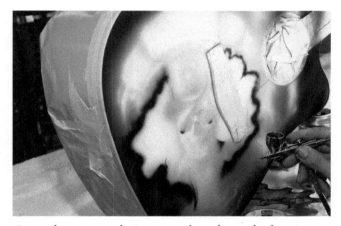

I use the same technique on the other side, keeping an eye on my reference sketch. Basically all I'm doing is creating my core shadows, which will anchor or tie together the entire design in the end.

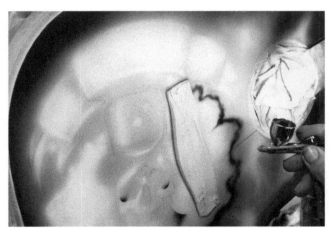

Using the same black mixture, I start to define the edge of the atomic blast. For this I use small irregular circles, varying the distance of my airbrush nozzle from the guitar surface.

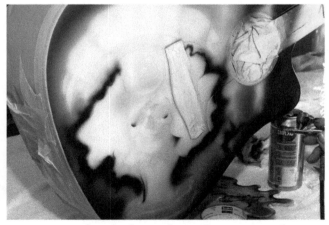

My mixture hasn't changed yet, I'm moving about the area on the bottom suggesting some of the smaller shapes and shadows.

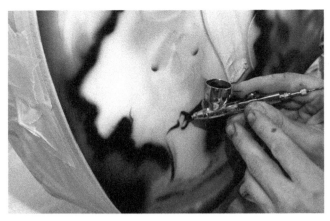

Using the same dark mixture, I move in closer to create some of the smaller shadows, or dark areas. Keep an eye on how far you hold your airbrush from the surface, as this determines the size of line.

I use a free-hand masking shield to create a sharp edge for the lower jaw. Then I fill in the drop shadow below it, being careful to control my over spray.

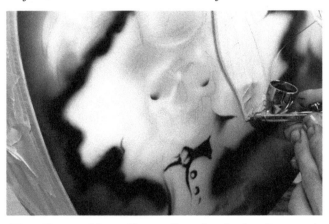

Here I'm working off one of my hard edges I created earlier, creating a nice transition into the areas of light. I do this by slowly moving my airbrush back, farther away from the surface, as I spray.

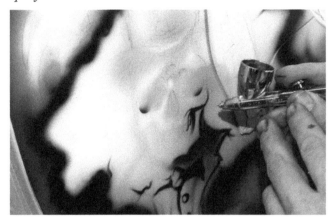

Here I'm doing the opposite of the last frame. I began with a free-hand line to suggest the shape of the cheekbone and eye socket. Then free-hand the smaller shapes and nuances, this gives softer edges to the shapes.

Moving back to 4 to 6 inches from the surface I create soft rings around the bottom of the atomic blast, keeping them fairly light gray as I don't want that area to be too dark in the end.

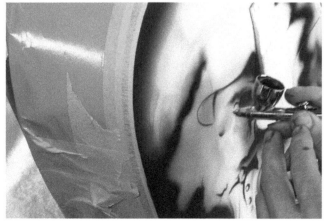

Using the same black mixture I continue in the same way using free-hand airbrushing where I want a softer edge, for example around the brim of the hat, and a mask for the hard edge on parts of the skull.

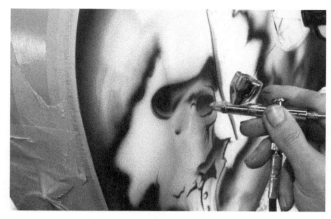

Using a straight shield (a business card), I cast a shadow above the left eye socket to suggest the light source is coming from the top right. Cast shadows have a hard edge as opposed to form shadows that are softer.

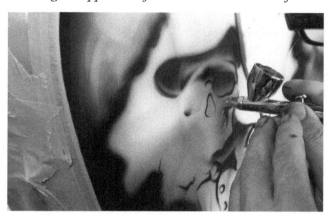

Moving closer and reducing my pressure, I start to suggest the triangular shape of the nasal cavity, keeping it in proportion to the eye socket.

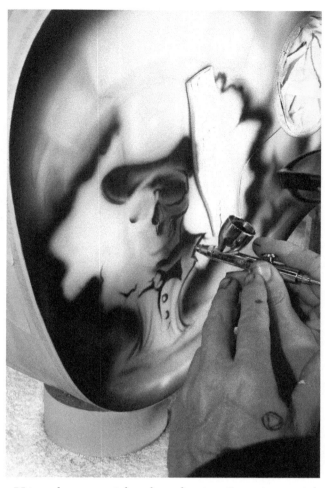

Using the same airbrush and paint, I continue with small corrections; in this case to the vest and collar.

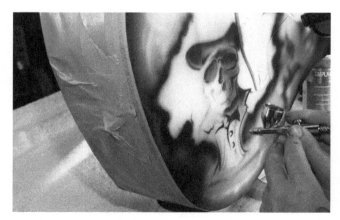

I now zip around the entire design doing a series of small corrections.

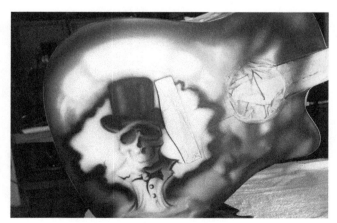

I'm now satisfied that I have the dark side where I want it, and decide to leave the light areas till later. That's the warm side and I use a whole different color for that.

Spraylat Sign Strip Blue is a latex mask. This product creates controversy, but I've never had a problem (it IS a good idea to practice first). It masks areas that are not flat, cuts out easily, and allows you to re-mask.

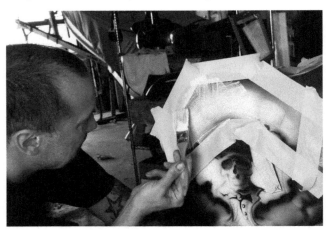

The trick is to be sure you apply enough material and that it's dry before you cut into it. You can spray it, and I do if it's a huge area. But by using a brush it's just as effective and there's a lot less to clean up.

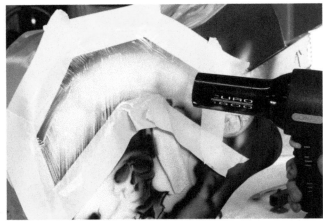

You will need a heat source to speed up the dry times between coats. Don't get too close, be cautious with plastic parts. I have tape around the outside, because I want to have good film thickness right to the edge.

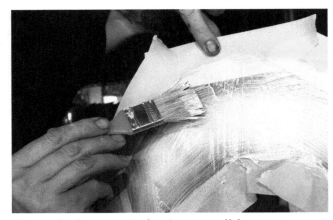

Once the first coat is dry (easy to tell because it becomes transparent) I'll apply 2 more coats, 3 three altogether to be safe. If the material is it's too thin it's hard to remove later.

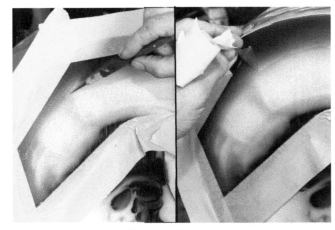

Once the masking fluid is dry (depends on shop temperature and use of a hair dryer) I lightly cut out my areas with a sharp X-Acto knife and peel off the latex. Here I'm reverse masking, note the next caption.

86

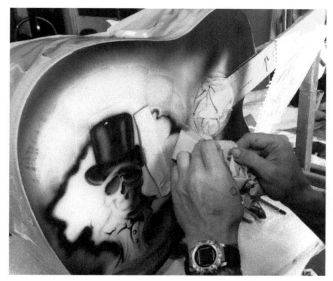

Reverse masking is masking the positive part of the shape and keeping the negative space around it open. So I cut out all around the edge of the next shape, which is the banner. I then remove the 2 inch tape around the edge because it's thick enough film and because we've painted up over the 2 inch tape ¼inch, it peels off like a breeze.

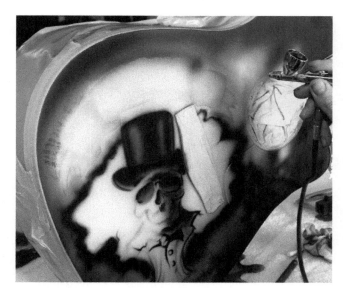

Still using the same dark mixture, I gently mist around the outside edge of my banner this will allow some sort of separation between my darks and lights. I know that phrase comes up very often but that's how I think about my visual creations. I'm describing their shapes amongst the influence of light and shadow. And how the light affects the colors and texture of those objects.

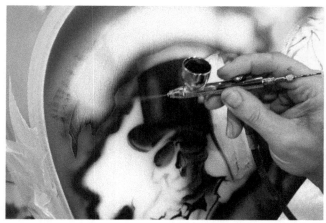

Now I move closer and with my dark mixture I add to the bottom shadow area. This will give me a good distinction between the light and dark as well, and suggest the shadow from underneath.

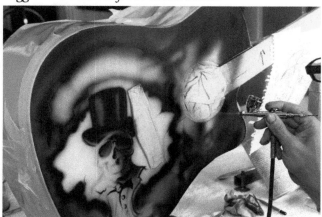

I've decided to darken some of the edges around the banner, as well as make a few changes here and there.

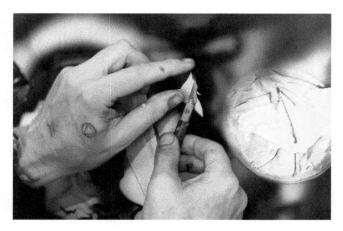

Using my X-Acto knife I gently begin to cut out masking from the areas on the outer edge of the banner. You don't have to press hard. If it's hard to peel it's not dry enough or you didn't have thick enough film.

Sometimes I use a free-hand shield in conjunction with my masking fluid to create the shapes I want. Basically, I detest taping things off, its time consuming and if I can avoid it I will - as long as it doesn't affect the final quality.

You can see the small bevel I created with the free-hand shield on the bottom of the ribbon. Now that I have the portion exposed I can the airbrush back a bit (because it's masked) to create a nice soft transition for my shadow. Once again keep in mind the dominant light source.

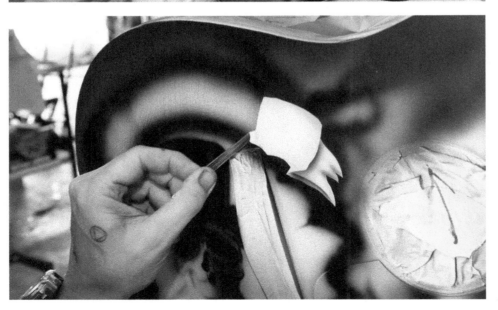

I repeat the process seen on the preceding page, gently cutting and weeding out the areas I want to expose in a sequence that allows me to spray the shadows I want. It's kind of like a puzzle, you need to think about it before you start.

Keeping the airbrush back about 4 inches, I repeat the process to create the small bevel on the bottom of the ribbon surface. The free-hand shield helps immensely. Imagine if I had to tape the area each time I did this.

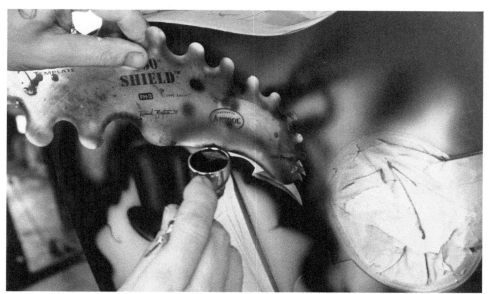

Once again, with my light source in mind, I model the shape of the banner to give it three dimensionality. I work back a distance to keep it smooth and soft looking. Thus the purpose of the mask, it allows me to create soft shade with hard edges.

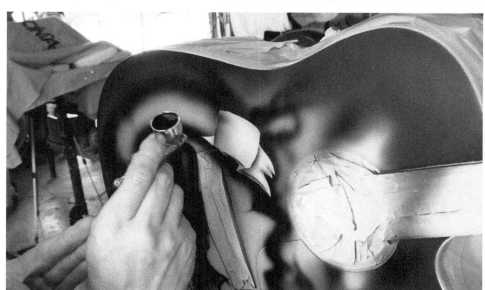

When peeling the masking fluid try to keep your fingers close to the edge of the masking fluid as you pull it off. This will reduce the problem of having it breaking off in little chunks.

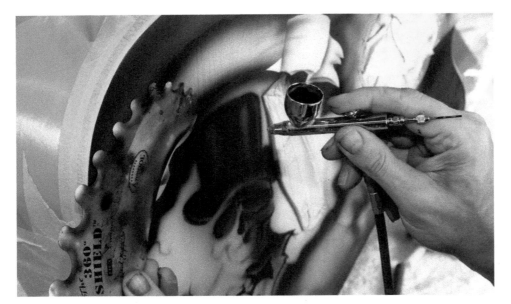

Here I'm still using the free-hand mask to create smaller shapes and nuances quickly and efficiently. By keeping the project moving it really helps me stay in my creative zone.

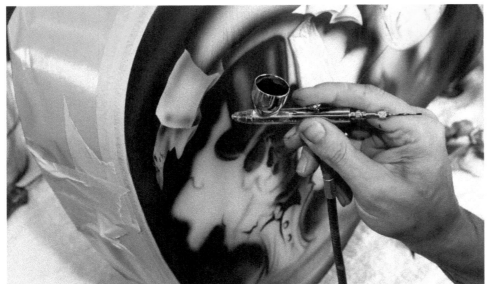

With the area shown exposed, I have the opportunity to create a form shadow that suggests the rolling edge of the banner. Be sure to give some thought to the fact that there will be some reflected light as well.

It's always nice when you peel away a big piece and you can see the intense contrast you wanted to achieve, taking form. Here you can see how this product peels off if applied correctly.

Now I have the large central portion of the banner open. I take my time modeling the illusion that this is a three dimensional banner. Let your mind leave and just follow your senses. I have to remind myself to be careful, it's very easy to overwork it at this point and lose all the contrast you work so hard to create.

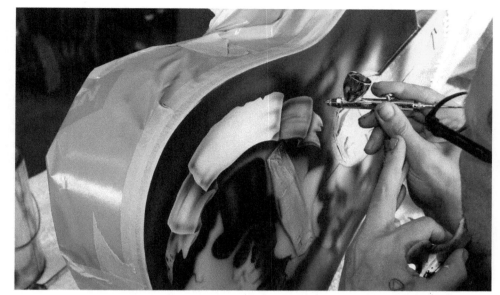

Now that you have the entire banner exposed it's much easier to evaluate the different tonal values, the lights and darks. Try to make everything consistent. all the shadows the same tone and the mid-tones likewise similar, with the light sources always coming from the same direction.

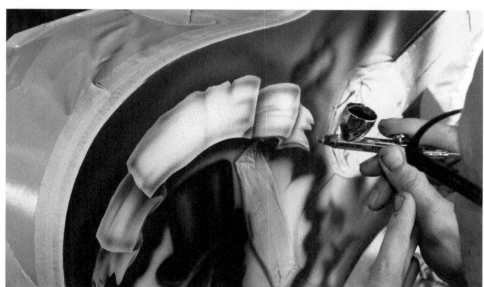

I've created another template out of mylar using my stencil burner. It's another excellent tool when I want to create interesting shapes on the fly. With practice you can make anything look gnarly.

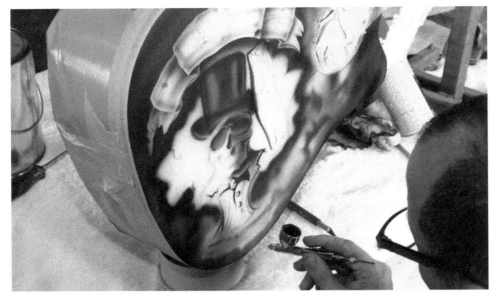

Time for some final adjustments with my dark black mixture, around the bottom of the shockwave from the nuclear blast. This will give me good contrast upon which to build some lighter shapes. Remember, you must have light to see dark and dark to see light.

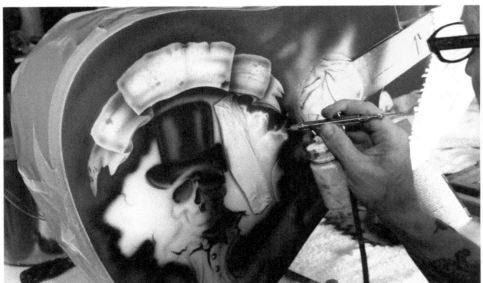

Switching to my suction-feed airbrush I put together a mixture of white with a few drops of transparent blue to cool it down - barely perceptible just a hint - I began to build upon the light colors I built up earlier. And because I didn't really saturate them completely and the fact that there was some over spray that crept over the original whites, I can start to suggest movement and shapes in the mushroom cloud.

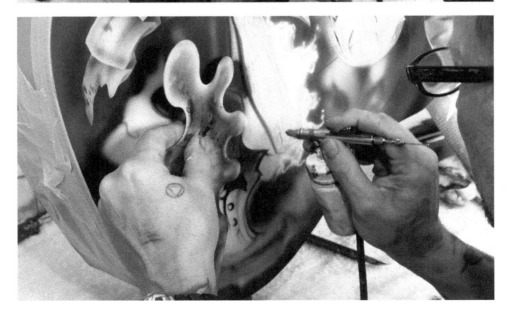

Using my free-hand masking tool I can now protect some of the dark areas from over spray and clean up the edges to create a sharp contrast. I have to be careful, I don't want to get too aggressive or I'll start carving into the dark shadows around the curve of the template.

Using white, I start to create and model the top edge of the shadows from the shockwave. I'm keeping it loose, the variety will give it a sense of movement.

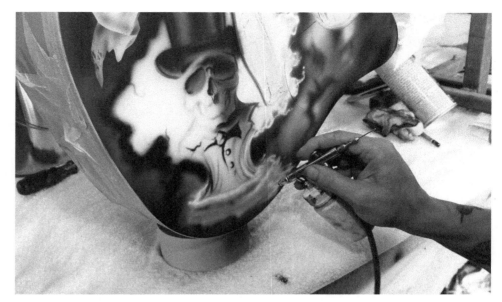

Here you can see the difference that the next layer of white made. It makes a very nice contrast on the outside of the blast You can definitely see the nice light and dark relationship we have maintained throughout the design.

If you hold your airbrush at an angle of approximately 45°(you will have to see what works best) you can create cool little meteorite thingys. I start by making a small white spot and turning the airbrush sideways and giving it a flicking motion, all while the paint is still coming out. You should maybe practice this before you try it on a project you care about.

93

Using the white again I move up the other side of the mushroom cloud and create the imagined havoc. I use the same erratic circular and zigzag motions I used earlier, with the airbrush at different distances from the guitar. You have to let your mind run wild - this will not look right if you're stiff as a British palace guard.

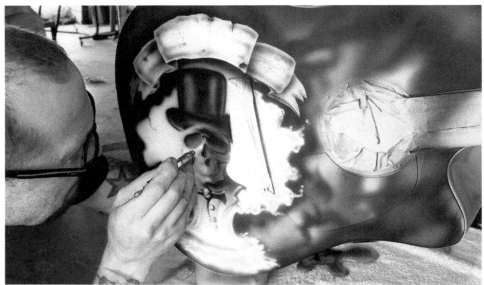

I now move into our bony friend's face, cleaning up and raising the opacity and brightness of the light areas that I want to be focal points. Time now to get really close with a good mixture that dries slow enough to allow me to get good detail. You could switch to a higher detail airbrush, like a Micron if you like for this part.

I want the overall theme to be a feeling of apocalypse or disaster, so why not have some lightning? I simply create jagged lines with the movement I like, and go back over them with a fine mist of white to give them that aura or glow. I thought I'd throw in a cloud structure to give the lightning a sense of origin.

What's an apocalypse without a few menacing skulls floating in the sky? It's also kind of my signature move ha ha. I just created these free-hand.

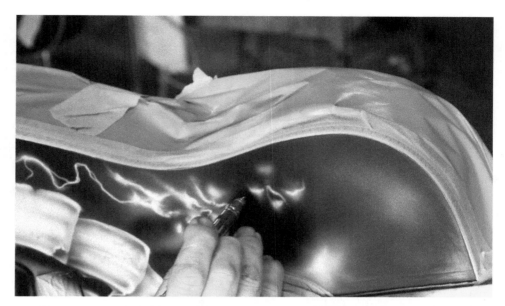

Well, I think that's pretty much it for my white under-painting. As I said, I think it adds movement as well as fills out some of the dead spots in the design.

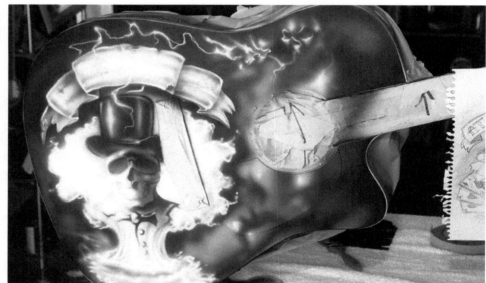

I'm switching to a transparent blue which is reduced quite thin so that I can build up the chroma and intensity slowly and with more control. Next, I move my airbrush back about 8 inches and lay down a fine mist or wash of color. Now you can see why you need the bright white underneath, without it the objects would be dull and uninteresting.

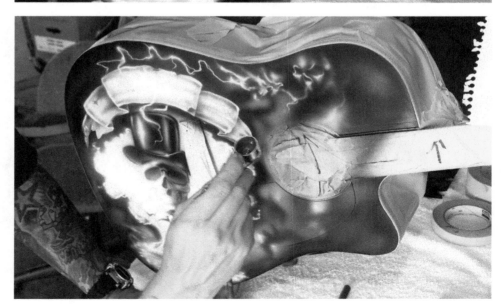

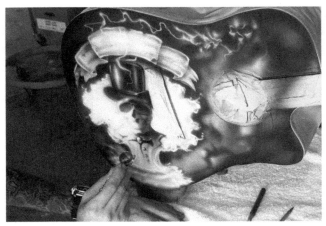

Using the same blue I add some coolness to the temperature of the right side, or shadow side, of my skull. I'm careful not to let it get into my light side or it will muddy up the colors I'm going to put their later.

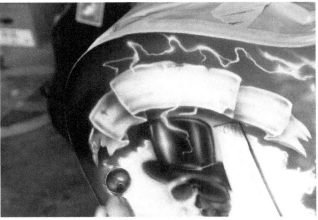

Here you can see how it works, the steps are black shadow, then your blue extension from the black, and then the purple or violet transition from there.

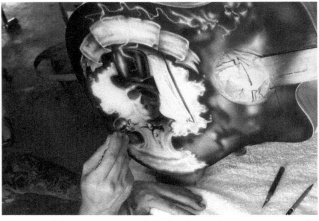

As usual I move my airbrush closer for more control over detail in the tight spots. I also add some blue to the banner shadows and in places where I thought it would add to the image.

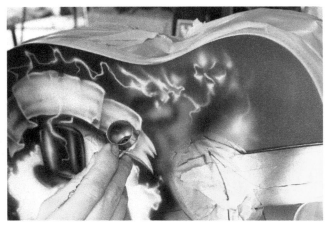

I add some pearl into my purple mixture, and start making the same transition from blue into the black background - this will look sick when the clearcoat goes on.

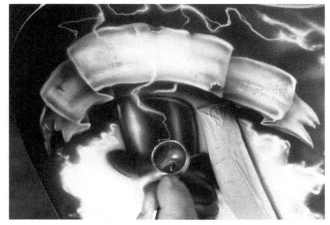

Now I choose the purple, (specifically a red blue secondary color) which is a warmer color than the blue and transitions toward a light source into my warmer callers that will be existing on the light side.

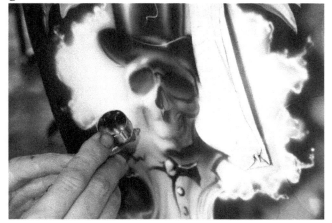

Here I carefully experiment with the first color I'm using to define my middle values. I want it dark enough to have some detail, but not so dark it ruins the contrast between the two universes of light and shade.

Now I think I found the right color, (orange with a drop of its complement purple or blue-ish) I lightly mist around the transition purple and I sprayed from the shadows as this gives me a bridge into the light without a muddy edge. Color theory is a very deep subject and would be a great topic to study with a few good books.

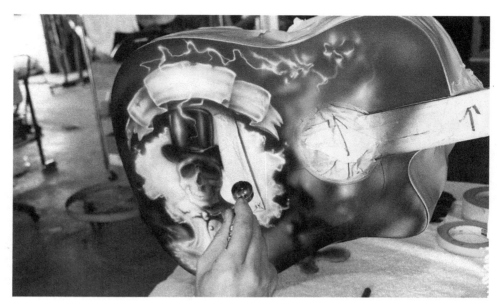

Here I start to gently mist the same transition mixture into the light side of the skull, and I leave out the highlight areas so they can remain the brightest brights in the light.

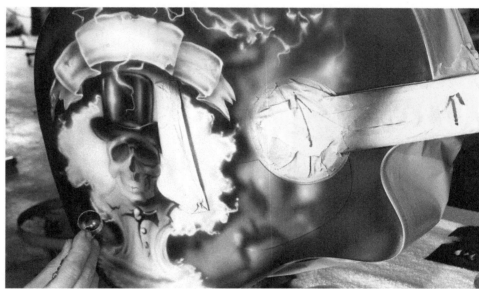

With a paper test panel, I try a darker mixture using a little purple, not much. That's why I use the paper, I want to get it just right and this way I can spray over the transition tone with a new darker tone and see how it's going to look prior to using it on the finished artwork.

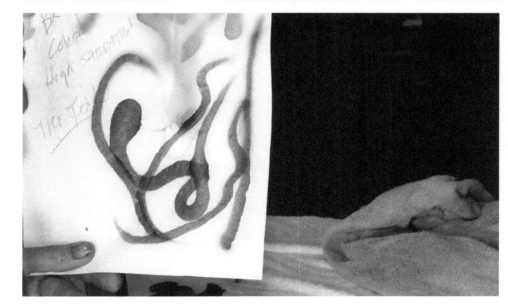

It's all in the details... I switch my airbrush to a Micron C. I need a little more control in these tight spots and I just find it much easier with the slower taper in the needle on the Microns. Now I get in close and start to define the shadow areas in the light side.

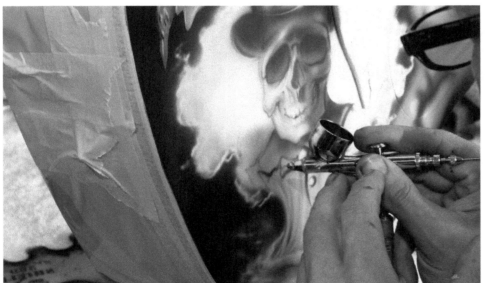

So just like we did on the shadow side, we are working from dark to light after laying down an initial ground color to work on. I see where it needs darkness and interpret how the light will affect the object the only difference is that we're using mid values instead of darker shade values.

Once I'm happy with the skull area, I know I'm going to need some dark areas in my blast. Even though the blast is going to remain quite light in value, we still need to make shadows, so to speak. In the light areas these are called halftones.

Now moving on to the lettering portion, I put some bright red, which is opaque, into my Micron and turn up the pressure as this paint is a little thicker and I want to cover faster. I freehand lettering as I see fit, I chose a kind of creepy loose, almost blood-like style (why not? it's an apocalypse).

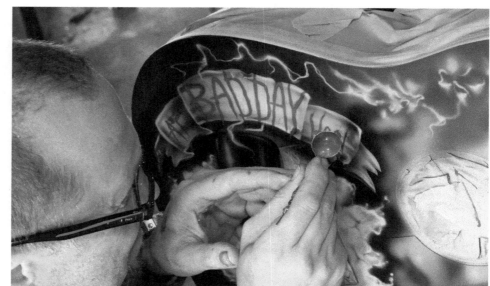

After finishing the red lettering I decided it didn't have enough contrast. In order to make the red stand out more I added a black border to the right side of each letter.

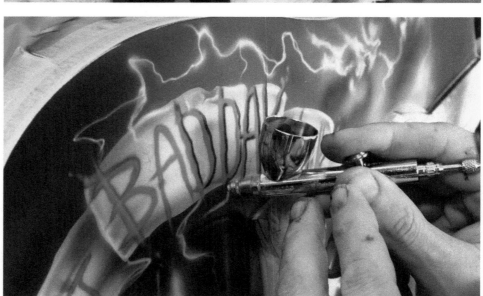

I finish the lettering and while I still have red loaded, I block out the shape of the roses and add a few finishing touches. I also get some transparent yellow the started to lay down a wash on the left side of my mushroom cloud.

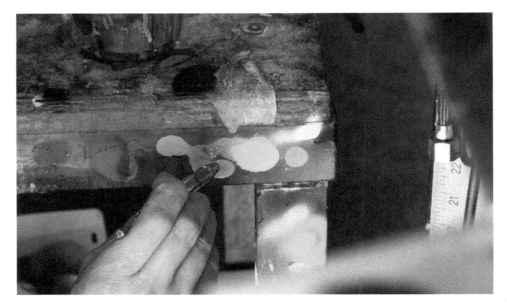

Increasing the chroma and intensity of my yellow, I experiment on my trusty work table - as you can see it's not the first time it's been abused.

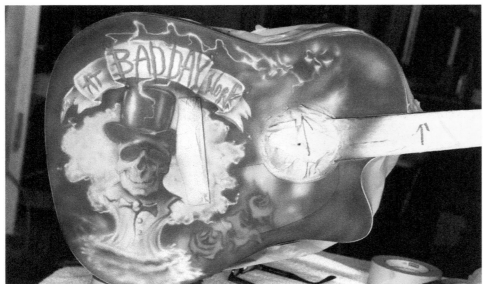

I step back again to inspect and make decisions as to what I'd like to change, or possibly add to the design.

Using artistic license I decided to add a little bit of blue-green to the area on the right side of the blast. I wanted to cool it down, and also wanted the viewer's brain to think of the green in association with the roses' foliage without actually painting it in. A bit of abstraction perhaps, but like I said we make decisions on how we want to express our ideas. I thought that the leaves would take your eye away from the main focal point.

100

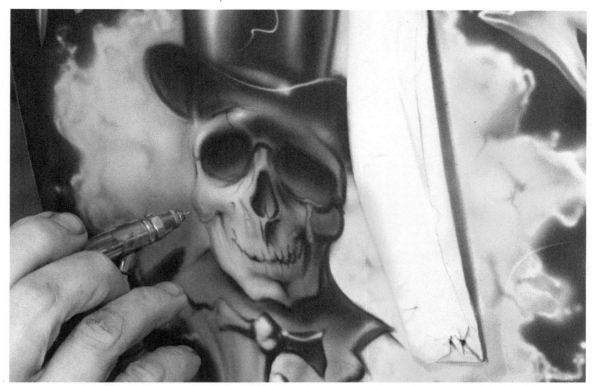

Finally, I make up some super bright yellow and heat up the edge of the transition into the white. Which gives me a very hot look that definitely contrasts with a cool blue and purples of the other side.

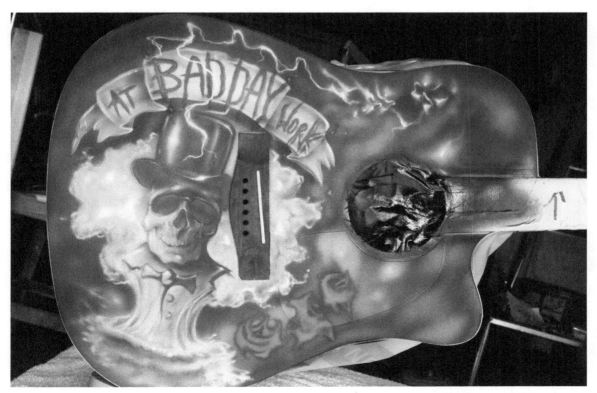

Here's the guitar before it heads off to the clearcoat booth. It's amazing how huge the difference is once the clearcoat is applied. Everything just merges together and the pearls blow your mind.

Chapter Ten

Horse Head "Night Mare"

An Elaborate Test Panel

This segment really illustrates the workflow of our procedure to investigate the content and style of an upcoming project and how important it is to get into our clients head. You need to find the style he or she is looking for when you paint their favorite subject on the metallic canvas that makes up one of their vehicles. This is especially important if your client is not local and comes from far away. In this case, thousands of miles separate my client from my studio.

Almost an abstraction, this horse head started out as a pretty simple image. In the end, it's the contrast in the colors and the bright highlights that makes the whole thing really pop.

Photos by Summer Goodeve

Thus I spent considerable time at the drawing table and on the phone and Internet. In the end I decided the best thing to do would be to paint up a mock sample on a large panel and send that to the client for approval. In this case I chose an aluminum sign blank as my test panel. From experience I find the amount of time you invest in discovering your clients expectations is never wasted. The time spent on this rather elaborate test panel far exceeds the frustration and time wasted on a job that's not successful due strictly to lack of communication at the beginning of the relationship. Not to mention the smudge to your reputation if the project crashes and you have a very unhappy (and sometimes very vocal) client.

I created a black and white pencil sketch to submit to the client based on our initial consultation. At this point I'm just looking for the major shapes and value separations.

Materials

Charcoal Mask
Aluminum sign blank.
Eclipse gravity feed airbrush and Iwata suction feed airbrush.
Standard pallet of colors in VOC compliant basecoat toners.
Transparent red, yellow, orange, blue, violet. Opaque red, orange, white, ivory, black.
Scotchbrite red/ brown abrasive pad.
Tack rag.
Basic drawing kit, pencils, paper etc.

I don't spend a lot of time on these initial drawings because I haven't nailed down exactly what the clients are looking for. You can waste a lot of time on details, save that for when you have a solid green light.

Here I'm using white basecoat and my suction-feed airbrush. I look for the light areas on my design and loosely lay them down, I like the suction feed for this part as I can change colors quickly and create more freely.

103

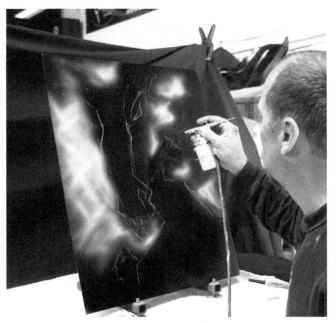 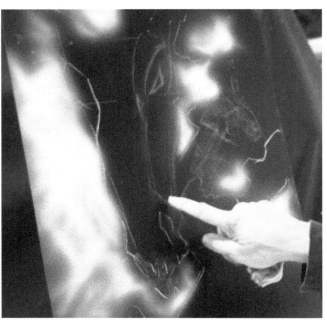

1. With my large suction feed eclipse I spray white basecoat in the areas that I want to use to push the black shapes forward. At this stage I like to think of everything as just two dimensional shapes like a puzzle. I'm not even thinking about their 3-dimensional qualities.

2. Note how lightly the chalk is on the surface of the basecoat, I can wipe it off with my finger - though I wouldn't suggest that. There is oil from your skin that can be transferred to your substrate. Keep everything off the surface and keep a clean rag between you and anything you're going to clearcoat later.

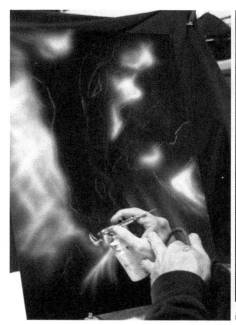  

3. At this stage I have no concrete ideas for color - I do know what I want to be light and dark so I work on that. Note how I hold my hands to steady them and keep the hose from flailing around.

4. Everything is very loose and noncommittal, but I do take care in keeping the shadow side separated from the light side. I talk a lot about light and dark shadow - because it's extremely important.

5. I have a lot of my background light mapped out. Here you can see I decided to have a direct light source from the top right area shining to the right of the nose. Here's another example of my hand position for steady brush and hand control.

Same white same airbrush. I continue to work my way around the image. I'm just under-painting the things that I'll make stand out with the transparent colors I'll be applying on top of them a little later.

At this point it's just a matter of pushing and pulling and adjusting things that just don't look right. I'd like to remind you that I'm usually wearing a charcoal mask, please do purchase and wear one.

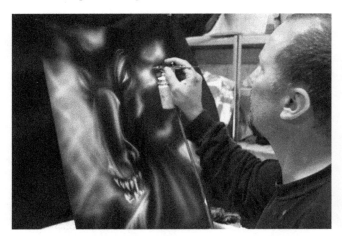

A lot of times I'll back way up and take a look at things from a distance, it's easy to get trapped into a focal jail - obsessed by one small area and forget about the big picture.

Here you can see basically the complete white under-painting complete. It really doesn't look like much now, but it gets the job done and separates the darks from the lights. This is a big reason I don't show customers their projects halfway along unless they're very visually literate or hip to the design process.

I switch to a transparent blue and mist all the areas that I want to remain in the background with coolish colors.

I pullback again to take a look at how things are coming along, checking my balance of light and dark values. I like to step back every 2 to 3 minutes for a good look.

Now I have a whole variation of blue and purple tones to work on top of. I begin the process again of pulling out highlights, this time thinking of the light source and making it a 3-D object.

This shows how bright your underpainting should be.

Note how I keep the centers of the blue areas very light - it gives the appearance of a glowing light source. The process can be worked back and forth until you achieve the desired effect.

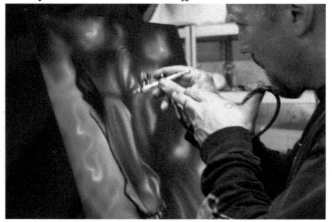

I carefully paint in the dark shapes that I envision to be silhouette or contour lines.

I carry on with the black-purple mixture, trying to keep in mind the style I'm going for, which is kind of street art. I guess simplicity is what I'm looking for more than anything.

This frame illustrates basic airbrush technique. I'm in close for detail with low-pressure, moving in small calculated strokes,

Here's the other basic airbrush technique for soft contours and edges - move back, increase the pressure and move with larger, gentler motions.

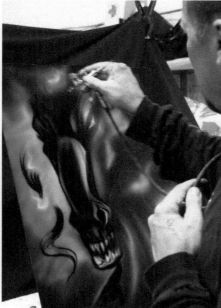

I remember this as the fun part - using the black-purple mixture in my more detail oriented gun, either SATA or Iwata. I start to imagine the forms that this demonic head should take on according to different light sources.

The really challenging part is to wind the different elements through the spaces that I left dark, and imagine what would show up to give the illusion of his body seen throughout the mist.

I think you can see what I'm going for with the hair. By winding in and out of the light and mist around his snout it give the viewer a now-you-see-it now-you-don't effect.

I'm so excited at the way my horse is turning out. I'm working with my dark color, and getting in a little tighter around the evil eye area.

With the airbrush positioned back about 12 - 16 inches, I'm spraying different layers of color and adjusting the values of the tones. Just let your mind tell you when it looks right, then stop.

At mid-point I'm starting to build up some of the lighter colors with white and blue layers to accentuate the form of the snout. I've put some blue pearl (more of a magenta) around the outside edges of the transparent blue color areas.

Here you can see the results of that detail from the last frame. The idea I have for this piece is to use simple line dynamics (thick and thin). I think less is more in some cases. A few confident lines in the right spot import just enough info for the viewer to happily play along and complete the idea in his own mind.

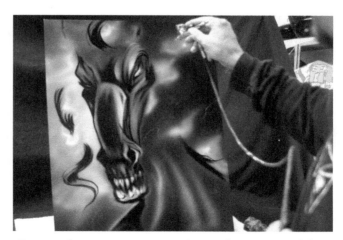

From a distance you can see I'm applying some of that magenta-purple color we talked about against the edges where the blue and black meet in a juxtaposition. It's a nice transition into the dark-black background.

108

This is a good angle to see what's going on. I never get too freaked out when I assess the progress. Each piece has its own personality and I never force it to be something else...

... now I turn my attention to some of the features in the face that are closest to us, sharpening up the edges and detailing the shadow forms.

By leaving some of front teeth lighter in color than the ones in the background it gives me the illusion of depth.

Here you can see the importance of good lighting in your studio. I run two types of lights, one warm and one cool, which balances out the light somewhat so it's like daytime even at night.

Using white in my airbrush I go back over the design and brighten up any areas that are too subdued due to overspray from my work on other parts of the painting.

I know that I'll be placing the fire element crossing over the nostrils and part of the face. So I'm blocking the area out with pure black to make it easier for me to freely express the fire forms.

With a piece of chalk I mark the shape of the fire. I tend to create blocks of fire rather than think of them as individual flames.

I'm carrying on with the exact same ideas, just in a different area. Now would be a good time to step back, and take a look at everything I've done to make sure that it all gels together.

Here's a picture of what the nose looked like before I blocked the area out. You can see that light colored fire going over the white would not contrast very much. It would look dull as opposed to going over the black that I created behind it in the last frame.

I step back and take a look to be sure I'm happy with the overall plan.

OOPS, it's the return of the green anomaly on our horse friend's chin. Oh well let's soldier on. Here you can definitely see my plan for the fire shapes, or heat or whatever you'd like to call it around the nose, the dark area where the fire will exit his nostril .

Light my fire... here I use a mixture of opaque orange and opaque yellow - free-hand shapes of what the fire might look like if it were spouting from the horse's nostril...

... I continue the process moving my hands freely, definitely without too many hard edges and not over-thinking the shapes, trying to move as if I were wind or air.

This is basically the same shot from a distance minus the green creepy thing under his chin. Neeeext ...

Note how beautifully the color theory is working. The delicious orange yellow vibrating against the blue violet is a real treat to my eyes. Color intrigues me and I will study it and enjoy it till my last breath.

Alternating airbrush-to-panel distance from about 3 inches to 8, I continue building fire shapes with the yellow orange mixture. Note how I like to make it a big shape, not lots of little shapes...

... next I mist a layer of red-orange very lightly over the original, leaving, some areas lighter some darker. Basically trying to create shapes that give the of feeling of radiating heat.

I carry on with the system of laying down the yellow-orange then the red-orange, building shapes as I go - continuously thinking about my design flow across the image.

Here's an example of getting in fairly close to make sharper lines. I move quickly and fluently with my forearm and hand, that way I don't get ragged or jerky lines.

A good look at the shapes in the fire - there are several different ways to do fire. I think your own method is OK as long as you get the feeling of illumination and heat movement. This example has a lot of impact, so it's good for street graphics.

112

More of the same technique here, I'm concentrating on which way the fire is moving, it's almost alive - with the wind being it's sculptor.

I like my flames, and if a little is good a lot is better - flames coming out of both nostrils.

I keep moving around with the same technique. All I can say is to keep shaking your paint mixture in the bottle, don't let it settle or you get an inconsistent result.

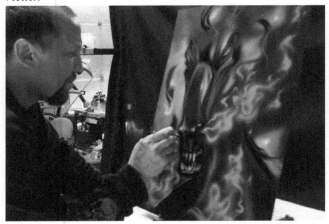

I tell students: "Let your imagination run wild ...there are no rules. Let the properties of light and shadow, shade and reflection, guide your decisions.

In the end my design looks like an S shape, but it fills up the panel without destroying the focal point. This panel is for submission to a customer, so I kinda went over the top with the content.

Another really fun aspect of this design is the reflected color from our secondary light source which is the fire itself. It can be put on any of the facing planes or sides of the object, just try not to overdo it.

Some places with the same yellow and no template. I freehand in the warm centers, creating that feeling of warmth emanating from the orange to light orange to yellow color banding.

I decided to tighten up some of the flames - bring the closer flames into focus, like landscape design where distant objects are fuzzy and closer objects are brighter and sharper.

Here's a nice shot depicting the work I've done so far with the fire, reflected lights and stylized black lines. Definitely quite a departure from my regular work.

Switching to a nice warm yellow in my airbrush, and my ellipse template, (store-bought) I create little warm highlights of varying shapes in amongst the orange flames that have already been laid down.

Above: Now in some areas, (up to 75% of the canvas) I spray a layer of transparent red over the work I've just done. This richens the fire quite drastically and knocks it back to more of the orange-red spectrum which is much more tasteful to me. It also plays well into the original double split harmony which I had in mind to begin with - orange-red to blue-violet.

Here I've tweaked a few more elements and added highlights into the darker fire. At this point I use orange tones, not yellow, to enhance the real-fire effect. I've also added some nice flow and movement to the mane as well.

This is a nice close-up shot of some of the details within the fire itself. I like to say: it can be anything you like as long as it goes from fuzzy to sharp, and from lower chroma red-orange to brighter orange-yellow hues.

115

More finishing touches, I add more heat and fire in the mouth area as well as additional teeth which are silhouetted against the bright fire - I like the silhouette better than showing the actual teeth.

I continue to adjust colors like the furrowed brow with the hair going over his nose and the drop shadow to give it depth. I also darkened some of the purples and blues a little to add contrast and harmonize better with the red-oranges at the bottom.

After closer inspection, and a day later, I walked back and decided that I couldn't leave well enough alone and decided I wanted more energy and contrast in the design...

... coming full circle I add some highlights to the blue reflective side of the horse. And there you go, with me looking a little worse for wear, but nonetheless finished and ready to send the image to the customer for approval.

Chapter Eleven

Devil's Playground

Machines with Heart

For this project I flew to Idaho where my friend, and client, Steve lives. I left home with nothing more than a pencil, which was promptly confiscated at customs. With no computer, vinyl cutter or other luxuries, I proceeded to embark on an adventure of hunter-gatherer. Securing equipment and product indigenous to the area became my first task. Next came the sketches and layouts of the complex graphics that would cover Steve's Big Truck from one end to the other.

Here's a finished shot of the driver's side door with our boney, gleeful jockey riding his hellish creation. The piece is approximately ten feet long, and I think it should make a great visual statement just sitting at a stop light.

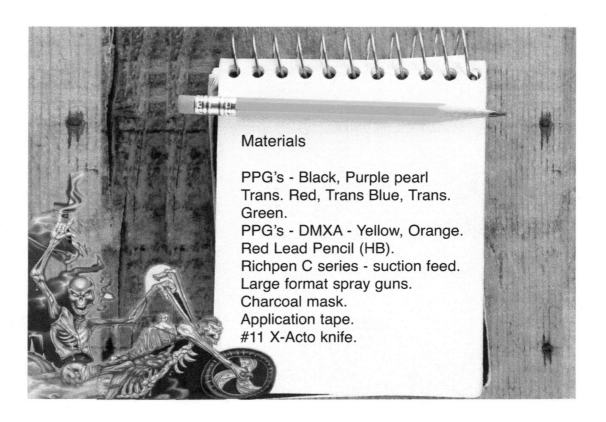

Materials

PPG's - Black, Purple pearl
Trans. Red, Trans Blue, Trans.
Green.
PPG's - DMXA - Yellow, Orange.
Red Lead Pencil (HB).
Richpen C series - suction feed.
Large format spray guns.
Charcoal mask.
Application tape.
#11 X-Acto knife.

Say hello to my little friend. This is how the vehicle came to me, ready to paint, thanks to Todd and his buds. At first the square footage of the project overwhelmed me, but I got over it. I start with the preliminary sketches and layouts, keeping them loose and basic as usual.

I start by taping the door jams and fender wells to avoid creeping overspray that could enter the clean black interior.

...be sure to use new blades and just cut the tape. The sharp blades give you more control.

Next, I lay down application tape, being sure to remove any loose spots or bubbles with a squeegee. Now I sketch my first element on the truck.

Here is a close-up of the cut-out application tape and masking, this will give us a nice crisp edge to work off of later.

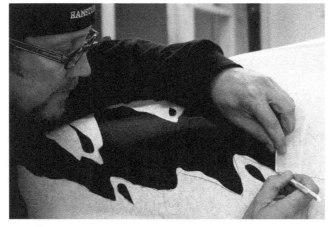

With a # 11 x-acto knife I carefully cut out the areas to be exposed. Don't hulk out and cut deep...

Using 24 inch body shop paper, I mask the outside edges to avoid overspray. Don't under estimate the creeping ability of overspray.

120

In this step I use bright orange from my mixing bank. It's opaque and builds color without the need for white underpainting. Using a French curve ellipse, I proceed to create random background texture to give the impression of depth.

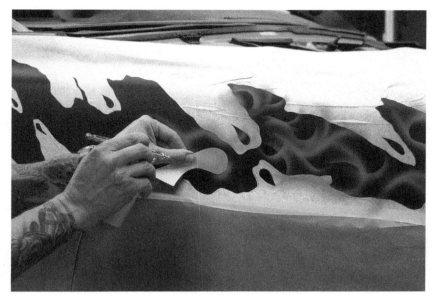

Next, I create a candy mixture using PPG's DMX series. It's a color concentrate that I add to a clear basecoat binder. I then reduce it to a spray-able viscosity of about 2:1. Next, I mist a light coat of this red candy on top of my orange shape, using a detail gun, to knock them back in intensity.

I repeat the process seen at the top of the page with the curve template and the same orange. The orange I applied earlier is darker now due to the candy red topcoat applied in the last step, so this application of orange will be lighter. I'm looking for dark areas, or windows, to fill up with the lighter tone for contrast.

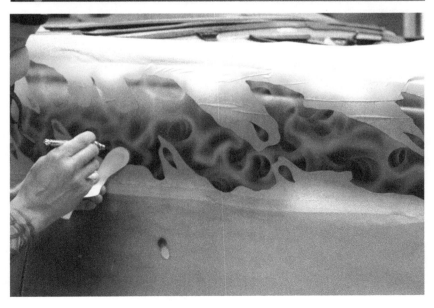

Using opaque yellow, I then add light-energy, winding the bright color into the tunnels I've created earlier with my template.

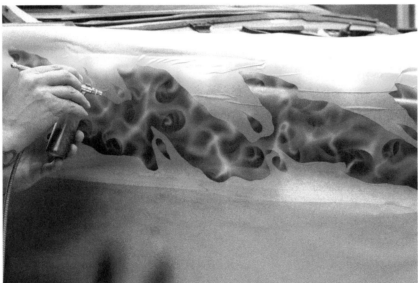

Another coat of DMX, but this time I use an orange mixture, reduced exactly like our earlier application of orange.

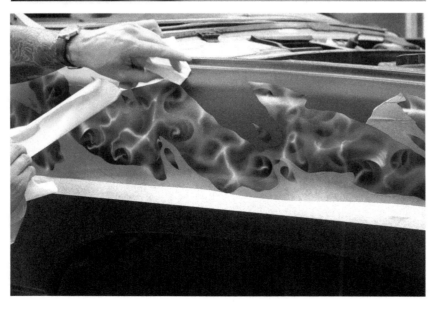

Now I remove the masking system. Always be sure to get any adhesive off the black basecoat by using PPG's DX 320 wipe.

Using pure white basecoat, I then air-brush the outer texture, the stretching sheet metal, bearing in mind a constant light source.

A transparent purple wash over the white I've just sprayed is a good compliment to the reds and oranges within the tear. The purple wash is just trans purple reduced with medium reducer, approximately 2 parts reducer to 1 part paint, sprayed from my airbrush at about 8 inches from the surface. I give all the white a nice, even mist coat.

Using pure black, I create the stitching that holds the tear together, I seldom use pure black, but in this case I wanted extreme contrast against the graphic. Going back in with white, I pop a few highlights onto the appropriate parts of the ripples in the sheet metal.

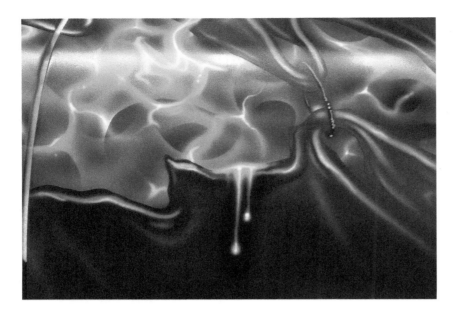

Here is a close up of some of the eye candy, namely the dripping lava and "stitching" that holds the tear together.

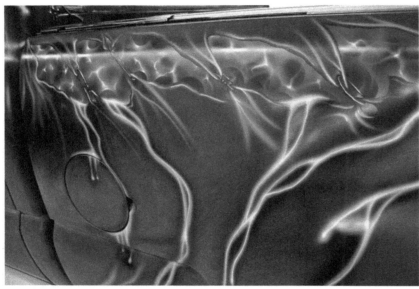

These are the lava flows that run down the quarter panels. I try to follow the contours of the vehicle just like liquid would flow naturally.

I also run the lava down the door pillars to sort of add some light and color to the top of the truck.

For composition sake I want to shoot, or project, my original sketch on the side of the truck in order to get a feel for placement and size. When projecting on black, the light is gobbled up and isn't conducive to a clean reference sketch. So I decide to use 24 inch application tape to mask the vicinity of my first image. This will allow me to better see the projected image (on the white tape), and also control the overspray created by the larger gun and the white basecoat.

I then use a large format spray gun and apply a medium dry coat of white to the area I've exposed in the application tape mask system. Wear your respirator!

Using a light lead pencil (HB), I draw in the major components of my sketch, leaving the fine detail to be created when I'm actually airbrushing. When doing work like this be careful not to contaminate the truck with the oil from your hands during the sketching. You can occasionally clean the areas of contact with DX 320 wipe.

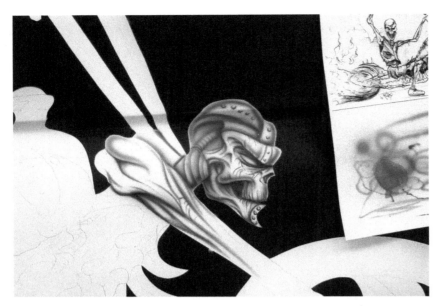

I start cutting out my malevolent friend with a transparent maroon mix (trans red 75%, purple 20% and green 5%). Working off my original design, I begin at the headlight. It seemed like a good place to start to get a feel for the color I chose.

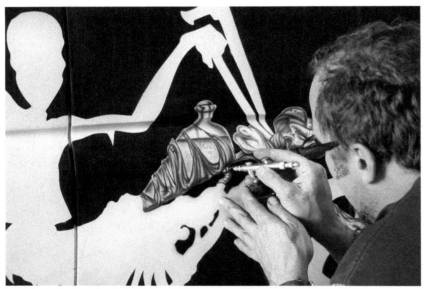

Here I am developing the texture of the sinew that holds the fuel tank. The beauty of not having a ton of reference lines is that you can create on the fly, designing as you go.

Our demonic daredevil emerges complete with fashion-conscious vest. I like the way his high five hand works in front of the other graphic elements.

A progression shot carrying on into the legs. I'm constantly flicking my concentration over the other areas I've worked, making adjustments for balance in the shadow areas.

This shows a progression into the rest of the image. Just remember, be sure to make up enough of your color to last through the whole project. You can remix, but it's a lot easier to roll through the entire painting without running out.

Using white, I under-paint the burnout and the cigar smoke. I then give the "Bike Thing" and the sinewy specter a case of heartburn by adding heat and lava in the key areas.

A close up of his head. The main thing to notice here is the highlight on the right side of the figure, it helps lose the cut out look, or masked white under painting.

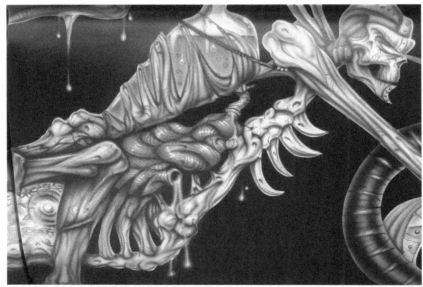

I use the lava to give a little life to the heart of the motor.

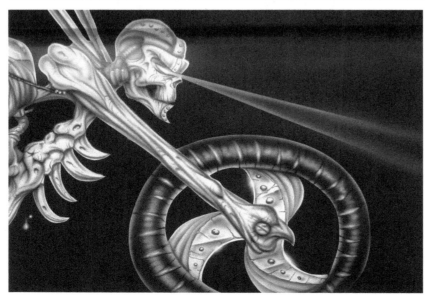

Slashed tires reveal a molten center. Just another point of interest that strays from the norm.

Using white basecoat, I suggest motion in the drive line and the rear tire.

Dark grey is used to indicate the misshapen tire, and the chain in motion.

I then heat up the burnout using the same principles mentioned at the beginning of the chapter.

Details are extremely important, note the little things like the flying, frying rubber.

Using darks and lights in silhouette, I create a hellish panorama. The strange rock formations contain personal hidden requests by the client.

A close up of the creatures hand against the torn or ripped graphic.

I use a series of loose templates, cut out of Bristol board, to create the mountains in the background, and the heat whick morphs into rivers and pools of lava.

By adding darks against lights, and lights against darks, behind my figure I'm able to lose the pasted look.

This is an overall shot of the passenger side. It's a lot of real estate to cover with an airbrush.

The next sequence reflects the client's fascination with flying up extremely steep mountains at high velocity on snow machines. Generally speaking, the color sequences are similar to the burnout, but with a few twists. So I begin by taping off with application tape.

Using a # 11 x-acto I cut out the outer edge of my design. Just be cautious not to press too hard and damage the underlying basecoat. Sharp blades are a must.

A medium-light basecoat of white will do the trick. I use a large format gun for this process. Be sure to pay attention to safety by wearing your respirator.

I remove the application tape revealing the silhouette of the image. Notice that I've left the back of the sled dark against a light background, this will really help with the illusion of heat later on in the project.

I develop the details in the same manner as the chopper in the previous steps using a mix (75% red 20% purple 5% green).

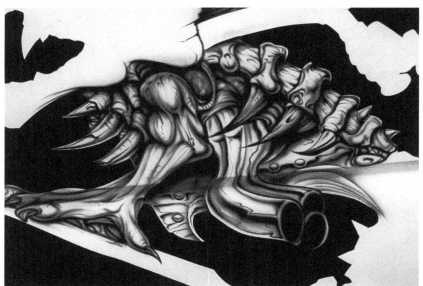

I continue airbrushing in all the details as I go, just letting my imagination run wild.

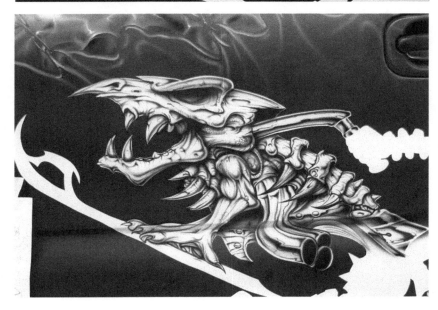

I add some more elements like guts and flesh to the sled, as well as heat in the exhaust pipes.

Using the transparent colors and sequences for heat seen in the previous segment, I create a wicked rooster tail to suggest horse power and movement.

To emphasize the heat from below, I create a secondary light source from above. I choose purple because it's a perfect complement to the oranges and reds. To do this I spray a trans purple wash to the top side of my design. I simply reduce trans purple basecoat and spray a light mist coat from about 8-12 inches, watching for even coverage.

Next, I complete the background in the same manner as the other side of the truck, with lava flows and strange scenery.

I then go in with white in my airbrush and enhance highlight areas through-out the image.

Next, is a simple matter of adding more eye candy such as lava drips, smoke and all the rest. I'm just kind of rounding everything out.

This is a great view of the driver's door. The look of glee is evident on our boney buddy as he hurtles forward from his fiery platform. Even though the figure is made of bone and sinew, you can still capture emotion through body language and position.

Here is a close up of our mobile monster showing the action as the beast rears out of the lava pool. Streaming liquid and other details add to this effect.

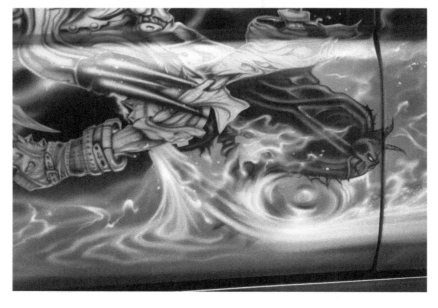

Little details like the crazy laced up motocross boots, as well as the torn pants, all add points of interest.

A bizarre 4x4 truck seemed to fit the bill. The iron-cage wheels filled with molten lava are a neat touch.

A custom built moto-cross bike doing a tail whip is always cool, especially when the frame is made of bones and the wheels are on fire.

Here is a shot of the tailgate prior to clearcoating. Somehow scantily clad ladies appear on the list of things to do in a lot of my commissions.

Speaking of which, I try to show some skin, but also try not to be too revealing. I guess you could say she's "HOT".

A sort of self portrait, as well as thanks to all the folks involved in the project.

Another part of the painting, a series of mountainous faces, each one dripping with hot lava.

The last thing I do is pinstripe the bumpers so they're not so bald. The rest of the stuff, like the hood and the tonneau cover, are shipped to my shop to be completed there. In fact, the air-brushing of the tonneau cover became the subject of a new Vince Goodeve DVD, now available through our web site or by calling our shop.

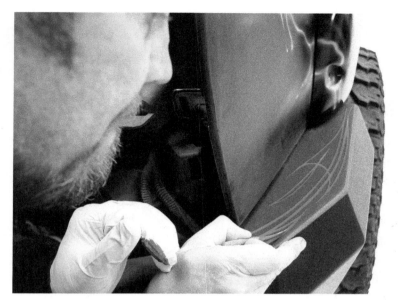

Running the striping brush over the slightly textured bumpers was initially weird, but once I found the groove, a matter of the right speed and consistency, it worked well.

Just a little color or stripping in the bumpers ties them into the rest of the truck. We also wanted to give it a moto-cross look that's showing up on some of the super cross bikes.

Books from Wolfgang Publications can be found at many book stores and numerous web sites.

Titles	ISBN	Price	# of pages
Advanced Airbrush Art	9781929133208	$27.95	144 pages
Advanced Custom Motorcycle Assembly & Fabrication	9781929133239	$27.95	144 pages
Advanced Custom Motorcycle Wiring - *Revised*	9781935828761	$27.95	144 pages
Advanced Pinstripe Art	9781929133321	$27.95	144 pages
Advanced Sheet Metal Fab	9781929133123	$27.95	144 pages
Advanced Tattoo Art - *Revised*	9781929133822	$27.95	144 pages
Airbrush How-To with Mickey Harris	9781929133505	$27.95	144 pages
Building Hot Rods	9781929133437	$27.95	144 pages
Colorful World of Tattoo Models	9781935828716	$34.95	144 pages
Composite Materials 1	9781929133765	$27.95	144 pages
Composite Materials 2	9781929133932	$27.95	144 pages
Composite Materials 3	9781935828662	$27.95	144 pages
Composite Materials Step by Step Projects	9781929133369	$27.95	144 pages
Cultura Tattoo Sketchbook	9781935828839	$32.95	284 pages
Custom Bike Building Basics	9781935828624	$24.95	144 pages
Custom Motorcycle Fabrication	9781935828792	$27.95	144 pages
Harley-Davidson Sportster Hop-Up & Customizing Guide	9781935828952	$27.95	144 pages
Harley-Davidson Sportser Buell Engine Hop-Up Guide	9781929133093	$24.95	144 pages
Harley-Davidson Twin Cam-Hop Up	9781929133697	29.95	144 pages
Harley-Davidson Evo Hop-Up/Build	9781941064337	29.95	144 pages
How Airbrushes Work	9781929133710	$24.95	144 pages
Honda Enthusiast Guide Motorcycles 1959-1985	9781935828853	$27.95	144 pages
Honda Mini Trail	9781941064320	29.95	144 pages
How-To Airbrush, Pinstripe & Goldleaf	9781935828693	$27.95	144 pages
How-To Build Old Skool Bobber - 2nd Edition	9781935828785	$27.95	144 pages

Books from Wolfgang Publications can be found at many book stores and numerous web sites.

Titles	ISBN	Price	# of pages
How-To Build a Cheap Chopper	9781929133178	$27.95	144 pages
How-To Build Cafe Racer	9781935828730	$27.95	144 pages
How-To Chop Tops	9781929133499	$24.95	144 pages
How-To Fix American V-Twin	9781929133727	$27.95	144 pages
How-To Paint Tractors & Trucks	9781929133475	$27.95	144 pages
Hot Rod Wiring	9781929133987	$27.95	144 pages
Hot Rod Chassis How-To	9781929133703	$29.95	144 Pages
Kosmoski's *New* Kustom Paint Secrets	9781929133833	$27.95	144 pages
Learning the English Wheel	9781935828891	$27.95	144 pages
Mini Ebooks - Butterfly and Roses	9781935828167	Ebook Only	
Mini Ebooks - Skulls & Hearts	9781935828198	Ebook Only	
Mini Ebooks - Lettering & Banners	9781935828204	Ebook Only	
Mini Ebooks - Tribal Stars	9781935828211	Ebook Only	
Power Hammers	9781929133604	29.95	144 pages
Pro Pinstripe	9781929133925	$27.95	144 pages
Sheet Metal Bible	9781929133901	$29.95	176 pages
Sheet Metal Fab Basics B&W	9781929133468	$24.95	144 pages
Sheet Metal Fab for Car Builders	9781929133383	$27.95	144 pages
SO-CAL Speed Shop, Hot Rod Chassis	9781935828860	$27.95	144 pages
Tattoo Bible #1	9781929133840	$27.95	144 pages
Tattoo Bible #2	9781929133857	$27.95	144 pages
Tattoo Bible #3	9781935828754	$27.95	144 pages
Tattoo Lettering Bible	9781935828921	$27.95	144 pages
Tattoo Sketchbook, Jim Watson	9781935828037	$32.95	112 pages
Triumph Restoration - Pre Unit	9781929133635	$29.95	144 pages
Triumph Restoration - Unit 650cc	9781929133420	$29.95	144 pages
Vintage Dirt Bikes - Enthusiast's Guide	9781929133314	$27.95	144 pages
Ultimate Sheet Metal Fab	9780964135895	$24.95	144 pages
Ultimate Triumph Collection	9781935828655	$49.95	144 pages

 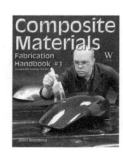   

Sources

Vince Goodeve
Mailing address:
55 River Street Tara, Ontario Canada N0H 2N0.
Shop address:
Springmount, Owensound, Ontario Canada
Phone: 519-934-3842
vincegoodeve@yahoo.com

BASF Automotive Refinish, contact for tech and
new product information
Internet: http://www.refinish.basf.us/
Phone: 800-532-7782

House of Kolor, find local distributors,
tech information, new product information.
Internet: www.houseofkolor.com/
Phone: 800-845-2500
Mailing: 1101 S. 3rd St.
      Minneapolis, MN 55415

Iwata, product information and tech help
Internet: www.iwata-medea.com/
      http://www.airbrush-iwata.com/

SATA, spray guns, airbrushes, painting
equipment and seminars
Internet: https://www.sata.com/
Mail: DAN-AM Company
      One Sata Drive
      P.O. Box 146
      Spring Valley, MN 55975
Phone: 800-533-8016

TCP Global, supplier of HOK paints as well as
spray guns, airbrushes, compressors and a huge
selection of body shop tools and supplies.
Internet www.tcpglobal.com/
Phone: 858-909-2110

CPSIA information can be obtained
at www.ICGtesting.com
Printed in the USA
BVHW021113121119
563577BV00014B/316/P